ANTHROPOLOGY AND CLIMATE CHANGE

ANTHROPOLOGY AND CLIMATE CHANGE

From Encounters to Actions

Editors

Susan A. Crate and Mark Nuttall

Walnut Creek, CA

Left Coast Press, Inc.
1630 North Main Street, #400
Walnut Creek, California 94596
http://www.lcoastpress.com

Hardback ISBN 978-1-59874-333-3
Paperback ISBN 978-1-59874-334-0

Library of Congress Cataloging-in-Publication Data

Anthropology and climate change: from encounters to actions/Susan A. Crate and Mark Nuttall, editors.
p. cm.
Includes bibliographical references.
ISBN 978-1-59874-333-3 (hardback: alk. paper)
ISBN 978-1-59874-334-0 (pbk.: alk. paper)
1. Climatic changes. 2. Ethnology. 3. Anthropology. I. Crate, Susan Alexandra. II. Nuttall, Mark.
QC981.8.C5A63 2009 304.2'5—dc22 2008042697

09 10 11 5 4 3 2 1

Printed in the United States of America

This book was printed on paper manufactured from 100% post-consumer fibers and processed chlorine free.

The paper used in this publication meets the minimum requirements of American National Standard for Information Sciences—Permanence of Paper for Printed Library Materials, ANSI/NISO Z39.48—1992.

Cover design by Andrew Brozyna

Cover Image: Several village herders gather with their cows at an oibon (water hole cut in river or lake ice) on the Viliui River in Elgeeii village, Suntar region, Sakha Republic, Russia.

Contents

Introduction: Anthropology and Climate Change

Susan A. Crate and Mark Nuttall

Increasingly, anthropologists are encountering the local effects and broader social, cultural, economic, and political issues of climate change[1] with their field partners. Wherever we go and work, we encounter people telling similar accounts of the changes they notice in the weather and climate. We work with Inuit in northern Canada, Alaska, and Greenland who talk about the sea ice thinning and disappearing, or about seeing insects for which they have no name in their language; we hear from reindeer herders in northern Fennoscandia about shifts in prevailing winds and the loss of precious pasture for their herds; we listen to horse- and cattle-breeding Sakha of northeastern Siberia, Russia explain how the warming of winter threatens to transform their ancestral cosmology; we research in sub-Saharan African villages facing increasing desertification; and we fail to find words of explanation as Quechua of the high Andes lament the disappearance of their glaciers, their main water source, and South Pacific islanders tell us of their fears of the rising tides that threaten to engulf their homes. Everywhere, from high-latitude taiga and tundra regions, to high-altitude mountain ecosystems, from tropical rain forests to near sea-level coastlines, there are compelling similarities in the narratives, accounts, and experiences of indigenous and local peoples who are already seeing and experiencing the effects of climate change. For them, climate change is not something that may happen in the near or far future but is an immediate, lived reality that they struggle to apprehend, negotiate, and respond to. The weather is increasingly unpredictable and people express concern that local landscapes, seascapes, and icescapes are irreversibly changing. We, with our field partners, are also encountering the local manifestations of this global phenomenon. And, like them, we are confronted with the challenge of comprehending and responding to it.

The global discourse on climate change has turned increasingly to adaptation as a priority for research and policy. Although many of us working on developing anthropological perspectives on climate change are versed in the frames of adaptive capacity and resilience, we nonetheless question whether these coping mechanisms are sufficient. Resilience, both social and ecological, is a crucial aspect of the sustainability of local livelihoods and resource utilization, but we lack sufficient understandings of how societies build adaptive capacity in the face of change. Furthermore, we suspect that environmental and cultural change, far beyond the reach of restoration, is occurring. Combined with institutional and legal barriers to adaptation, the ability to respond to climate change is severely constrained for many people around the globe. Some of us feel we are in an emergency state as field researchers and struggle to design conceptual architecture sturdy enough to withstand the storminess of the intellectual and practical challenges before us. We are confronted with an ethical and moral issue.

And we are left with a flurry of questions, directly related to our age-old struggle as academics to reconcile anthropology's applied, public, and activist roots (Lassiter 2005, 84). What is our proper response and what is our responsibility to our research partners in these revelations? How do we translate, advocate, educate, and mediate? What are the theoretical frames that inform our queries? What insights can we gain and use from the work being done where communities are the hardest hit—where climate change is already having profound effects, for example in the Arctic, Africa, South Pacific islands, and other low-lying lands? What are the challenges faced by the current scientific models as we try to bring research to bear in a meaningful way? How do we understand the complexity of everyday life in relation to climate change? How can we transform knowledge into action, vulnerability to learning to cope and to be responsible? How do we link our expertise to this arena in such a way that we are not part of the problem but part of the solution? How do we negotiate and communicate anthropological insights effectively to influential policy makers? How do we make claims for greater participation in global discussions on adaptation to climate change?

Given the increasing effect that climate change is having on local populations across the globe and the highly charged geopolitical arena in which action must be taken, understanding anthropologists' roles in the field as we encounter, communicate, and act in response is paramount. These questions and more are considered, explored, sometimes answered, and other times left unanswered, by the contributors to this volume.

The Moving Target of Climate Change

Although there have been voices alerting the world about climate change for several decades, the new millennium has ushered in an increasingly potent stirring up of public, private, and international attention to and, in some cases, action on the issue. Numerous scientific reports compiled by international teams of experts confirm that climate change is not only happening but is very likely caused by human activity, a point clarified by *Climate Change 2007*, the Fourth Assessment Report of the Intergovernmental Panel on Climate Change (IPCC 2007). Politicians (and would-be politicians) have brought climate change into their campaign vocabulary, framing it as one of the most urgent issues of our times. Climate change consistently makes front page stories of mainstream magazines and newspapers. The media portrayal of climate change, the excessive dramatizing of apocalyptic events, or the downplaying of scientific evidence and the critique of scientific motive are critically important for us to understand, for they have significance for how climate change is defined, understood, and legitimated. Public awareness has risen to the extent that climate change is not just a topic of conversation but a call to action to make major changes in consumer lifestyles. Popular magazines are touting green tips. Major box stores have reprioritized to embrace sustainability. Many of us ponder to what ends these actions will lead and if these efforts are merely a feel-good exercise while the earth continues to heat up and public opinion oscillates between climate change as fact or fiction. We sense a deeper calling, one that compels us to focus the anthropological gaze on vistas beyond the immediate locality and investigate the multifaceted roots of the climate crisis to understand how genuinely to address the issue.

If we frame our inquiry on the scale of global geopolitics, we see the causes and effects of climate change to be about people and power, ethics and morals, environmental costs and justice, and cultural and spiritual survival. Scholars are beginning to address the equity and justice implications of climate change (Thomas and Twyman 2005). On a temporal scale, the effects of climate change are the indirect costs of imperialism and colonization—the "non-point" fall-out for peoples who have been largely ignored. These are the same peoples whose territories that have long been a dumping ground for uranium, industrial societies' trash heaps, and transboundary pollutants. Climate change is environmental colonialism at its fullest development—its ultimate scale—with far-reaching social and cultural implications. Climate change is the result of global processes that were neither caused nor can be mitigated by the inhabitants of the majority of climate-sensitive world regions now experiencing the most unprecedented change. Thus indigenous peoples and other place-based peoples find themselves at the mercy of—and having to adapt to—changes far beyond their control. Yet climate change is a threat multiplier. It magnifies and exacerbates existing social, economic, political, and environmental trends, problems, issues, tensions, and challenges.

Environmental problems can, it is widely acknowledged, become problems for national and international security. Barnett (2003) has argued that, despite climate change being the most profound of global environmental change problems, it has received little systematic attention as a security issue. In 2007 the United Nations Security Council held its first debate on the implications of climate change for international security. The UN also estimated that all but one of its emergency appeals for humanitarian aid that year were climate related. The European Council has also considered the impact of climate change on international security. One challenge for anthropology is to address the security dimensions of climate change. The most obvious issues include humanitarian aspects, political and security risks, conflicts over resources, border disputes, tensions over energy supply, migration, political radicalization, structural violence, and tensions between different ethnic and religious groups. However, the anthropological gaze needs to also settle on the governance of national and international security issues, and the tools, instruments, and institutions utilized by states within a broader context of policy and practice.

The Cultural Implications of Climate Change and Indigenous Peoples

From an anthropological perspective, climate change is ultimately about culture, for in its wake, more and more of the intimate human-environment relations, integral to the world's cultural diversity, lose place. For indigenous peoples around the world, climate change brings different kinds of risks and opportunities, threatens cultural survival and undermines indigenous human rights. The consequences of ecosystem changes have implications for the use, protection, and management of wildlife, fisheries, and forests, affecting the customary uses of culturally and economically important species and resources. The effects of climate change are not just about communities' or populations' capacity to adapt and exercise their resilience in the face of unprecedented change. Climate change is also about the relocations of human, animal, and plant populations to adjust to change and to cope with its implications. Such relocations, both actual and projected, entail a loss of intimate human-environment relationships that not only ground and substantiate indigenous worldviews, but also work to maintain and steward local landscapes. In some cases, moves will also result in the loss of mythological symbols, meteorological orientation and even the very totem and mainstay plants and animals that ground a culture.

Indigenous peoples themselves may argue that, despite having contributed the least to greenhouse gas emissions, they are the ones most at risk from its consequences due to their dependence upon and close relationship with the environment and its resources. Their livelihood systems are often vulnerable to environmental degradation and climate change, especially as many inhabit economically and politically marginal areas in fragile ecosystems in the countries likely to be worst affected by climate change. Massive changes in ecosystems are occurring and have in many cases been accompanied by opportunistic and often environmentally devastating resource exploitation. To indigenous peoples this means that climate change is not something that comes in isolation; it magnifies already existing problems of poverty, deterritoriality, marginalization, and noninclusion in national and international policy-making processes and discourses.

We need to be careful not to be too overconfident about our research partners' feelings about their capacity to adapt. As anthropologist Piers Vitebsky (2006, 10) remarked, "Eveny are highly adaptive. Sometimes they joke, 'This is our home. If the climate gets too hot, we'll just stay and herd camels.'" Although it seems completely plausible that such highly adaptive cultures as the reindeer-herding Eveny of northeastern Siberia will find ways to feed themselves even if their reindeer cannot survive the projected climactic shifts, as anthropologists we need to grapple with the implications of reindeer-centered cultures (or other groups dependent on species that may not adapt) losing the animals and plants that are central to their daily subsistence practices, cycles of annual events, and sacred cosmologies. The cultural implications could be analogous to the disorientation, alienation, and loss of meaning in life that happens when any people are removed from their environment of origin, like Native Americans moved onto reservations (Castile and Bee 1992; Prucha 1985; White 1983). The only difference is that the communities experiencing the effects of climate change are not the ones moving—their environment is.[2] As the earth literally changes beneath their feet, it is vital to understand the cognitive reverberations and cultural implications to a people's sense of homeland and place.

If we agree, as Keith Basso convincingly argues, that human existence is irrevocably situated in time and space, that social life is everywhere accomplished through an exchange of symbolic forms, and that wisdom "sits in places" (1996, 53), then we need to grapple with the extent to which climate change will increasingly transform these spaces, symbolic forms, and places (Crate 2008). It follows that the result will be great loss—of wisdom, of the physical make-up of cosmologies and worldviews, and of the very human-environment interactions that are a culture's core (Netting 1968, 1993; Steward 1955). Climate change leads us to ponder both literal and figurative transformations of cultures and environments and even a future of unmanageable change. As anthropologists, we need to look closely at the cultural implications of the changes global warming has and is bringing.

Despite the impact of climate change on indigenous peoples and their traditional knowledge, international experts and policy makers most often overlook the rights of indigenous peoples as well as the potentially invaluable contributions from indigenous peoples' traditional knowledge, innovations, and practices in the global search for climate change solutions. And since adaptation to climate change is something that primarily takes place at the local level, it is paramount that indigenous peoples and place-based societies themselves define the risks related to rapid change. An exemplary text, *Voices from the Bay*, published in 1997, reported on an indigenous knowledge project that originated with the Hudson Bay Traditional Ecological Knowledge and Management Systems (TEKMS) study in northern Canada. From the perspective of its indigenous Inuit and Cree participants from communities around Hudson Bay, one aim of the study was to document their knowledge of the environment so that it could be used and incorporated into environmental assessments and policies. Another was to communicate that knowledge effectively to scientists, and to indigenous youth of the participating communities. It was one of the first attempts by indigenous peoples to put a human face on climate change, to record indigenous observations and knowledge of environmental change, and to argue the right for the inclusion of traditional ecological knowledge as an integral part of the decision-making process for the Hudson Bay bioregion. In its approach and methodology, it emphasized the active participation of indigenous communities living in the region. It was community based and community driven, involving indigenous people in all aspects of the research process: design, development, compilation, synthesis, and the production of results.

Voices from the Bay served to illustrate that local people can contribute to identifying and understanding the ecological processes and dynamics of their ecosystem. Scientists became aware of how weaker currents are changing sea-ice regimes, of the departure of belugas whales from river mouths that have become too shallow for molting, of the sensitivity of sturgeon to changing water quality and river diversions, and of the type of damage caused by freshwater diversions. As a model for how the situated and narrative dimensions of local knowledge can be combined with expert scientific knowledge, *Voices from the Bay* was cited in discussions during the early stages of the Arctic Council's Arctic Climate Impact Assessment (ACIA 2005) to ensure the recognition of the validity of ecological insight inherent in indigenous knowledge. Although it emerged from a project in Canada's north, its relevance is global.

CLIMATE CHANGE, HUMAN RIGHTS, AND THE INTERPLAY OF MULTIPLE STRESSORS

Communities differ in the way they perceive risk, in the ways they utilize strategies for mitigating negative change, and in the effectiveness of local adaptive capacity. In our field contexts we see that the effects of climate change are prompting the adoption of different subsistence and local economic strategies to suit new ecosystem regimes or, with more rapid change, the displacement and resettlement of peoples who risk losing their homeland to environmental change. Policy responses need to be informed by a greater understanding of how potential impacts of climate change are distributed across different regions and populations.

The second part of the Intergovernmental Panel on Climate Change (IPCC) fourth assessment shows that the world's poor, already struggling to achieve their basic needs of food, water, and health, will suffer the worst effects of climate change: "Poor communities can be especially vulnerable, in particular those concentrated in high-risk areas. They tend to have more limited adaptive capacities, and are more dependent on climate-sensitive resources such as local water and food supplies" (IPCC 2007, 9). In the same context, an important sector of the poor has been repeatedly absent—that of women (including within that IPCC summary document). Women make up 70 percent of the world's inhabitants living below the poverty line (Röhr 2006). Women, in their roles as the primary managers of family, food, water, and health, are hit the hardest and must deal very directly when the impacts of climate change are brought home (Wisner et al. 2007).

Take an example to illustrate: When a poor community in Mexico's Yucatan Peninsula is hurricane struck, there would be three to four women dead for every man (Aguilar 2008). The reason for the disparity is that women, due to the culturally specific evaluation of their gender, face different vulnerabilities and many live in conditions of social exclusion. Examples include exclusion from survival skill learning, such as tree climbing and swimming, which help during floods; restrictions on women's movement in times of crisis, including dress codes requiring lengthy garments and prohibiting women from leaving the home without a male's permission; and unequal allocation of food resources to girls and women, rendering them physically weaker in time of evacuation and crisis (Aguilar 2008).

Climate change brings additional vulnerabilities for indigenous peoples, which add to existing challenges, including political and economic marginalization, land and resource encroachments, human rights violations, and discrimination. Because climate change has effects on the myriad of rights necessary to lead a productive and healthy existence, including subsistence rights, economic rights, cultural rights, intellectual property rights, and the like, it is implicitly a human rights issue. Framed within the context of legal and human rights challenges, we can argue that maintaining cultural diversity and recognizing indigenous livelihood rights are prerequisites for successful adaptation in the face of change. Although most major international treaties concerned with human rights contain provisions concerned with the environment, universal recognition and protection of environmental or ecological rights continues to be a significant challenge for international human rights law. The potential threat of climate change to their very existence combined with various legal and institutional barriers, including conservation and wildlife management regimes, resource-use quota systems, trade barriers, and so on, which affect their ability to cope with and adapt to climate change, makes climate change an issue of human rights and inequality to indigenous peoples.

In parts of the world there has been a swell of advocacy by affected communities in response to the local impacts of climate change. One example is the petition to the United States by the Inuit Circumpolar Council (ICC) to consider the human rights issues of climate change in the Arctic and their intrinsic role in reducing greenhouse gases as a way to mitigate. Past chair of ICC Sheila Watt-Cloutier's testimony explicitly posits climate change as a human rights issue: "Inuit are taking the bold step of seeking accountability for a problem in which it is difficult to pin responsibility on any one actor. However, Inuit believe there is sufficient evidence to demonstrate that the failure to take remedial action by those nations most responsible for the problem does constitute a violation of their human rights—specifically the rights to life, health, culture, means of subsistence, and property"(Watt-Cloutier 2004). A second example is the yearly restatement of the 2000 declaration by indigenous peoples at the annual COP United Nations Framework Convention on Climate Change conventions. Here is an excerpt from the 2004 convention in Buenos Aires:

> We request the urgent need to continue to raise awareness about the impact of climate change and approaches of climate mitigation and adaptation measures on Indigenous peoples and request a High-Level Segment on "Indigenous Peoples and Climate Change" be held during the 11th session of the Conference of the Parties. Panelists on this High-Level Segment shall include representatives of the UN Permanent Forum on Indigenous Issues.[3]

Clearly many indigenous groups are actively expressing their concern that the local effects of climate change may surmount their adaptive capacity and threatens survival for their communities.

Climate change issues, of course, have to be approached, understood, and resolved within a context of the interplay of multiple stressors. Human activities, industrial development, consumerism, resource-use regulations, and global economic processes have far-reaching consequences for the environment and on indigenous and local livelihoods. Indigenous and local economies are not self-reliant closed systems, and although their involvement in global networks of production and consumption may provide avenues to strengthen and extend their possibilities, it also introduces greater elements of risk and perhaps makes people and their livelihoods less resilient to coping with and adapting to climate impacts. For some people, climate change may not be the most immediately pressing issue facing them. Social, cultural, and economic change often has more immediate effects (Nuttall et al. 2005).

By exploring these field contexts, one aim for anthropology is to show the complex and multisited ways that indigenous and other marginalized peoples are shaping and being shaped by being within the world system, in order to highlight both the tenacity and the susceptibility of their cultural survival—which itself depends on an intimate knowledge of and connection to the natural world—the very relationship that substantiates their utility for adaptation and resilience. Through the investigation of local capacities for adaptation and resilience in the face of climate change, anthropology can tease out how larger-scale processes including industrial development, resource use regulations, global economic flows, and related human activities affect local environments and tend to magnify the impacts of weather and climate variations on indigenous and local livelihoods. Not only research but policy responses should also recognize climate change impacts within the broader context of rapid social, cultural, and economic change and, in their implementation, should underscore the reality that climate change is but one of several problems affecting people and their livelihoods in many parts of the world today.

In the reflexive context, when as anthropologists we engage in climate issues, we need to consider our response when our consultants' accounts and testimonies enter our story vis-à-vis one of the main drivers of climate change being Western consumer culture. It is imminent that we radically transform our consumer culture into a culture of environmental sustainability. Our awareness and actions demand a multisited analysis to expand the context from a "committed localism" and explore the complex interactions of the larger world system that shape the local (Marcus 1998, 83). With this in mind, how then can we translate our field experience of climate change into compelling global messages of what is to come as warming proceeds and as a means towards reevaluating consumer lifestyles and moving towards a carbon-free, sustainable society? Similarly, how can the research methods we develop in the process of working with our consultants inform potential research paradigms within our home communities to begin a similar transformation?

What's Being Written on Climate Change in the New Millennium[4]

The new millennium has also ushered in a steady stream of edited volumes and several monographs addressing the sociocultural, human-ecological, and ethical issues of climate change. Many of these publications interweave a variety of interdisciplinary fields, including anthropology, geography, environmental science, public policy, business studies, economics, communication, psychology, and the like to lay the groundwork for the vital multidisciplinary dialogue we must initiate and act upon to comprehensively address climate change.

McIntosh, Tainter, and McIntosh's (2000) volume assembles papers representing the broad geographical, cultural, and temporal coverage of global change in history and prehistory and focuses on the need to bridge the social and biophysical sciences—to build "a common language to appreciate the full history and prehistory embedded in human responses to climate". The authors emphasize understanding our species' symbolic past, retained in social memory, or "the long term communal understanding of landscape and biocultural dynamics that preserve pertinent experience and intergenerational transmission; the source of metaphors, symbols, legends and attitudes that crystallize social action." Analysis of social memory clarifies how communities curated and transmitted past environmental states and responded to them, and thereby can inform the present. The editors argue that economists and policy makers need to understand how social memory works because it

a) is relevant to small-scale subsistence producers who are most vulnerable to climate change but most often left out of macro policy decisions and their effects, and b) implicates cultural conservation to the extent that indigenous social memory is a great repository of human experience and therefore a vital resource for resilience and adaptation in our rapidly changing contemporary global context. The editors emphasize the need to tap this great reservoir of human experience for legitimate, appropriate, and economically and culturally sustainable responses to climate change.

In *Weather, Climate and Culture*, Strauss and Orlove (2003) integrate understandings of two camps within anthropology, the materially grounded ecologists and the meaning-centered symbolic anthropologists, to examine how humans think about and respond to meteorological phenomena. They structure their volume around two key issues of weather and climate: how humans experience weather and climate in differing time frames and how people use language in response to weather and climate, to emphasize how both human perceptions and reactions to these phenomena are ultimately shaped by culture. In context of the worldwide concern for climate change, the volume shows how time and talk are central to both how an understanding of unprecedented climate change is increasing and to the cultural patterns centered on language that underpin international climate policy negotiations. Contributors to the volume forefront the value of using the "anthropological eye and ear" to understand how societies perceive and act in response to climate perturbations.

There are also several noteworthy volumes that focus at a regional/ecosystem level. In *The Earth is Faster Now*, Krupnik and Jolly (2002) have assembled reports of long-term research and collaborative efforts from across northern Canada and Alaska to reflect how indigenous people are seeing changes and what they are saying about those changes. The editors intended the volume to both inform public policy, through its translation of local, place-based perceptions of change, and also to balance scientific research focused on futuristic models of what might happen with local experience that shows what is happening. Contributions describe projects that are in many ways exploratory by incorporating both new methods and tools for learning and sharing about data based in indigenous knowledge and collaborative forms of research that engage multiple stakeholders and go far beyond "scientific informing."

The volume *A Change in the Weather: Climate and Culture in Australia* explores the cultural space between weather and climate and how the very climatic feature that first attracted whites to Australia, that "the sun offered a new source of light and energy to escape the gloom of Britain," is reframed in the context of climate change to serve as a reminder to its people of global responsibilities (Sherratt, Griffiths, and Robin 2005). Even the flood and drought cycles so distinctive to Australian experience and identity are now shared with the rest of the world. The editors argue that this loss of identity and, in many ways, of innocence for the weather that once surrounded inhabitants in familiar and reassuring ways now serves as a bellwether for action from the local to the global.

Considering the prominent role that adaptation plays now and as climate change proceeds and the disparity in who it affects, Adger, Arnell, and Thompkins (2006) discuss comprehensively about the justice implications of climate change and how its unprecedented forces raise moral and ethical questions about vulnerability, inequality, fairness, and equity. Contributors emphasize that issues of justice are essential considerations in international negotiations where power relations between the rich and powerful and the marginalized desperately need to be brought into question and shifted. Fairness in adaptation is key in the developing world where the most vulnerable—the old, young, poor, and those dependent on climate-sensitive resources—can either benefit or become more vulnerable as a result of international action. This volume challenges us to think about the higher purpose to politics and law and the need to recognize and act upon the justice and equity issues of the causes of and responses to climate change.

Darkening Peaks: Glacier Retreat, Science and Society compiles essays addressing the nature, history, and consequences of the recent unprecedented retreat of the world's mountain glaciers (Orlove, Wiegandt, and Luckman 2008). The editors frame their topic using five main terms to capture each of the major aspects of glacier retreat: *perception*, *observation*, *trends*, *impacts*, and *responses*. By weaving these concepts, the authors show how human responses to glacier retreat are shaped as much by cultural attachment as by economic issues. The editors argue that, in the end, it is the cultural and iconic power of glaciers for humans who both inhabit their presence and who live afar that will be the force to motivate action on climate change.

In Creating a Climate for Change: Communicating Climate Change and Facilitating Social Change, Moser and Dilling (2007) forefront stories of success in communication and social action on climate change. They predicate the volume on the (to date) ineffectiveness of communication that has prevented the general public from understanding and taking action against climate change to bring to light effective modes of communication to quicken those processes. They argue that effective communication that mobilizes action must engage relevant social and cognitive characteristics. The volume distills the scholarship of both practitioners and an interdisciplinary research group to offer improvement on current communication strategies that empower individuals and communities to act in response to climate change.

Several recent anthropological monographs provide insights to our present climate change crisis. Archaeologist Arlene Miller Rosen (2007) gives an in-depth understanding of how different Terminal Pleistocene through Late Holocene communities were impacted by and responded to historical climate change. Her emphasis on working within the medium to small temporal scales of environmental change to understand the growth, development, and mutual relations of ancient societies of the Near East highlights the contemporary importance of the local scale to appreciate how, for example, small shifts in rainfall can affect people living without benefit of world market systems and international aid programs. She argues that societies do not interact directly with their environment but with their *perceptions* of that environment. The environment is only one of many actors in determining social change and plays a less important role than perceptions of nature. Accordingly, societies overcome environmental shifts in a diversity of ways and failure to do so signals a breakdown in one or more social and political subsystems. Success or failure is most often related to internal factors: social organization, technology, and the perception of environmental change.

In *Do Glaciers Listen: Local Knowledge, Colonial Encounters, and Social Imagination*, Julie Cruikshank (2005) draws some important lessons for our pursuit of defining anthropology's role in climate change. Specifically she emphasizes how glaciers, previously considered eternally frozen, largely inert, and safely distant, gain new meaning in the context of contemporary climate change concern. Glaciers of late are understood as a new endangered species as they melt and fail to reproduce themselves to serve as a "cryospheric weather vane for potential natural and social upheaval." Furthermore, Cruikshank stresses how glaciers undergoing rapid environmental change are interpreted differently. In the Andes, a long-held ritual practice involving pilgrims carrying pieces of glacier away has recently stopped to prevent the glacier's diminishing. In Asia, local men "plant" ice in opposite-gender ice fields in order to grow the glacier and prevent drought. And in Peru's Cordillera Blanca, campesinos take scientists' measuring devices because they believe those instruments are what are causing drought. In short, there are contrasting approaches to understanding climate change and they all deserve attention and consideration as we approach strategies and solutions.

In contrast to the above publications and others coming to and in press at this time, this present volume represents a novel attempt to understand what roles anthropologists have in response to climate change. The intent is to open dialogue among anthropologists to questions concerning the extent of our roles as advocates, communicators, educators, practitioners, and activists.

FROM ANTHROPOLOGICAL ENCOUNTERS TO ACTIONS: THE CHAPTERS

This edited volume comprises twenty-four chapters, organized into three sections to explore anthropology's role(s) in climate change issues and with the objective of locating theoretical frames, research approaches, and applied practices that can move us from impartial witnesses of our field collaborators' experiences of unprecedented environmental change into the realm of action-oriented researchers and effective academics and practitioners.

To these ends, the volume has three parts:

Part 1: Climate and Culture

This section lays the groundwork for the rest of the volume by establishing a) climate change's ultimate and direct interrelationship with human culture (historically and now); and b) anthropology's privileged position to investigate that relationship. Climate change is having impacts on culture, ways of life, spirituality, and in other arenas that are not "obvious." Anthropologists are finding evidence/effects of climate change in "unexpected" places. There are both theoretical/epistemological reasons (we see climate and culture/environment and society as inextricably linked, as opposed to the traditional natural science approach that does not deal with humans), and methodological reasons (ethnography, participant observation, collaborative research, community-based research, decolonizing methodologies, etc). This section paints a vivid picture of not only what we know, but also how/why anthropologists are witnessing dimensions of climate change that other researchers have not seen or are not seeing, or just have not prioritized.

In Chapter 1 Fekri Hassan looks at climate change and culture, emphasizing the importance of human agency from an archaeological perspective. While, he argues, we have some considerable knowledge of climate change in historical times, we lack a clear understanding of its scale and its magnitude, as well as its frequency. Nor are we certain of the causes of particular paleoclimatic events. And, significantly, we cannot say with any degree of certainty what kind of impact climate change has had on human history. Without this clear understanding of past impacts, how can we claim to devise reliable scenarios for the future? While this points to lack of specific paleoclimatical data, it also highlights the inadequacies of current methodologies and interpretation. Hassan cautions us to avoid simplistic notions of determinacy and indeterminacy, arguing that the impact of any climatic event depends on the local social and ecological setting. He advocates that it is only through long-term archaeological and historical analysis, as well as detailed examination of local and regional social dynamics that we can begin to pinpoint the differential impact of the same climatic event. Hassan reinforces the argument that we need to be attentive to considerations of human agency and different responses and strategies in any one particular locality, as well as to local and regional scales, organizational complexity, ideology, technology, and social and cultural values of local populations within an interregional context.

In Chapter 2 Nicole Petersen and Kenneth Broad argue that, based on a deeper understanding of how climate influences, shapes, and organizes our lives, we are moving to organize this understanding in epistemologically complex ways. They survey the shifting terrain of research on weather and climate in anthropology, examine some of the narratives surrounding climate and climate change in anthropology, and provide a context for current climate anthropology. Their chapter considers the shift in theoretical and practical interests for anthropologists concerned with climate and climate change research. While earlier interests may have focused anthropological attention on local weather, seasonal variations, extreme events, and local cosmologies, current discussions about climate and weather now differ because of the attention given to both the global nature of anthropogenic climate change and the production and distribution of forms of scientific knowledge. Anthropologists may have previously regarded a discussion of climate to be necessary to frame their narratives of local ethnographic specificity, but a reaction to environmental determinism as a driver of cultural variation made some tread carefully when even thinking about the significance of climate for the people and environments they studied. Current anthropological engagement with climate and weather can perhaps be traced to the emergence of several areas of research interest, including work dealing with the social and ecological aspects of disasters, research on vulnerability and adaptation, ethnoecology, the idea of a global climate and its implications for how anthropologists carry out place-based research, uncertainty in climate modeling and forecasting, the emergence of climate as a global issue that influences shifting ideological positions, and climate change as an area of human rights and social justice.

In Chapter 3, Carla Roncoli, Todd Crane, and Ben Orlove present a robust overview of anthropology's last decade of active involvement in climate change research to highlight both the field's recent surge of interest and projects and the field's effective epistemological and methodological approaches. They begin their exploration by showing how "being there" in the context of ethnographic fieldwork provides a distinctive lens into the dynamics of climate and culture. The chapter next explores the four levels of interpretations relevant to climate-culture interactions, including perception, knowledge, valuation, and response. The authors further encourage anthropologists to enlist a diversity of intellectual traditions to interpret and illuminate how local communities are perceiving, understanding, valuing, and responding in order to be more effective in local to global scales, as facilitators of community adaptation, and as an authoritative voice in climate change debates. This advice also comes with an important warning not to conform to value systems or compromise on our disciplines's core ideals of cultural sensitivity and social equity, in the process of reaching out into other disciplinary and institutional contexts. In conclusion the authors emphasize that the climate research and policy communities are in fact recognizing anthropology's critical role in moving forward both the policy setting and the adaptive capacity–building agendas needed to forestall climate change—now it is a matter of how anthropologists step out to meet that call.

In Chapter 4 Anthony Oliver-Smith provides a detailed analysis of anthropology's long history of and privileged role in working with human migration in response to both short- and long-term environmental change. He reminds us that throughout history humans have adapted to environmental change, and one major way they have done so is through migration, be it permanent or temporary. In the recent period, we are observing what were once temporary migrations turning into permanent ones due to both the forces of unprecedented climate change and inadequate government response. Not only do displacements due to climate and governments exacerbate local, regional, and national issues but they are also a prime cause for cross-border and international conflicts. He emphasizes the critical role of anthropologists, as climate change increasingly uproots large numbers of people, being that of action-oriented interpreters who can bring to light the complex nature of local contexts. Anthropologists must recognize that displaced communities all need to mobilize social and cultural resources to reestablish viable social groups that underpin human well-being and material life. In this era of increasingly displaced communities, Oliver-Smith calls on anthropologists to work with other social scientists to develop sound theory and methods in response.

Part 2: Anthropological Encounters

This section includes eleven case studies that work to illustrate climate change's direct interrelationship with human culture and anthropology's privileged position to investigate that relationship. Authors tell what they are encountering, how the people they are studying are responding, and how their subsistence, culture, spiritual orientation, and the like are being affected. Cases are drawn across global latitudes, longitudes, and altitudes from the Arctic, Africa, South America, the Pacific Islands, Australia, North America, and Europe. Each author discusses how they became involved in the study of climate with their field consultants, what impact they are having and with what outcomes.

In Chapter 5 Susan Crate discusses the significance of climate change for the Turkic-speaking Sakha horse and cattle breeders of northeastern Siberia. Sakha personify winter in the form of the white Bull of Winter, and stories told about it seemingly offer explanatory accounts of the temperature events of the extreme sub-Arctic winters that characterize this part of the Russian Federation. Yet as Crate argues, as they reflect on their observations of climate change, some Sakha are pondering whether the story may well become one about something that used to be, rather than an account of what is. Crate reports on local observations of change that have compelling similarities to what anthropologists working elsewhere in the circumpolar North also say people are seeing. Noticeable amongst these changes are warmer winters with more snow, and cooler summers with more rain. One of the implications of this seasonal change, argues Crate, is that people say they can no longer read the weather. And worryingly for the Sakha, food harvests are being affected and horses and cattle are facing greater difficulty finding fodder. Crate teases out local understandings of the causes of climate change, showing how Sakha attempt to relate their perceptions of natural variability and change with their perceptions of anthropogenic influences, including technological change in Soviet and post-Soviet Siberia. In her discussion Crate focuses on climate change as being ultimately about culture. Its root causes, she argues, are in the multiple drivers of global consumer culture; its impacts are evident in the way it transforms subsistence cultures (such as that of the Sakha); and it can only be forestalled, she suggests, via a cultural transformation from degenerative to regenerative consumer behavior. For Crate, anthropologists are strategically well placed to interpret, facilitate, translate, communicate, and advocate (both in the field and at home) in response to conditions giving rise to climate change, to the cultural implications facing communities as they cope with change, and for the actions needed in response to climate change.

In Chapter 6 Anne Henshaw takes us to Inuit communities on southwest Baffin Island, Nunavut, Canada, with whom she has worked on a long-term environmental knowledge project for the last ten years, to understand how climate change can be both a threat and a welcome opportunity for place-based peoples. The chapter focuses on how sea ice, which dominates the Inuit seascape for most of the year and is a critical element of Inuit travel and hunting, has become a critical barometer of environmental change especially in terms of recent climate change. She explores this dynamic in both local and global perspectives, through the careful analysis of interviews with community members and historical records. On a local level the loss of sea ice both poses threats to travel and hunting and brings new subsistence resources. Globally, Inuit have been outspoken about how the rapid loss of sea ice and other unprecedented changes represent threats to their basic human rights. Henshaw concludes that anthropologists have a role in the climate change issue, but that often indigenous groups are their own best advocates. Anthropologists are better suited to facilitate collaborative, community-based projects that can work to mediate the complex and rapidly changing social and political environment in which climate change is taking place.

In Chapter 7 Sarah Strauss considers climate change impacts as they affect a village and its locality in the Swiss Alps. Building on her work on understanding the social lives of water and the weather, Strauss broadened the scope of her study of one local area to include consideration of climate variability and modeling in the wider context of the Swiss Alps. Glacier stories are of particular importance to her work—glaciers have historical social, cultural, and economic importance for people of the Swiss Alps, providing water for drinking and power generation, as well as ice before refrigeration. They continue to be powerful attractions for tourists, and they act as markers of environmental change. Strauss also describes how glaciers have local significance as repositories of lost souls. Siren tales are told of entrapped souls that call out to the living, reminding people to stay close to home and not wander too far into the mountains. The cautionary nature of these tales reveals local perspectives on the dangerous nature of glaciers, as well as the importance of following local religious traditions. While glaciers have been an integral part of culture and social life in the Alps, Strauss discusses how concern over their retreat and demise is forcing people to think about the implications of an ice-free future. Local people are considering what may happen if their glaciers disappear. While some are not too perturbed about a glacier-free scenario, others are anxious about water supplies and electricity generation, a decline in tourism and its consequences for the local economy, and hazards from flooding and avalanches. While locally specific, these concerns highlight issues that also have implications for people who live far beyond this narrow Alpine valley.

Tim Finan's aim in Chapter 8 is to consider what the role of anthropology is—and indeed can be—in understanding what he calls the twisting trail of the impacts of climate change. While acknowledging the sophistication of the science that informs climate change reports and impact assessments, most notably the IPCC reports, he nonetheless criticizes these reports for failing to address the impacts on human systems. The IPCC exemplifies the scientific tradition of portraying climate change without a human face. Anthropology, he argues, provides an appropriate lens that allows a sharper focus on how best to assess adaptation, vulnerability, and resilience, and provides both the theoretical concepts and the methodologies needed to shift the focus to the dynamic interface of natural and human systems under change. Finan draws on his work on shrimp aquaculture livelihoods and sea-level rise in coastal Bangladesh to illustrate his argument that anthropology can assemble the necessary toolkit to understand how changes in the natural system will revise current terms of engagement at the level of communities and households. In Bangladesh, moderate flooding is, from a livelihood perspective, important for fertility and the replenishment of freshwater fish stocks. Excessive flooding, however, is catastrophic, causing death and disruption to livelihoods. Finan argues that the natural and human systems have negotiated an uneasy balance. People's livelihoods are dependent upon the annual renewal of the resource base, yet extreme events can be devastating. Finan shows how livelihood stress introduced by global warming is manifest in severe cyclonic events, storm surges, and excessive flooding. However, the vulnerability of coastal livelihoods in this deltaic system is not determined by climate change alone, but by a complex interplay of environmental, political, social, and economic systems distributed across a vulnerable landscape.

Benedict Colombi's concern in Chapter 9 is with the cultural history of human interactions with the Columbia River basin in the northwest United States. He discusses Nez Perce relations with the environment, particularly the significance of salmon for cultural and ecological sustainability. Fish and water are of central importance to Nez Perce social and cultural life, with salmon bringing the energy of the ocean inland to plants, animals, and people. Without the salmon, say local people, the river would die. Salmon are a keystone species, argues Colombi, but also fundamental to the ideological and material foundations of Nez Perce culture. Within the discourse of climate change, they are also cast in the role as an indicator species of the health of the Columbia River and its ecosystems. Colombi discusses local concern over what the effects of climate change on ecosystem regimes will mean to cultural and subsistence rights. Climate change is not only perceived as a threat to salmon—and thus to local and regional biodiversity—but also to indigenous rights. The effects of climate change already experienced, and the worries about what may happen, are prompting local discussion about adaptive strategies and the building of sustainable indigenous economies within a context of the intricate relations between people, environment, and salmon. But while it is feared that climate change will contribute to the transformation of people's relationships to place, Colombi draws attention to the need to understand this with reference to two centuries of social and ecological change brought about by settlement, commercial growth, urban and agricultural development, and the impacts that timber cutting, grazing, fire suppression, and hydroelectric development have had on the ecosystems of the northwest.

In Chapter 10 Jerry Jacka documents the effects of climate change for Porgerans, several thousand Ipili and Enga speakers in the western central highlands of Papua New Guinea. The author illuminates the importance of considering medium-term climatic trends, in this case more frequent El Niño events, and how local peoples perceive how these trends affect the moral, agricultural, environmental, and cosmological dimensions of their livelihoods. Through his careful anthropological analysis of development, migration, and land cover change, Jacka shows how the last twenty years of gold mining development, which has brought increases in population and deforestation to the Porgeran homelands, has affected indigenous subsistence practices and, perhaps more profoundly, his consultants' cosmological orientations. Daily the Porgera people comment about how the physical manifestations of environmental change are signs of the world's end, an understanding based upon an enduring cosmological belief system. Through his analysis, Jacka clarifies the importance of taking into account local perceptions and understandings of what the local effects of climate change mean, in this case, the "societal breakdown between native Porgerans and the rituals oriented toward more powerful spirits that control the cosmos."

Beth Marino and Peter Schweitzer explore in Chapter 11 how anthropologists conducting climate change research can either make or break their investigation depending on whether they use the term *climate change* explicitly in their field inquiry or not. Working in five Inupiaq villages of northwestern Alaska, the authors show how the use of the phrase *climate change* alters local-level patterns of speech. In part this phenomenon is a result of the world's overdocumentation of and myopic focus on climate change centering in the Arctic, and especially in Alaska and Canada. As a result of this overemphasis, not only do most global inhabitants directly identify climate change as something going on only in the Arctic, but the Alaskan and Canadian Arctic is inundated with what Marino and Schweitzer refer to as the "photo snatchers," who, in addition to snatching the world's images of climate change, also leave behind certain specific ideas and concepts about the global process for the local inhabitants. The authors found that when they used the term *climate change*, consultants responded by giving summaries of information they had gleaned from scientific and other media outlets. However, when they asked about change in the context of the local environment, consultants shared their personal experiences based on daily and seasonal activities. Marino and Schweitzer conclude their piece by suggesting that perhaps the anthropological investigation of climate change will proceed much farther if "we stop talking about it."

In Chapter 12 Donna Green documents how historically highly adaptive and resilient cultures are often rendered highly vulnerable in the face of climate change due to the interplay of an accelerated rate of environmental change and profound government mismanagement and neglect. Working with Torres Straits Islanders, living between mainland Australia and Papua New Guinea, who are seeing the most immediate effects of climate change due to sea-level rise and inundation of their islands, Green explores the extent to which traditional environmental knowledge (TEK) can both guide appropriate local-level adaptation strategies and provide needed historical observations for climate scientists. The author discusses how Torres Straits Islanders are adapting by building houses on stilts and away from lower areas—but that it seems that eventually relocation will be inevitable. She emphasizes the need to fund more projects focusing on TEK and climate change because, based on her experience, when people have information and see change occurring around them, they act. Although the government funded the initial studies, they are not forthcoming with more funds to continue the work.

Inge Bolin takes us to the high Peruvian Andes to the rapidly changing world of the Quechua people in Chapter 13. She reveals how highly adaptive and resourceful people are unequivocally challenged by the impacts of climate change, in this case, due to the unprecedented retreat of glaciers that supply all their drinking and agricultural water. Issues of water scarcity are not new to the Quechua but are an issue that humans of the Andean Cordilleras have found ways to adapt to since the time of the first inhabitants. Myths, legends, and both historical and contemporary spiritual practices show this continuity. Bolin explores how water scarcity due to the rapid melting of mountain glaciers affects not only contemporary Quechua physical survival but also their spiritual practices. She describes how locals reinstate ancient Inca water-conservation practices to some success. These include a) the Inca practice of terraced gardens, both preexisting and newly built, to prevent water run-off and erosion; b) the pre-Columbian practice of conservation tillage, which also lessens water loss; and c) the revival of ancient Andean subterranean water channels to transport water to dry areas. Despite these and other local and regional efforts, Bolin argues that much more must be done and largely on an international level to ensure that the Quechua, and others who make up the one-sixth of the world's population that relies on glaciers and seasonal snow pack for their water supply, will be able to continue to inhabit, much less survive, in their ancestral lands.

Heather Lazrus draws on rich ethnographic material from her research in Tuvalu and illustrates in Chapter 14 how people perceive, understand, interpret, and cope with changes they notice and experience in their South Pacific island homes. Lazrus argues that while Tuvaluans draw on locally based traditional knowledge to understand atmospheric and climatic disturbances, their understandings and responses to climate change are, despite their apparent remote geographical location, nonetheless also informed and enriched by different forms of knowledge derived from their participation in broader transnational networks. Local knowledge about ocean currents, wind, and precipitation intersects with scientific and universal ways of knowing about anthropogenic influences on the atmosphere. The scientific consensus is that climate change will dramatically transform this small South Pacific country, a mix of atolls and table reef islands. Local concern is with sea-level rise, increases in sea surface and subsurface temperatures, acidification of the ocean and coral bleaching, coastal erosion, an increased intensity (but decreased frequency) of rainfall, and increased frequency of extreme weather events including drought. Lazrus places the discussion of climate change in Tuvalu within a context of anthropological engagement with global debate about the avoidance and amelioration of problems associated with it. She identifies three areas in which anthropology can play a role in this debate: by contributing to the understanding of how impacts of climate change are constructed, how agency is retained, and how governance can promote autonomy and sovereignty as necessary for dealing with climate change and its impacts.

Based on his extensive work in the Kalahari Desert, Robert Hitchcock discusses in Chapter 15 how the San of Botswana have slipped further into poverty as a result of economic and environmental changes that have occurred over time in southern Africa. To meet the challenges arising from these changes, San have employed a range of often ingenious strategies, which include diversifying their subsistence bases, utilizing water control techniques, and depending on other groups or, in some cases, governments and international agencies for food and support. The periodic but all-too-frequent experience of hunger and privation arising from climatic events, the infestation of pests, and the outbreak of diseases such as malaria demonstrates the vulnerabilities to which San livelihoods are exposed in this vast "sand-sea" ecosystem. Nearly all the San with whom Hitchcock and his colleagues have worked over the years recognize that environmental change is occurring. For some, environmental change is nothing new, but its character is now different than many remember. While some San report that environmental and socioeconomic conditions are getting worse in the Kalahari, these changes are exacerbated by the fact that the options San have to deal with them are increasingly limited, and that the strategies traditionally employed to cope with environmental stress are no longer effective. Hitchcock's chapter reveals a diversity of understandings and explanations that San have for why the environment is changing. Most San seem to agree that they are witnesses to seasonal shifts and uncertainty in the environment, yet not all attribute this to climate warming and some argue that the local areas in which they live are actually getting cooler. Yet all agree that the changes are affecting their well-being and that the biggest challenge they face is environmental variability and risk.

Part 3: Anthropological Actions

This section includes nine chapters exploring what anthropologists are doing and can/need to do both in their field research context, in the larger anthropology community, and to reach out to wider audiences from the local to the global. The authors explore various research approaches, conceptual dilemmas, practical actions, and professional orientations that anthropologists can take, including appropriating field research agendas in order to effectively address climate change, finding and using effective methods of communicating at home and abroad about the issues and necessary actions to forestall climate change, actions to transform consumer culture, engaging campuses in sustainability initiatives, building decentralized resilient communities, affecting policy, working with and within business, and the like. Granted, these themes overlap, as they rightly should. Perhaps the biggest challenge is how we can act to make change, and how to take on so big a change.

Richard Wilk contributes a thought-provoking piece questioning anthropology's role in interpreting and addressing the issue of consumer culture as it pertains to climate change. In Chapter 16 he begins by critiquing the commonly accepted perspective that Western consumers just need to "tighten up their belts" to effectively counteract climate change. His argument lies in the fact that, contrary to popular belief, the majority of greenhouse gas emissions are not due to individual consumption but rather to pivotal political and historical decisions. In the end, they are rooted in the variety of "cultural ideals about justice, comfort, needs, and the future." In this light, anthropology, as a discipline that has always considered these factors in its research program, is uniquely situated. In the same breath the author critiques anthropologists' lack of engagement with other anthropologists, unions that Wilk claims could be pivotal in developing a comprehensive and sophisticated theory of consumer culture, useful to policy makers and laypersons alike. Wilk concludes by encouraging anthropologists to play a more proactive role in the work of creating a more sustainable economy by first and foremost creating new models of comparison and collaboration so that the whole truly may be more than the sum of the parts.

Shirley Fiske's aim in Chapter 17 is to explore the opportunities for how anthropologists can engage with the policy concepts in US national legislation, as well as internationally, that are influencing and shaping the implementation of strategies for global carbon markets and emissions trading. She discusses policy debates in the US Congress, highlights policy questions, and provides examples of how anthropologists have analyzed the link between local communities and policy instruments, such as carbon offsets and sequestration. She draws on unique experience of climate change policy gained from her positions as observer and participant in its discussion and development in legislative and executive branches of the US government. She describes her own roles, but also considers the changing nature of dialogue about climate change and various attempts to make it a legislative priority. Her experience has given her fascinating insight into the shifting terrain of the negotiation of policy questions about mitigating and controlling carbon emissions. Here is fertile ground for the anthropologist, she argues, and she urges us to take a closer look at domestic policy debates, raise questions about the process, comment on legislative provisions, and focus on (and deconstruct) the language and careful phrasing of policy concepts, such as cap and trade. The domestic discourse in the US, Fiske argues, revolves around reducing emissions, cap and trade, carbon taxes, credit allocation, carbon reduction technologies, and so on. Framed and viewed as a problem of and for technology, rarely is it discussed as something that has consequences for people, families, and communities.

In Chapter 18 Mark Nuttall drives home the importance of grappling with the interplay of climatic, social, economic, and political factors when anthropologists work in indigenous communities confronting climate change, or how the "regional texture to climate change means [that] changing environments are perceived and experienced differently." Based on his long-term research in Greenland, Nuttall illustrates just how climate change means different things for different people. In the context of Greenland's Home Rule government, politicians see climate change as an opportunity for mining and hydrocarbon development that for them translates to greater political and economic independence from Denmark. Conversely, Greenlandic peoples interpret climate change not as a change to some environment outside themselves but as a change to their personhood, illustrating how climate and culture in many indigenous worldviews are inextricably linked. Although these same communities have a historical precedence for adaptation to environmental change, Greenlanders' capacity to adapt to change is highly dependent on the strength of their sense of community, kinship, and close social associations. Similarly, the last one hundred years of changes in marine resources, both from climate change and the politics of resource rights, have affected the Greenlandic fishing industry and translated into varying degrees of social and economic consequences for communities. Nuttall concludes by emphasizing that communities' ability to adapt to climate change has more to do with issues of autonomy—their capacity to make decisions on their own and to continue to exercise their ancestral way of *becoming* with the environment around them in order to adapt and be flexible in coping with climate variability and change.

In Chapter 19 Pamela Puntenney helps us to focus on the complexities and complications inherent in developing and implementing the multilateral agreements and other policies needed to tackle the climate change issue in the global policy arena. The author argues that the three most serious threats to our world ecosystem—climate change, loss of biodiversity, and urbanization—can only be addressed through the creation of comprehensive agreements founded on principles of sustainability, equity, and justice. She discusses the growing concern and sense of urgency to bring the human-cultural dimensions of climate change to the policy forefront and then turns to how social scientists can affect decision-making. The most vulnerable countries need action now. She reminds us that with our anthropological training comes the irrevocable responsibility to assume our "anthropological footprint" and engage ourselves in discussions and directives on international policy making, debates, policies, decision-making, the building of sustainable enterprises, and plans of action, each of which is increasingly focusing on how issues of culture are central.

Gregory Button and Kristina Peterson discuss, in Chapter 20, the efforts of the people of the community of Grand Bayou in Louisiana to forge collaborative partnerships with social scientists and physical scientists in the wake of Hurricane Katrina. Their analysis focuses on a participatory research project and the importance of such an approach for the validation of lay knowledge, a necessary prerequisite for the synthesis of local ways of knowing with science. This, they demonstrate, not only increased the community's understanding of environmental disasters and climate change, but also contributed to local people's empowerment and enhanced community approaches to adaptation to dramatic environmental change. While Button and Peterson discuss the importance of understanding the environmental setting of Grand Bayou as a way of assessing vulnerability to environmental disasters, they draw attention to the sociopolitical and economic relations that also play a part in making a community vulnerable to risk. Grand Bayou residents have been witness to the dramatic topographical transformation of their locality over the past few decades, a transformation attributable in part to the digging of the Mississippi Gulf outlet, increased coastal erosion resulting from clear-cutting of cypress, the building of transportation canals for the petrochemical industry, and oil and gas development. Environmental change and socioeconomic and geographical marginalization not only exposed people to the impact of Hurricane Katrina—it also hid them from the sight of federal officials offering aid for debris cleanup afterwards. The essence of Button and Peterson's chapter is to show that vulnerability to disasters can be reduced if communities are empowered, and if they are including in the planning process for disaster preparedness, response, and mitigation.

In Chapter 21 Noel Broadbent and Patrik Lantto discuss climate change from an Arctic cultural perspective. The Saami people of Sweden are facing similar dilemmas as other indigenous peoples not just elsewhere in the circumpolar North, but elsewhere globally. However, Broadbent and Lantto argue that the Saami are boxed in by national narratives and welfare-state policies that have redefined their rights and identities. Among the problems they face is the fact that the Swedish government has not recognized them as indigenous people in accordance with United Nations policies. In recent court cases Saami land-use rights have been challenged and lost. The Swedish nation-building process incorporated myths about the origin and the nature of both Swedish and Saami identities, and these cultural and political myths still influence environmental and cultural policies. Archaeological evidence is nevertheless opening the door to new interpretations regarding Saami territories and identities, and yet recent government policy reports do not allow archaeological evidence to be used in court cases. Broadbent and Lantto discuss the role of anthropology in helping to contextualize contrasting narratives and unraveling the complex webs of discourse emerging from them. At the national level, they argue, historical narratives and social policies can either enhance or limit societal responses to climate change, particularly when indigenous minorities are involved. Based on this, they conclude that the Saami may well survive as fully acculturated Swedish citizens but face major political obstacles to sustaining land-use practices and traditions that make them part of the indigenous world.

In Chapter 22 Peggy Barlett and Benjamin Stewart ask how anthropologists employed in institutions of higher education can contribute effectively to educating about climate change. Their interest, however, is not just with making students aware of climate change and its causes and consequences, but with expanding awareness and galvanizing action within the life and culture of the institution. They draw on their work concerned with institutional change and sustainability and focus on efforts at Emory University to promote cultural change on campus as part of climate change teaching strategies. Their starting point for discussion is that, while institutions of higher education have a major role in contributing expertise for climate change research, debate, and policy, colleges and universities are huge operations that also leave their own large carbon footprints. Higher education institutions, and the people who study and work in them, are consumers of energy. Even an academic who favors the use of chalk instead of high-tech equipment in classrooms is still nonetheless implicated in transnational networks of production and consumption. Barlett and Stewart discuss strategies for teaching and educating students about climate change, but also about human agency and engagement with the local, regional, and global challenges associated with it. But they also stress the importance of emphasizing the personal moral dimensions of climate change. How much do we care, they ask, about biodiversity and polar bears? Do we care enough to sacrifice the daily comforts of heating or air conditioning? The critical rethinking of priorities and values—those we hold as individuals, but also those of institutions—is a fundamentally important response to climate change.

In Chapter 23 Lenora Bohren explores anthropology's role in "car culture," based upon her own experience as an anthropologist working as director of the National Center for Vehicle Emissions Control and Safety (NCVECS) at the University of Colorado. She first chronicles the late-nineteenth-century development of the car, with its beginnings in Europe and its standing as an expensive proposition and mainly a status symbol and high-ticket item for the wealthy. It was in the twentieth century, when Ford designed a car for the average American, that the industry took off. Despite the fact that the rise of the car lessened the issues of waste and housing for horses, it also brought with it a myriad of resulting environmental and cultural problems. These included the need for proper roads, traffic congestion, exhaust problems, disruption of community life, and the increased costs to car owners for insurance, fuel, repair, maintenance, and health issues. These issues prompted a series of emissions regulations in the US beginning in the late 1960s and up to the 2007 Executive Order directing the EPA to develop regulations for CO_2 emissions. Bohren next illustrates how cars are a cultural phenomenon, represented and understood differently in different countries. It is to this extent that anthropology has a role. By highlighting the interface between culture, technology, and environment, anthropologists can facilitate culturally appropriate and long lasting change.

In Chapter 24 Nicole Stuckenberger discusses her work curating the museum exhibition Thin Ice at Dartmouth College's Hood Museum, focusing on communicating the diversity of local understandings and perceptions of climate change to an audience oriented to a largely Western scientific knowledge base. Because the development of effective policies depends upon the engagement of a multiplicity of stakeholders, including local residents, regional and national governments, scientists, policymakers, and indigenous groups, the exhibit serves to underscore the importance of cultural perceptions and understandings by exploring past and present Inuit perceptions of environmental change. In order to convey the holistic Inuit perspective to museum visitors, the exhibit avoids displaying each domain of life in isolation but rather integrates various elements via the use of the indigenous teaching method of storytelling. Stuckenberger describes the challenges of designing an exhibit to communicate indigenous ways of knowing and worldviews to a mostly Western audience. In the end the author emphasized the important role of anthropologists in cultivating this and other visual and experiential perspectives on climate change in order to enrich and inform larger-scale perspectives and reach a greater public.

Together these three parts comprise a whole that is by no means to be considered a final assessment of the state of anthropology and climate change at the beginning of the twenty-first century, but rather a preliminary reading of a novel, rapidly expanding, and crucial dialectic and an invitation for increased dialogue and collaboration. From the first roots of this book project, in Susan Crate's conversations with Jennifer Collier at the 2006 Society for Applied Anthropology meetings, through several double paper panels and policy forums at SfAA and AAA annual meetings on the topic, it is clear that anthropology has a multitude of roles to play and offers a privileged set of ways of knowing, keys to understanding, and avenues of action related to our affinity for issues of culture. It is also clear that we have much to learn and explore across the scope of this volume, from encounters to actions.

Acknowledgments

As editors, our first acknowledgements go to all the people with whom we have worked as anthropologists in the field, especially the people of the Viliui regions of western Sakha, Russia, for it was in the context of Susan's work encountering the changes to their physical and cultural worlds that the seed for this book was planted. Thanks go to the National Science Foundation (NSF) and Anna Kertulla de Echave, Office of Polar Programs, Arctic Social Sciences Division for funding support. Tremendous thanks go to Jennifer Collier, our editor at Left Coast Press, for her unequivocal enthusiasm for the book idea from its inception in early 2006 and her steadfast dedication to making it a reality. The book also would not have been possible without all the panelists who participated in panels and policy forums at the 2007 American Anthropological Association and 2007 Society for Applied Anthropology meetings based on the book theme and organized by Susan—including most of the authors of the book and others who could not contribute to this edition. In the same breath, we also thank each of the thirty contributors to this volume. We also thank Barbara Rose Johnston, Luke Eric Lassiter, Bonnie McCay, Gail Osherenko, Michael Sheridan, Bruce Winterhalder, and Oran Young. Susan lovingly dedicates this book to her daughter, Kathryn Tuyaara Yegorov-Crate, and Mark to his son, Rohan Nuttall, and we only hope it contributes in some way to making their world, and that of all the generations to come, a humanly habitable ecosphere.

Notes

1. We are using the term *climate change* throughout this introduction to refer to the contemporary phenomenon of anthropogenic global climate change, as distinct from natural climate variability. Whereas *climate variability* indicates changes in climatic conditions that scientists consider to be due to natural mechanisms and processes, and as such are entirely unrelated to human activities, climate change is defined as a variation in climatic parameters and is attributed directly or indirectly to human activities (Lange 2005, 365). Such variations in climatic parameters occur in addition to or despite natural climate variability. *Climate change* and *global warming* (the enhanced greenhouse effect and the trend in the increase of mean global surface temperatures) are often used interchangeably, particularly by the media, resulting in a confused understanding of the terms, but climate change can result in the cooling as well as the warming of the earth's near-surface atmosphere.
2. We take poetic license here by saying that "the environment moves." It works well within the analogy. We fully acknowledge that the environment cannot move but that it changes.
3. http://www.tebtebba.org/tebtebba_files/susdev/cc_energy/buenosaires.html. For the original indigenous declaration on climate change, see http://www.treatycouncil.org/new_page_5211.htm (accessed June 1, 2006).
4. By the time this book is in the readers hands, there will invariably be as many new volumes and manuscripts on the topic.

References

Adger, W. N., N. Arnell, and E. Thompkins. 2005. Adapting to climate change: Perspectives across scales. *Global Environmental Change* 15: 75–76.

Aguilar, L. 2008. Acknowledging the linkages gender and climate change. Paper presented at World Bank Workshop, "The Social Dimensions of Climate Change," March 5–6, 2008, Washington DC. http://siteresources.worldbank.org/EXTSOCIALDEVELOPMENT/Resources/244362–1170428243464/3408356–1170428261889/34083591202746084138/ Gender_Presentation022808.pdf

ACIA. 2005 *Arctic Climate Impact Assessment: Scientific report*. Cambridge: Cambridge University Press.

Barnett, J. 2003. Security and climate change. *Global Environmental Change*, 13(1): 7–17.

Basso, K. 1996. Wisdom sits in places: Notes on a western Apache landscape. In *Senses of Place*, eds. K. Basso and S. Feld, 53–90. Santa Fe, NM: School of American Research Press.

Castile, G. P. and R. L. Bee, eds. 1992. *State and reservation: New perspectives on federal Indian policy*. Tucson: University of Arizona Press.

Crate, S. 2008. Gone the bull of winter? Grappling with the cultural implications of and anthropology's role(s) in global climate change. *Current Anthropology. In press.*

Cruikshank, J. 2005. *Do glaciers listen? Local knowledge, colonial encounters, and social imagination*. Vancouver: UBC Press.

ICC. 2005. Inuit petition inter-American commission on human rights to oppose climate change caused by the United States of America. http://www.inuitcircumpolar.com/index.php?ID=316&Lang=En (accessed April 10, 2008).

IPCC. 2007. *Climate change 2007: Impacts, adaptation and vulnerability. Working Group II Summary for Policymakers*. Geneva: IPCC Secretariat. http://www.gtp89.dial.pipex.com/spm.pdf (accessed April 10, 2008).

Krupnik, Igor and Dyanna Jolly, eds. 2002. *The earth is faster now: Indigenous observations of Arctic environment change*. Frontiers in Polar Science. Arctic Research

Consortium of the United States/Smithsonian Institution—Arctic Studies Center. Arctic Research Consortium of the United States, Fairbanks, Alaska, USA.

Lange, M. A. 2005. Climate change. In *Encyclopedia of the Arctic*, ed. Mark Nuttall, 365–73. New York and London: Routledge.

Lassiter, L. E. 2005. Collaborative ethnography and public anthropology. *Current Anthropology*, 46(1): 83–107.

Marcus, G. 1998. *Ethnography through thick and thin*. Princeton, NJ: Princeton University Press.

McIntosh, R., J. Tainter, and S. Keech McIntosh, eds. 2000. *The way the wind blows: Climate, history, and human action*. New York: Columbia University Press.

Moser, S. and L. Dilling, eds. 2007. *Creating a climate for change: Communicating climate change and facilitating social change*. Cambridge: Cambridge University Press.

Netting, R. M. 1968. *Hill farmers of Nigeria: Cultural ecology of the Kofyar of the Jos Plateau*. Seattle: University of Washington Press.

———. 1993. *Smallholders, householders: Farm families and the ecology of intensive, sustainable agriculture*. Stanford, CA: Stanford University Press.

Nuttall, M., F. Berkes, B. Forbes, G. Kofinas, T. Vlassova, and G. Wenzel. 2005. Hunting, herding, fishing and gathering: Indigenous peoples and renewable resource use in the Arctic. In ACIA, *Arctic Climate Impact Assessment: scientific report*, 649–90. Cambridge: Cambridge University Press.

Orlove, B., E. Wiegandt, and B. Luckman, eds. 2008. *Darkening peaks: Glacier retreat, science and society*. Berkeley: University of California Press.

Prucha, F. P. 1985. *The Indians in American society: From the Revolutionary War to the present*. Berkeley: University of California Press.

Röhr, U. 2006. Gender and climate change. *Tiempo* 59(April): 3–7. http://www.tiempocyberclimate.org/portal/archive/pdf/tiempo59high.pdf.

Rosen, A. M. 2007. *Civilizing climate: Social responses to climate change in the ancient near East*. Walnut Creek: AltaMira Press.

Sherratt, T., T. Griffiths, and L. Robin. 2005. *A Change in the weather: Climate and culture in Australia*. Canberra: National Museum of Australia Press.

Steward, J. H. 1955. *Theory of culture change*. Urbana: University of Illinois Press.

Strauss, S. and B. Orlove, eds. 2003. *Weather, climate, culture*. New York: Berg Publishers.

Thomas, D. and C. Twyman. 2005. Equity and justice in climate change adaptation amongst natural-resource-dependent societies. *Global Environmental Change* 15: 115–24.

Vitebsky, P. 2006. Reply in Letters. *Natural History* 115(2): 10.

Watt-Cloutier, Sheila. 2004. "Climate change and human rights" in Carnegie Institute's *Human Rights Dialogue* special issue on "Environmental rights" http://www.cceia.org/viewMedia.php/prmTemplateID/8/prmID/4445

White, R. 1983. *The roots of dependency: Subsistence, environment, and social change among the Choctaws, Pawnees, and Navajos*. Lincoln: University of Nebraska Press.

Wisner, B., M. Fordham, I. Kelman, B. R. Johnston, D. Simon, A. Lavell, H. G. Brauch, U. Oswald Spring, G. Wilches-Chaux, M. Moench, and D. Weiner. 2007. Climate change and human security. http://www.radixonline.org/cchs.html (accessed 4/10/08).

PART 1
Climate and Culture

Chapter 1

Human Agency, Climate Change, and Culture: An Archaeological Perspective

Fekri A. Hassan

Introduction

In 1914, the geologist and explorer J. W. Gregory inquired if the earth was drying up. He surveyed various sources of data from Scandinavia to China and concluded that there was, in general, no evidence of climate change in historical times. Furthermore, Gregory noted that there was a great deal of controversy concerning paleoclimatic interpretations and the probable causes of climate change. Today, our knowledge of climate change in historical times has improved immensely (Mayewski et al. 2004; Shulmeister et al. 2006), yet we are still not clear about the magnitude, scale, timing and frequency of climatic changes, and are unable to provide conclusive evidence of the causes of certain paleoclimatic events (for news on current research on past global changes see the PAGES project [Past Global Changes]).[1] We are also far from certain of the probable impact of climate change on the trajectory of human history. H. H. Lamb (1982), in his seminal volume *Climate History and the Modern World*, reviews in detail both the lack of sufficient paleoclimatological data and, more importantly, the limitations of our methodologies and our interpretative strategies.

Climate change is no longer a matter solely of academic interest. People throughout the world are now aware of the impact of climate change on contemporary societies. For example, when the Sahelian droughts in the 1970s caused severe famines threatening the lives of millions of Africans, most scholars agreed that the droughts were, in part, caused by human activities, such as land use, overgrazing, overfarming, inappropriate irrigation, and inadequate governmental policies (Hulme 2001; Reynolds and Smith 2001; Zeng 2003). This spurred debate over the exact role of human activities in land degradation and droughts and ushered in an increasing appreciation of human agency in environmental change.

This chapter elucidates the intractable and complex interrelationship between climate and human societies with a plea to overcome simplistic notions of determinism and indeterminacy. Climate, as I show through case studies in Southwest Asia and North Africa, played a major role in the origins of agriculture, the emergence of state societies, and the temporary breakdown of the centralized organization of complex state societies. However, the impact of any climatic event depends on the local ecological setting and the organizational complexity, scale, ideology, technology, and social values of the local population. It is only through long-term archaeological and historical analysis, as well as detailed examination of the social dynamics on local and regional scales within an interregional framework, that we can begin to detect the differential impact of the same climatic event.

I wish here to emphasize the rigor needed to make any assertion of the causal role of climate. It is misleading, for example, to list the frequency of radiocarbon age determinations as a proxy to climate change, or to cull a selection of a few archaeological cases from different regional and temporal contexts to underscore the role of climate in the "collapse" of civilizations on the basis of dubious chronological determinations, climatic reconstructions, and cultural interpretations. Frequencies of radiocarbon age measurements of archaeological occurrences are not directly a function of intensity of human occupation, but more so of a series of formation processes as well as the intensity of archaeological research and dating objectives. It is important to infer climatic events from well-dated geological and environmental proxies. Archaeologists also need to pay attention to the screening of radiocarbon data to ensure that they accurately date the event under consideration. Precision in this can be improved by the use of multiple measurements and statistical analysis. In this chapter, I report events based on radiocarbon age determinations as calibrated radiocarbon years before present (cal BP).

At present, new drilling methods to obtain samples, dating, and analytical techniques have improved our current understanding of climate change. Data retrieved over the last two decades have shown that climate change is not always gradual or cyclical, but that dramatic shifts in climate can take place rapidly (Allen and Anderson 1993; Street-Perrott and Perrott 1990). Such rapid events occur within less than a century and the change can be quite severe (Adams, Maslin, and Thomas 1999; Rahmstorf 2001; Taylor 1999). One of the most remarkable transitions, which had a major impact on the history of humankind, was the end of the last major cold spell, a period known as the Younger Dryas. A large part of this global switch from cold to warm climate took less than twenty years: "There was no warning. A threshold was crossed, and the climate in much of the world shifted abruptly from cold to warm" (Taylor 1999, 323). Abrupt climatic events are not just rapid; they are also breaking points, thresholds resulting from strongly nonlinear responses to "forcing"[2] (Rahmstorf 2001, 2). Rahmstorf singles out an event, dated to 8,200 years ago during the Holocene that shows a spike in arctic ice cores, which affected the North Atlantic. He also recognizes the abrupt desertification of the Sahara 5,500 years ago with evidence from Atlantic sediments off northeastern Africa that show a sudden and dramatic step-function increase in wind-blown dust, indicative of the drying of the African continent. Similarly, the last ice age, like previous ice ages, was punctuated by abrupt climatic transitions.[3]

I also wish to emphasize the role of human agency. Different groups guided by the decisions of all or key figures in the community seem to pursue different strategies in coping with a climatic crisis in their region. However, it seems that the choice is limited to a few options at any one time, and that certain options seem to be favored by the majority of the population. Further, interaction within and between regions seems to provide different groups with the choice of reconsidering their position. In some cases they may adopt or adapt certain innovations that they deem to be advantageous. However, people inevitably make decisions on the basis of subjective probabilities, biased information derived from anecdotal evidence, and prejudicial positions based on their values and worldviews. Additionally, the long-term consequences of any decision are not, in most cases, apparent within the lifetime of a person or a couple of generations. It is thanks to human ingenuity, not climate change, that in responding to environmental crises or endogenous cultural perturbations (Hassan 1993) people tend to make adjustments to sustain their modes of life. In so doing, they unwittingly set the stage for new problems that future generations have to cope with.

Endowed with the vision of a "good life" and the capacity to modify habitats, people have introduced changes that ultimately lead to the emergence of new conditions that expand their capability of modifying their environment. Today we have reached the point at which our human impact now extends to a modification of the forces that regulate the climate of our planet (Goudie 1993; Hassan 1992; Redman 2004). Our environmental impact has increased with our continued expansion and intensification of food production and industrial products, caused by adapting agriculture, developing complex managerial organizations (state societies), expanding the size of the labor force, and utilizing progressively more powerful sources of energy beginning with draft animals and ending with nuclear power. In the meantime, our numbers have soared from a few million during the Pleistocene to more than six billion today, with unprecedented rates of population increase over the last one hundred years (Lutz, Sanderson, and Scherbov 2004).

Clearly our modern crises are not caused solely by climate change. But today, as in the past, societies become vulnerable to millennial or centennial climatic changes because they do not recall or anticipate abrupt, severe climatic events that are outside the range of human reckoning and collective memory (Hassan 2000b). People and societies, as my analysis of agricultural origins in this chapter shows, are conservative. They tend to guard and cling to the paradigms, values, and institutions that have proved to be successful in their own past based on coping with decadal fluctuations in climate change or multigenerational social perturbations. As a result they are reluctant to undertake corrective actions that go against their social grain—for example, moving from a mobile subsistence economy to a settled economy (a process that took more than five thousand years in the Levant). The veneration for old habits, which may become encoded in religious precepts, is in most cases advantageous, but it is harmful if adhered to dogmatically in all cases given endogenous cultural developments and novel external conditions. Today certain interest groups in many countries are still reluctant to accept the impact of climate change on current and future human affairs. In so doing they miss not only the lessons from world history, but also the opportunity to examine the social dimensions of climate change critically.

Also, because modern industrial nations overvalue industry and science, current efforts to cope with climate change focus on reducing emissions or the use of new alternate technologies instead of considering the social dimensions of climate change. Studies of climate change are still dominated by climatologists and environmental scientists, though the last few years have witnessed some overtures for collaboration among social and environmental scientists (see Costanza, Graumlich, and Steffan 2007; Costanza et al. 2007; Dearing, Cromer, and Kiefer 2007).

The threat of climate change may divert our attention from the explosive nature of our postindustrial world, with its inequities and mass poverty, expanding population, progressive organizational complexification, and spiraling demands for critical resources. These factors compound the danger from climate change. We should recall that climate change, as in the case of the Classic Maya civilization, may be a catalyst hastening collapse.

In the following sections I discuss the problem of confusing correlation and causation and the role of human agency in recognizing and responding to climate change with reference to the limitation of the human scale. I then review the main millennial climatic events that influenced the course of human history over the last 13,000 years to illustrate the frequency of abrupt, severe events that humans have had to cope with since the emergence of humankind. I subsequently deal briefly with the role of climatic events in the peopling of the world during the last 100,000 years, followed by a summary of the role of climate change in the origins of food production, with a primary focus on Southwest Asia and North Africa. This discussion reveals the importance of regional expressions of climate change and the role of independent innovations that converge toward similar solutions following a similar sequence in historical developments. I then deal with the impact of climate change on the spread of food production to other parts of the world and the beginning of significant cultural differentiation under varying environmental conditions. The last part of this chapter considers the origins and collapse of civilizations in different parts of the world to illustrate that the impact of severe abrupt climatic events is not necessarily negative. This should serve as an antidote to sensationalist accounts of the role of climate in destroying civilizations.

Correlation, Causation, and Oversimplification

Archaeologists have for a long time debated the role of climate change in human affairs. Gordon Childe (1928, 1934) made one of the most influential contributions with his contention that desiccation by the end of the glacial period led to the emergence of agriculture, as people congregated along river courses and within the orbit of oases, where conditions were appropriate for the domestication of plants and animals. The relationship between climate change and the origins of food production was revisited in the 1960s, as Binford (1968) linked post-Pleistocene climatic events with agricultural origins in arid regions. He hypothesized that a worldwide change in sea level in post-Pleistocene times led to a greater exploitation of fish and other aquatic resources, which prompted sedentary settlements and rapid population increase where food resources were abundant. Excess population from such optimal zones was forced into less productive, marginal habitats where food production was initiated out of necessity.

More recently, Weiss highlighted the causal role of climate in cultural evolution (Weiss et al. 1993). Together with Bradley et al. (2001, 610), by making use of high-resolution paleoclimatic data they argued that climate change has been a primary agent in repeated social collapse. However, such generalizations fail to acknowledge the intricate relationship between human agency and the impact of climate change on societies. Archaeologists have been hampered in the past by crude data on climate change and poorly dated environmental and cultural events. Explanations based on climatic forcing depended upon rough correlations, which were often hastily translated to causal linkages with climate playing a determining role.

Correlations are uncertain because climatologists still do not agree on the mechanisms and details of climate change during the last 10,000 years (the Holocene) in spite of attempts since 1983 to using radiocarbon dating (Bond et al. 1997). In addition, some climatic events are global, while others are regional or local. In some cases, in the absence of well-dated climatic records in an area, correlation with global events may be erroneous when local evidence is lacking. Moreover, some global events are time transgressive and may not be of the same amplitude, duration, or even direction (e.g., wet episodes may be synchronous with dry events) in different regions.

Even when adequate records of climate change and high-resolution archaeological data exist for a given region, other appropriate information—for example, on rates of depopulation and relocation and the extent of famines, morbidities, or land use—needed to explain the causal role of climate change may be lacking. This may be due to the selective nature of the archaeological record or lack of sufficient archaeological research. There are also only a few special cases when there are high-resolution and appropriate data both for climate change and cultural developments to allow for credible assertions on the causal links between climate and culture. To establish such causal links, it is imperative to show how climate change influences human activities through its impact on surface and groundwater resources, availability and quality of food, settlements, transportation, industry, trade, or any other cultural variables influenced directly or indirectly by changes in climatic parameters. That a drought may have led to the "collapse" of a civilization or to a dramatic cultural transformation may be intuitively appealing, but that is not sufficient to validate any such claims, since droughts are common and climate is constantly changing.

Today, the realizations not only that climate changes over millennia, but also that drastic shifts in climate can and did occur within a century or less, make it possible to comprehend how abrupt climatic events could have been influential in the course of human history. People are likely to respond to climatic signals if they are within the scope of their perceptual span. Climatic events that are not perceptible or too distant in the past are not likely to be effective in the way people react to their environment (Hassan 2000b).

Human Agency, Scale, and Adaptability

An appreciation of the role of human agency is essential if we are to make any sense of the impact of climate change on human societies. Societies vary in their size, scale of organization, social differentiation, subsistence activities, productive strategies, ideologies, and worldviews. They are maintained, reproduced, and transformed as a result of the impact of day-to-day practices by individuals who are constrained by their perceptions, beliefs, norms, values, and mind-sets (cognitive schemata). They are likely to maintain or alter their practices and ideas subject to what they perceive to be positive or negative outcomes of their actions (Craik 1972). Neither the past nor the future is too far from the present when it comes to making decisions, because people consciously or routinely select from among alternatives based on their past experiences and future plans and possibilities.

Perceptions of climate change can only be assessed through tangible proxies within the purview of the culturally conditioned canons of perception and comprehension. In a hunting-gathering society, the group is small and differentiation of occupations is minimal. In addition, such groups are intimately familiar with the seasonality, condition, and changes in the location of game and wild plants. They are also familiar with different habitats over a large region. Within that context they are likely to detect environmental change that may ultimately be directly or indirectly related to climate change, but might also be due to changes in endogenous landscape responses or to overhunting or overexploitation of certain resources.

Climatic change can only be inferred from proxies. Among all potential proxies, those that will be observed are those that are pertinent and relevant to the life support system, comfort, and sense of aesthetics among other natural phenomena that an individual or a group consider of interest and value. In addition to perceptions of changes in amount of rainfall, seasonality, frost, wind, heat, and humidity, changes in climate are likely to lead to noticeable changes in surface water distribution, vegetation, animals, and geomorphological and sedimentary processes such as soil erosion, deflation, or invasion by sand dunes (Bridges 1997; Derbyshire 1997; Lawlor 1997; Mannion 1997; Stonehouse 1997; Whitehead 1997). In dealing with climate change and human responses it is crucial to employ both an ecological approach and a focus on landscape ecologies (see Burel and Baudry 2003; Cordova 2007). It is now becoming clear that ecological responses to climate change over the last fifty years were associated with major shifts in the distribution of species (Walther et al. 2002) and that abrupt climate change can have catastrophic ecological consequences (Davis and Shaw 2001; Scheffer et al. 2001).

Information gained from individual experiences is likely to be shared by members of hunting bands, as well as among other bands that habitually come in contact with each other. It is also likely that practices that ensure survival in the face of food shortage, cold nights, or lack of critical resources will be adopted and shared within and between groups. Given that climate changes on millennial, centennial, and decadal scales, people are not likely to recall millennial or centennial events, and are more likely to recall extreme events that have been experienced in their lifetime or by others in the community who often span no more than three generations, or roughly sixty years (Hassan 2000b). More distant memories are likely to fade exponentially with a selective retention of those that have caused severe hardships and loss of life. Given interannual variability in climatic conditions, people are most likely to respond to such "routine," predictable change by regulating the size of a group, mobility, and other means of coping. They are also likely to have set responses for dealing with decadal changes, but are not likely to have a stock of actions in response to widely spaced centennial to millennial events.

Smithers and Smit (1997) provide a useful guide for interpreting the impact of climatic perturbations on social "systems" and the probable adaptive responses. The intensity of the impact depends on the magnitude, areal extent, frequency, duration, and suddenness of the climatic disturbance. The system will respond depending on its stability, resilience, vulnerability, flexibility, and scale. As a result the system will either fail (collapse) if people are incapable of overcoming the negative impact of climate change in time or undergo remedial actions if people do not or cannot act in time through effective coping mechanisms.

In the following section, I discuss briefly one of the well-recorded climatic events and its implications for cultural responses.

Global Climatic Events—the "Medieval Warming Period"

Analysis of the series of Nile floods dating back to the seventh century AD, which serves as a proxy to climate changes below the millennial scale, reveal decadal variations in the amount of Nile flood discharge in the range of

40 years (from 20 to 50 years) with four episodes lasting from 120 to 170 years and an average of 140 years. Over a period of 1,220 years, historical accounts highlight the recurrence of stressful events from the tenth to fourteenth centuries when severe famines, civil disorder, depopulation, and plagues were attributed to the vagaries of Nile flood discharge. Recent analysis of the frequency of extremely low or excessively high Nile floods revealed that these famines coincided with an increase in the frequency of extreme events, and a reduction in their spacing (Hassan 2007b). Such episodes linger and become a part of collective memory, triggering human responses that lead to new practices or ideas that may then become normative and precipitate structural changes in social organization, ideology, or productive technologies.

It is clearly not coincidental that the extreme fluctuations in Nile floods and associated famine were synchronous with dramatic climatic changes in Europe during the period that came to be known as the Medieval Warm Period, described by H. H. Lamb in his pioneering work *Climate, History and the Modern World* (1982). This period has become known as a time when Norse farmsteads were established in Greenland, vineyards flourished in sunny central England, and barley and wheat were grown in south Iceland. This was also a time that "saw a great expansion of European civilization" (Street-Perrot 1994, 518). Although the magnitude, duration, and geographical extent of this period have been debated, it is becoming clear that it was a global event with different manifestations in different regions. A study of relict tree stumps rooted in present-day lakes, marshes, and streams in California (Stine 1994) revealed that sustained droughts occurred in two clearly separable intervals in AD 892–1112 and 1209–1350. Similarly, recent analysis of Nile floods reveals that the period from the ninth to fifteenth centuries was not uniform but was characterized by critical flip-flop transitions from one climatic regime to another at AD 900, 1010, 1070, 1180, 1350, and 1400 (plus or minus twenty years).

Recent reexaminations of the Medieval Warm Period (MWP) narrow this period to AD 890–1170, when warmth is indicated across a number of temperature proxy data from the Northern Hemisphere (Osborn and Briffa 2006). Hughes and Diaz (1994) observed also that the period from the ninth to fourteenth centuries AD was warmer for some regions, particularly during the summer, than those that prevailed until the recent decades of the twentieth century. These regions included Scandinavia, China, the Sierra Nevada in California, the Canadian Rockies, and Tasmania. However, these warm episodes were not strongly synchronous. Other regions such as southwest United States, southern Europe along the Mediterranean, and parts of South America do not reveal such warming conditions. Similarly, the medieval climate was clearly unusual in some areas, but studies of large-scale climate variations reveal that some regions do not follow the global or hemispheric trend (Bradley et al. 2001). Further, the warmest medieval temperature was not synchronous around the globe, and the High Medieval period (AD 1100–1200) was not warmer than the late twentieth century (Bradley, Hughes, and Diaz 2003).

The so-called Medieval Warm Period does not in fact represent warming conditions throughout the world. It was a complex phenomenon that probably was expressed differently in different regions, with different impacts on local environmental conditions. This period may thus serve as an analogue for other case studies in which high-resolution climatic and historical data are missing.

Archaeologists should be aware of the ambiguities and uncertainties of paleoclimatic reconstructions before embarking on oversimplified generalizations. On the other hand, it is certainly possible that certain global events that were sufficiently severe and extensive have had an impact on the course of human societies. Global extreme events occur at a millennial scale and can have an indelible impact on human life support systems. The effect of such events, however, depends on the scale of societies, their ability to rebound without significant structural changes, and the potential to take timely remedial actions to sustain viable populations. In many cases, cultural developments in response to climate change in one area can influence adjacent populations. They are also likely to lead to further social changes that can, in the long run, lead to a transformation of the overall identity of society, thus marking a transition from one cultural state to another.

Climatic Events and Human Dispersal: The Peopling of the World

Before agriculture, the effect of severe climatic events led mostly to dispersal into more favorable regions, extinctions, or small incremental change in technology and land use. This was probably because early humans were on a foraging track with an emphasis on the improvement of hunting and food collection technologies, including hunting implements and strategies. The number of foragers was small; their organization was minimal to allow them to disband and regroup as needed in the face of food shortages and in order to roam over large areas. Throughout a history of numerous glaciations and warm periods, early humans spread out of Africa to populate most of the inhabited world by the end of the last glacial maximum. The option of dispersal was congruent with a mobile mode of subsistence. It was also unhindered by fixed territorial ethos or a strong affiliation to a permanent group identity. Dispersal and adaptation to new habitats led, by the end of the last glacial, to a world of foragers in very different world biomes with advanced stone technologies well tuned to their habitats (Hassan 1981).

Human dispersal out of Africa took place between 130 and 190 kyr (130,000 and 90,000 years ago) and may have coincided with the global climatic changes associated with the last major glaciation.[4] Modern humans appeared in Palestine ca. 90 kyr, and remains of early humans in Southeast Asia date to ca. 75 kyr (Burroughs 2005; Hoffecker, Powers, and Goebel 1993; Mellars 2002; Schurr 2004; Turner et al. 2001).[5] This suggests that this phase of dispersal may have been associated with warmer interstadials that cluster in the period ca. 85–75 kyr.

Evidence of a third wave of dispersal is from Nieah Cave and Wadjiak in Southwest Asia 50–40 kyr ago and in northern Europe ca. 40 kyr ago, apparently triggered by severe cold conditions ca. 50 kyr (Heinrich event 5). The distances involved were not very long by comparison with the second wave. These waves were separated by approximately 25,000 years, suggesting that the dispersal was rapid and not as gradual as what happens during warmer periods.

The next wave, which included the penetration of Australia 38–30 kyr ago, coincided with a period of frequent millennial changes in climate starting before 40 kyr until 36 kyr (Heinrich event 4). The subsequent wave is indicated by the penetration of northeastern Siberia ca. 20 kyr during the Last Glacial Maximum (LGM), perhaps in response to episodic amelioration in climate during that cold phase. The movement into North America began perhaps before ca. 16 kyr because postglacial warming conditions would have made it difficult to cross the Bering Strait (see Bryan and Gruhn 2003). However, another wave could have crossed during the Younger Dryas 13,000–11,600 years ago. The movement into hitherto uninhabited territory was rapid, reaching Patagonia in South America by 11,000 cal BP.[6]

The Transition to Agriculture: Climate, Regionality, and Innovation

The transformation from the peripatetic life of foragers into sedentary farming communities approximately ten thousand years ago was a momentous development in the history of humankind. From a sparsely populated planet exploited by loosely organized bands, the world population shot from less than ten to fifty million in a few thousand years (Hassan 1983). Novel social, religious, and political foundations attending this transformation provided the basis for the rise of complex chiefdoms and state societies. Explanation of the causes for this transformation, aptly called the *Neolithic Revolution* by Gordon Childe (1936), has become one of the most intractable puzzles in contemporary archaeology (e.g., Asouti 2006; Bar-Yosef 1998; Bellwood 2005; Cappers and Bottema 2002; Garcea 2004; Harris 1996, 1998; Hassan 2000a; Issar and Zohar 2004; Kuijit and Goring-Morris 2002; Reed 1977; Rosen 2006; Simmons 2007). At a time when it was believed that the end of the Ice Age led to desiccation in the Near East, Childe (1936) hypothesized that domestication of animals and plants began as people and animals took refuge in well-water locations on the banks of rivers and in desert oases. Our understanding of the effect of glacial conditions and the subsequent warmer conditions in Southwest Asia and North Africa provides a more informed evaluation of the role of climate change in the origins and spread of food production in southwest Asia and North Africa.

Although most archaeologists thought that they could determine whether climatic events "caused" agricultural origins by examining proxies of climate change at the time domesticates appear, it is likely that domestication and the appearance of an agricultural mode of life was an intractable, multi-stage, long-term transformation that has to be traced back to the climatic and environmental transition from the last glacial to the period of postglacial warming (Hassan 1977, 604; 1981, 219). My contention has been that "the onset of the Holocene and the retreat of glaciers marking the termination of the last major glaciation, and the possible impact of such changes on wild resources in climatically unstable areas such as semiarid and subtropical regions, seem to explain the independent emergence of food production in several places of the world beginning with the Holocene. The change in subsistence patterns that ultimately led to agriculture, however, must be sought in the impact of climatic fluctuations associated with the Terminal Pleistocene on cultural systems that were receptive for the transition in areas where domesticable plants were available." (Hassan 1981, 219)

Human dispersal during the last glacial out of Africa led to the peopling of the world; regionalization; growth of local populations; and a series of innovations in tool making, hunting gear, foraging strategies, social organization, communication, symbolism, and art. Populations built up in the midlatitude areas and, during the last glacial maximum, in well watered areas along riverbanks, grasslands and woodlands.

The impact of global climatic warming on world populations beginning ca. 16,000 cal BP, as well as subsequent severe climatic fluctuations—particularly the droughts associated with the Younger Dryas (13,000–11,600 cal BP)—varied widely depending on the local expression of the global climatic events and the local environmental settings. The origins of food production were first of all a matter of a sequence of social transformations brought about by changes in the mode of subsistence in response to climatic fluctuations involving population situated in vulnerable subarid ecotonal habitats. In North Africa and Southwest Asia (the NASWA region), these habitats included the northern Levant, the southern Levant and the Sinai, the Nile Valley, the African Sahel, the Mediterranean coastal region of North Africa, and the series of massifs in the Sahara. In these habitats, people responded to local conditions by making the most appropriate decisions given local opportunities, available resources, and their perceptions of the food potentials of specific subsistence modes. I argue that in all areas, sedentary life was resisted as highly risky under unstable climatic regimes and that similar steps were undertaken to intensively utilize cereals and probably "keep" or "manage" animals.

In Africa, the period 16–8.2 kyr witnessed the invention of pottery, the invention and spread of innovative hunting equipment such as the bow and arrow, the use of grinding stones to process cereals and tubers, the harvesting of wild sorghum and millet, increased reliance on fishing, the "management" of Barbary sheep, the domestication of cattle, and probably the preservation of fish (Garcea 2004). During this period, in some cases base camps in favored spots were repeatedly visited. A mobile strategy and an ethos of sharing and exogamy, as is the case among foragers, made it possible for some items and ideas to be passed from one end of North Africa to the other through corridors connecting one region with the other.

In the Levant a similar adaptive response was taking place, but differences included a lack of pottery (Simmons 2007), less emphasis on fish resources, and a more intensive utilization of wild wheat and barely. Here too, and in spite of a precocious attempt to restrain mobility from 14,500 to 13,000 cal BP (the Early Natufian), it was not until much later that mobility was constrained. The switch to more sedentary conditions was due to the onset of stable wet climatic conditions during the Holocene Wet Maximum from 11.6 to 8.2 kyr. It was during this period that successive generations of foragers, who took advantage of the well-watered habitats in the Levantine corridor, had enough confidence to settle down and progressively increase their dependence on homegrown cereals and legumes. This option was not open to, or welcome by, neighbors who lived in less favorable desert areas such as the Negev and the Sinai, despite other shared cultural traits. Within the span of two millennia some of the farming hamlets in the Levantine corridor developed into "mega-villages" with elaborate social organizations, ritual, and art. During 10,600–8,800 cal BP, known as the Pre-Pottery Neolithic Period (PPNB), naked six-row barley and free-threshing bread and hard wheat were cultivated, as were pulses such as peas and lentils, and sheep and goats were domesticated. In northern Levant and central Anatolia, cattle were domesticated by 8,800 cal BP (Bar-Yosef 1998), and in Greece domesticated cattle appear ca. 9,000 cal BP and pigs and goats ca. 10,000 cal BP (Halstead 1996).

The adoption of farming and herding during the PPNB was associated with the development of sedentary populations in villages of substantial size with sophisticated architectural features, including nonresidential structures with communal or ritual significance, and elaborate nonresidential sites (Simmons 2007). It is not clear if the farming villages constituted egalitarian or hierarchical societies, suggesting that social organization was not strongly hierarchical and may have consisted of communal management of village affairs. Mortuary "skull cults" and evidence of rich symbolism suggests that ritual and religion might have played a major role in group dynamics as a means of alleviating conflicts and promoting solidarity. It is noteworthy that some groups opted to pursue hunting and foraging, while others continued to lead a nomadic life with the adoption of domesticates, especially animals, eventually evolving into pastoralists (Simmons 2007, 167).

In the succeeding period 8,800–8,200 cal BP, the end of the PPNB, populations in the southern Levant shifted east, particularly to highland Jordan, where they established a series of very large settlements or mega-villages. Such sites are also recorded outside the southern Levant, including Çatalhöyük. The emergence of large settlement aggregates does not seem to be related solely to climate change, but rather to a combination of factors. It appears that sedentary farming communities overexploited their surroundings and that relocation to more favorable, previously unused habitats was one of the responses (Rollefson 1992). Aggregation of communities from neighboring villages to form compound villages was apparently facilitated by social and kinship ties.

By 8,200 cal BP, the mega-villages were suddenly depopulated. The PPNB came to a sudden end. The 8,200 cal BP event was apparently responsible, at least in part, for the collapse of the Final Pre-Pottery Neolithic (Bar-Yosef 1998). After favorable climatic conditions for close to a millennium (Kohler-Rollefson 1988, 88), these early Neolithic societies overextended (thus reducing their margin of safety) and exposed themselves to the hazard of unanticipated climatic turns. The collapse of the PPNB was not a local event; this pronounced, millennial, global, abrupt cold event (Alley et al. 1997) was a severe "environmental crisis" (Cordova 2007).

The period 8,200–6,800 cal BP was characterized by unstable climate with a shift from forest to maquis[7] in the northern Levant and more olives in the southern Levant (Rosen 2006). It was under this climatic regime that village communities were established with greater reliance on a mixed farming-herding economy that established the foundation of the Mediterranean agrosystem (Butzer 1996). Dairying apparently led to the widespread use of pottery for keeping and serving milk, butter, yogurt, cheese, and other milk products, and thus this phase is characterized as the Pottery Neolithic. The Neolithic communities in the Levant appeared to have maintained a heterarchical social organization based on aggregation of extended families within a communal organization with an emphasis on religion and ritual as a means of group solidarity (Simmons 2007).

Climatic Instability and the Spread of Farming, 8,200–6,800 Cal BP

One of the main aspects of the period postdating the 8,200 cal BP event is the spread of farming activities into coastal areas in Syria and Lebanon and eastward into Mesopotamia and Iran. This led to regional differentiation that was to become the foundation of further cultural trajectories: a transition to agrarian state societies occurred in Mesopotamia and the Nile Valley, while the southern Levant, restricted by its limited resources, remained at a stage of relatively small-scale social complexity. The onset of drier conditions detected at 8,200 cal BP and the perpetuation of unsettled climatic conditions until 7,800 cal BP, when a period of greater rainfall became evident, were in part responsible for a simultaneous dispersal of cattle from the Eastern Sahara toward the Central Sahara and the introduction of caprines from the Levant to the Red Sea coast in Egypt and the Eastern Sahara. All these developments came in the wake of the "collapse" of the Levantine PPNB farming communities ca. 8,200 cal BP.

In short, the populations taking advantage of the changes in vegetation and animals as a result of the postglacial warming commencing ca. 16,000 years ago also developed significant new subsistence and settlement strategies over a period of three thousand years. The period 14,500–13,000 cal BP was particularly wet and warm. However, the "good life" under centuries of relatively lush vegetation and abundant resources came to an abrupt end with the onset of the Younger Dryas (13,000–11,600), a cold interval of 1,400 years, which in turn led to strategic responses and innovations. This period, however, also came to an abrupt end with the return of warmer and wetter conditions. The appearance of sedentary farming communities in the Levantine corridor during the tenth millennium cal BP coincided with the practice of cattle keeping and intensive utilization of sorghum in the Eastern Sahara. Both developments were synchronous with fairly stable wet conditions that lasted, with minor interruptions, until the transition to cold climatic conditions ca. 8,200 cal BP, followed by inclement weather conditions until ca. 7,800 cal BP. The 8,200 cal BP (8.2 kyr) event hastened the collapse of PPNB communities in the Levant and led to a dispersal of populations from the Levant into North Africa, the Arabian Peninsula, and Europe. The movements were rapid. The displaced individuals and groups were not demographically viable, and it was advantageous to intermarry with the indigenous local foraging populations, fusing multiple subsistence and social traditions and adapting new farming and animal-keeping technologies to the local ecological settings.

The 8,200 cal BP event must be regarded as one of the most important climatic events of the Holocene. It led to a wave of dispersal and the transformation of many parts of the world as local populations began to adopt, adapt, and modify their ways of life in response to innovative methods and technologies of food production.

Climate Change and Saharan Populations in 7,800–6,800 cal BP

Although wet conditions recurred, the period 7,800–6,800 cal BP was characterized by frequent climatic oscillations and droughts. Climatic conditions became progressively drier and led to the onset of severe aridity by 5,500–5,300 cal BP, signaling the prevalence of desert conditions over most of North Africa. The succession of wet and dry episodes seems to have encouraged successive movements of cattle and then ovicaprids (sheep and/or goats) westward following the better-watered range and basin areas associated with Saharan highlands, such as the Ennedi, Tibesti, Tassili, and Hoggar massifs (Hassan 2000a, 2002b).

A millennium after the initial spread of cattle out of the Eastern Sahara and the introduction of sheep and goats from the Levant into the Eastern Sahara, a number of independent events occurred simultaneously ca. 6,800 cal BP, such as the first appearance of domesticated cattle in the Sudan along the banks of the Nile and at Merimde farther north in the Nile Delta ca. 6,800 cal BP, and in the Central Sahara ca. 6,900 cal BP. These events were a response to the onset of dry conditions in the Eastern Sahara, which is well documented by the disappearance of the temporary lakes that once filled the depressions where oases are located. The disappearance of those lakes and the onset of a freezing cold desert climate are attested at Farafra Oasis at that time (Hassan et al. 2001; Hassan 2003). In the Central Sahara, local conditions in the massif range areas ca. 7,800–5,200 cal BP were favorable for the emergence of varieties of livestock keeping and pastoralism, since large basins were fed by great wadis (desert water courses) that drained from the massif ranges. Seasonal variations in rainfall between north and south might have allowed transhumant pastoralism.

Meanwhile the populations that remained in the Eastern Sahara, which receives less rain, were becoming increasingly isolated. Their livelihoods were precarious and those that remained depended more on occasional springs than on rainfall. Sheep and goats were incorporated into a subsistence regime that also emphasized hunting gazelle and other desert game. However, there is no evidence that the adoption of sheep or goats originating in Southwest Asia was accompanied with the introduction of the cultivation of domesticated wheat or barley.

Climate and the Origins of the Egyptian Civilization

Desert dwellers apparently struggled during 7,800–6,800 cal BP under uncertain climatic conditions and frequent spells of aridity. By 6,800 cal BP many of the inhabitants of the Sahara adjacent to the Nile Valley apparently decided to risk settling along the banks of the Nile. Until then they probably had not regarded the area as an ideal habitat because its peculiar resources were not concordant with their preferred mode of life as desert hunters and gatherers. It is precisely at this time that we begin to see the first evidence for sheep/goats and domestic cereals, introduced earlier from the Levant into the Sahara, in the Nile Valley (Hassan 1988; Wetterstrom 1998). The early Neolithic sites of the Nile Valley contain artifacts analogous to those from the Eastern Sahara dating from 7,600 to 7,300 cal BP (Hassan 1988, 2003).

For perhaps as long as 6,800–5,800 cal BP, the Nile Valley south of the Delta and the Faiyum were inhabited by a mélange of communities with various subsistence strategies dominated by hunting, fishing, or herding. During that time, the main farming communities were located along the margins of the Delta, in contrast to Upper Egypt where farming villages become a major feature of the cultural landscape by 5,800 cal BP (Naqada I).

This suggests that the inhabitants of Merimde Beni Salama were the result of the fusion of farmers who relocated to the edge of the Delta and Delta dwellers. It took the southerners a longer time to adopt the new way of life, perhaps as a result of increasing contacts or incentives to enter in economic exchanges for goods provided by the northerners.

This transition to a predominantly agrarian mode of subsistence by 5,800 cal BP marked the emergence of the economic foundation of the Egyptian civilization. Within the relatively short span from 5,800 to 5,300 cal BP, Egyptian farming communities made a transition to a state society that was fast by comparison to Mesopotamia, where this transition is estimated to have taken two millennia (ca. 7,500–ca. 5,500 cal BP). In both regions, the cultural developments that led to the rise of state societies, altering the cultural map of North Africa and Southwest Asia, were due to the vast fertile plains of great rivers. In Egypt, however, the flood plain in Upper Egypt is narrow and linear with the Nile as a single transport artery. Political transformations may have been attempts to counteract the decadal variations in Nile flood discharge (Hassan 1988). The recognition of a global climatic cold event leading to a reduction of Nile floods is likely to have hastened the process of unification as it became advantageous to expand the size of the groups participating in a shared economy as a buffer against crop failures. The frequent depiction of boats suggests their use for food transport. There is also evidence for expansion of trade and elaboration of rituals.

Core and Periphery: A Divided World

By contrast to the agrarian civilizations of Egypt and Mesopotamia, the rest of North Africa and Southwest Asia, which lacked the fertile plains and flood discharge of great rivers, was constrained by the limited productive potential of rain-fed agriculture and pastoralism (see Barker and Gilbertson 2000). Improvement in agrarian productivity there depended on a suite of methods of water harvesting such as check-dams, terracing, cisterns, and wells (Hassan 2007a). In many areas, farming was risky if the interannual variability was high, as it is today in the Mediterranean coast of Egypt. In addition to herding and pastoralism, desert nomads exchanged food and goods with settled farmers, and on occasion benefited from raids. Subject to the vagaries of decadal droughts, the nomads had recourse to either a pattern of episodic raiding or a flexible mode of farming and herding.

The emergence of farming in the Sahel appears not to date earlier than 3,300–3,100 cal BP in northeast Nigeria and Burkina Faso (Breunig and Neumann 2002; Wetterstrom 1998) and 3,800–3,700 cal BP in Mauritania (Ambelard 1996). It came soon after the appearance of the first pastoralists in the Sahel of northeast Nigeria as the culmination of the movement westward that began from the Eastern Sahara around 7,800–6,800 cal BP. Within the Sudano-Sahelian zone of Africa, people today practice nomadic pastoralism, settled agriculture, and a variety of different agropastoral strategies.

In this zone, rainfall is unpredictable and subject to frequent oscillation of the intertropical convergence zone. Because the livestock is rain fed and the grain (sorghum) fields are not irrigated, rainfall is the determinant of the year-to-year variation a family experiences in both crop and livestock yields.

Using a stochastic modeling technique whereby a decision is made to cultivate, keep family wealth in the form of livestock or stored grain, or some mixture of the two, Mace (1993) concluded that pastoralists are likely to increase their chances of long-term survival by taking up agropastoralism if their wealth declines below a certain level. Inversely, when agropastoralists become wealthy, they will usually do better to give up cultivating and devote their effort to their herds. Her conclusions are confirmed by ethnographic observations. However, the model does not deal with the consequences of severe, prolonged droughts that are likely to lead both to the failure of crops and the decimation of herds and stored grain, which may force pastoralists to raid settled farmers, become settled farmers if possible, or emigrate. They may also buffer the stochastic variations in the yield of crops and herds through trade. This can be a source of wealth, prestige, and power, but it ultimately depends on the viability of the agrarian communities. Lees and Bates (1974) present a model for the origins of another type of specialized nomadic pastoralism within a cultural setting characterized by specialized irrigation. In this case, the origin of pastoralism is predicated upon the increased labor requirements of irrigation agriculture and the consequent conflict of interest related to land use and labor allocation in alluvial zones, as well as the hazards of irrigation agriculture through time. Being mobile, pastoralists respond to climatic events in a way that differs from that of sedentary farming communities.

Agrarian Communities and Pastoral Nomads

The rise of the agrarian state and pastoralism created a new cultural dynamic that became a major force in the history of world civilizations from Egypt to China. Invasions by pastoral nomads who may be unified under charismatic leaders, especially at times of droughts, may allow them to overrun agrarian-based states and to establish their own dynasties. In Egypt and Mesopotamia, the nomads were an integral force in the history of the two regions. It would be useful to examine the history of nomadic invasions in connection with climatic upheavals and droughts.

In Arabia, the earliest evidence of herders dates to 7,700–7,000 cal BP, even though moist conditions prevailed as in the Eastern Sahara and the Levant from 9,800 onward. The introduction of herding into the Arabian Peninsula thus seems to have been caused by the dispersal of nomadic herders or hunter-herders from the Levant after 7,800 cal BP simultaneously with their movement into North Africa. Pastoralism thrived in Arabia, supplemented by the hunting of gazelle and ostriches. Cattle herding is attested at 6,700–6,300 cal BP in Yemen and along the Tihama coastal plain

(Edens and Wilkenson 1998). By 4,400 cal BP, a millennium after the end of the Holocene wet phase (McClure 1976), early cattle pastoralism begins to disappear in favor of sheep and goat herding, which becomes dominant ca. 2,700 cal BP. Camels were apparently domesticated sometime prior to 3,000 cal BP. In Eastern Africa, goats and sheep were introduced into the Sudan from the Sahara by 6,800 cal BP, and reached southern Kenya and northern Tanzania by 4,500 cal BP (Gifford-Gonzalez 2000), where the emergence of specialized pastoralism dates to 4,000 cal BP (Gifford-Gonzalez 2000; Marshall 1994).

In sum, the initial phase of food production in the Levant was followed by a period of dispersal into Arabia and North Africa from 8,200 to 7,800 cal BP. This led to a variety of regional developments depending on local ecological settings and past cultural histories. The onset of desertification by ca. 5,300 cal BP and a series of hyperarid spells at 4,200 and 3,300 cal BP led to further dispersal from North Africa toward West and East Africa. Outside areas where extensive fertile land and water persisted, sustainable economy under repeated decadal and centennial droughts and high interannual unpredictability of rainfall led to ingenious water harvesting and management technologies. Eventually some of these societies developed into agropastoral communities who used raiding and trade to buffer their survival in climatically unstable areas.

The inherent instability of agropastoral and agrarian economies is further compounded by unpredictable centennial and millennial variations in climate (Allen et al. 2000). Such climatic variability also played a key role in the rise and collapse of some early civilizations depending on their local environmental and cultural circumstances.

Climate and the Collapse of Civilization

Perhaps in the rush to dramatize the impact of climate change in the past, as a warning or a premonition, some attention has been given to the role of climate in the breakdown of complex state societies, especially during the third millennium BC (de Menocal 2001; Weiss 2000). A climatic deterioration ca. 4,200 has been detected in many areas (Dalfes, Kukla, and Weiss 1997; Drysdale 2006) and is considered a cause for the breakdown of complex societies at that time. Although I agree that we can make a case for the role played by millennial abrupt severe climatic events on the temporary descaling of social complexity and the breakdown of centralized authority, I do not think that climate change should be blamed for what is in fact a problem in the social management of natural resources. This is abundantly clear from the thorough discussion of the topic of the collapse of complex societies by Tainter (1988). It is also invalid to restrict the role of climate change to the unmaking of civilizations.

The rise of social complexity appears in the first place as a mechanism to buffer shortages in agrarian productivity within an ecological context that has the potential to accommodate a very large population. The initial success of linking villages in a corporate society that pools its resources, and thus evens out the chance loss of crop yield by a single household or a village, leads to expanding this strategy to more and more communities. Since food security is enhanced as the number of participants increase, the temptation to expand the number of joint villages also increases. However, this process requires specialized managers and specialists to enhance solidarity and minimize conflicts through ritual among other means, and thus a complex social organization characterized by administrative and religious elites who wield power in order to execute their assigned tasks. This implies an increase in the number of non-food-producing elite and the rate of consumption by the elite, who reveal and consolidate their power through lavish displays of wealth, possession of exotic goods, and monumental palaces and temples. The elite also carry their duties with the help of an ever-expanding cast of clerks, guards, entertainers, servants, and craft specialists. Such dramatic increase in the non-food-producing sector places a premium on progressively higher crop yields. Meeting the rising demands may be accomplished through conquest, slavery, overtaxation, or encouraging population growth. Since most capital outlays are not for increasing productivity through capital investments in productive projects, the maintenance of the expanding system relies on pacifying a hard-working population living close to a level of misery, conquest, or expanding into progressively less fertile lands.

Complex agrarian societies thus evolve through time into metastable organizations that are vulnerable to internal or external perturbations. Internal perturbations may result from mismanagement, rivalries among elite groups, or revolts, while external perturbations may be due to adverse unforeseen consequences of farming (e.g., salinization, erosion, pollution, depletion of key resources), attacks by foreign enemies, or climate change undermining the life-support systems. The perturbations may not lead to a "collapse" in the sense of the breakdown of the organization, but may serve as a means to develop better tactics and strategies to increase the resilience and robustness of the organization, or to place it on a new level (of complexity) where productivity, at least in the foreseeable future, does not exceed the cost of managing and maintaining the complex organization. The effect of the perturbation may also be temporary. The system may recover due to the reimplementation of the previous level of complexity. In the case of climate change, the impact of the perturbation is also a function of the rate of environmental recovery.

In the case of Egypt (Hassan 1993, 2007c; O'Connor 1974), the sudden, unanticipated series of reduced Nile flood discharges led to a breakdown of centralized authority for a few decades. This breakdown, associated with famines, violence, and civil disorder, was primarily due to a social system that exploited farmers to support a non-food-producing elite, who siphoned labor and resources for the construction of monumental pyramids and to lead a life of luxury and leisure. Within decades, bountiful Nile floods were resumed, and a new set of "managers" emerged to reunify the country and reestablish the monarchy. Within two centuries and following internal conflicts between rivals for the throne, all of Egypt was again governed by monarchs who restructured the government and promulgated a new code of ethics and royal responsibilities.

China and Central America: Agriculture, Climate, and Collapse

Examples from Central America and China also show marked differences in the pace and timing of cultural developments in response to global climatic events. In Central and South America, wet conditions from 14,600 cal BP with a maximum from 13,000 to 8,800 cal BP led to the formation of lakes, attracting hunters and foragers (Messerli et al. 2000). In central Chile, however, Paleo-Indian occupation ended as a result of rapidly increasing aridity ca. 11,000 cal BP that caused mega faunal extinction (Núñez, Grosjean, and Cartajena 2001). As a result of the droughts Paleo-Indians retreated to ecological refuges around lakes (see also Zárate, Neme, and Gil 2004)[8]. In the Atacama Desert (and Altiplano of northern Chile), the first period of human occupation came to an end by ca. 9,000–8,800 cal BP. Hyperarid conditions from 8,800 until 4,000 cal BP made the region inhospitable (Messerli et al. 2000). Extreme events at Querbrada Puripica from 8,800 to 5,500 cal BP contributed to the transformation of Archaic hunters into the complex Late Archaic cultural tradition (Grosjean et al. 1997).

The few areas blessed with flowing water (e.g., Puripica on the western slopes of the High Andes) provide evidence of domestication by 5,500 cal BP and a shift to farming with channel irrigation and terracing after 3,300 cal BP. Since the shift from hyperarid to more humid conditions with a return of rain-fed lakes occurred ca. 4,000 cal BP, domestication would have been achieved under hyperarid conditions, and farming under more moist conditions. However, the lack of detailed, high-resolution environmental data hampers any attempt to make reliable interpretations. The shift to farming coincided, however, with a large-scale reoccupation of the Atacama area (by 3,400 cal BP). Data on climatic variability over the past 3,500 years in the Yucatan Peninsula (Mexico) (Curtis, Hodell, and Brenner 1996; Hodell, Curtis, and Brenner 1995) reveal that the period 3,500–1,800 cal BP was wet. Afterwards exceptionally arid events occurred at AD 862, 986, and 1051.

The period AD 980–1050 also correlates with the first phase of the Medieval Warm Period, which is manifest in Europe, North Africa, and North America. Curtis et al. (1996) attribute what they refer to as the Classic Maya Collapse (AD 750–830) to droughts. They also note that other cultures in South and Mesoamerica experienced declines at or near the time the Classic Maya collapse, citing the abandonment of Teotihuacán, Mexico, around AD 750–800, and the collapse of the Andean Tiwanaku at about AD 1000. Shimada et al. (1991) also suggest that the prominent pre-Hispanic culture of Mochica (also known as Moche), with its heavy dependence on large-scale irrigation, experienced an upheaval as a result of a series of severe sixth-century droughts, including one of the severest droughts of the past 1,500 years in AD 562–594.

There are, however, reasons not to single out climate change as the main cause of the collapse of the Classic Maya. For example, the Classic Maya civilization developed under arid conditions and is thus likely to have developed mechanisms to cope with droughts. Also, the collapse alternatively may be explained by a shift in organizational strategy, population movements, or the cessation of monumental carving in Mayan cities. Moreover, the facts that the hydrogeological situation is rather complex and that we lack a model of the sociodynamics of the collapse cast serious doubt on the primary role of climate as a forcing factor. Scarborough (2003) does provide a sociodynamic interpretation, arguing that the collapse was primarily caused by what may be called *cultural gigantism*. The Classic Maya depended on an elaborate system of waterworks and control by elites who continued to aggrandize their civic centers, creating a superstate in a fragile, dry subtropical environment. Their growth was unsustainable, and severe droughts would have only hastened the civilization's demise.

Some of the most salient global climatic events that have had a significant impact on the course of human civilization appear to have been connected with cold events associated with equatorward shifts of the Intertropical Convergence Zone (ITCZ). This, in turn, causes droughts in a transcontinental belt that extends from Central Mexico and the Andes to China (Hassan 2002a). In China, Cohen places the initial period of rice domestication at 14,600 cal BP in the south at Hamadong, 16,000–13,000 in the Xia Ren Phase in northern Jiangxi (Xianrendong and Diatonghuan sites) and 13,000–10,800 during the Wang Phase (Cohen 1998). A subsequent shift from garden to paddy rice agriculture occurred in the Jiangxi phase (11,000–6,800 cal BP). Liu (2004) links the emergence of sedentary villages with pottery, dogs, and pigs to 11,000–9,700 cal BP in the central plains with the advance of monsoonal rainfall northward into the present arid and semiarid regions. It appears that the transition to the intensive collection of rice was linked to the postglacial warming commencing 16,000 years ago. Additionally, the Younger Dryas (13,000–11,600 cal BP) was followed by the beginnings of full domestication, paddy farming, and the establishment of sedentary villages during the period 11,600–8,200 with warm, wetter conditions. Recently, a five-year resolution, absolute-dated oxygen isotope record from Dongge Cave, southern China, provided evidence of a warm, wet season during the summer months when the Intertropical Convergence Zone (ITCZ)[9] shifts northward and monsoonal convective rainfall reaches its maximum.

During a period of warm and wet climate from 9,800 to 4,500 cal BP, early Neolithic cultures were established in the Yellow River valley from 9,000 to 7,000 cal BP. Under more humid and warmer conditions from 7,000 to 5,000 cal BP, populations increased rapidly. According to Huang and Zhang (2000) pollen, phytoliths,[10] and carbonized grains reveal that cultivated varieties of rice occurred slowly from 7,000 to 6,300 cal BP, with maximum variation from 6,300 to 5,500 cal BP (late Neolithic) due to effective artificial selection. Millet also appears with sedentary villages at 10,000–7,000 cal BP. During the period 6,000–5,500 cal BP, the number of agrarian settlements increased dramatically.

According to Stanley and Chen (1996), Neolithic settlements in the southern Yangtze delta plain dating from 7,500 to 4,400 cal BP began within five hundred years of delta formation as a result of sea rise. However, continued sea level rise led to a rise in groundwater level and poor drainage, resulting in a shift of Neolithic settlements eastward toward higher, more restricted areas of the Yangtze delta chenier plain. The transition to fortified settlements in the middle reaches of the Yangtze in 6,400–6,100 cal BP (Yasuda et al. 2004) coincided with the onset of arid conditions as a result of decreased monsoon activity and cold climate, as in North Africa (see also Wang et al. 2005). Rapid urbanization at 5,300 cal BP also coincided with a period of drier conditions in the Eastern Sahara. A period of abandonment of settlements at 4,000 cal BP apparently was caused by the 4,200 abrupt cooling event (Yasuda et al. 2004). Liu (2004) recognizes the multiple roles of population increase, shifts in the course of the Yellow River, soil erosion from farming, and climatic fluctuations in the rise of hierarchical complex societies in China as a result of conflicts leading to political integration. He also concludes that the Longshan culture in the Yellow River valley declined around 4,000 cal BP. Wenxiang and Tungsheng (2004) similarly recognize a collapse of Neolithic cultures around Central China at the time of the 4,000 cal BP cooling event.

CONCLUSION

Climate changes at different scales. Societies are often "adapted" or better "tuned" to multidecadal fluctuations, but are not immune from experiencing environmental stresses caused by unanticipated multicentennial and millennial severe, abrupt climatic events. This is mostly a function of the limited temporal range of collective memory, which is often restricted to a few generations. However, even with memory aids, such as historical records, predictability of climatic events far into the future is virtually impossible and too distant to matter within the operational range and concerns of human societies. Accordingly, an abrupt climatic event consisting of a series of unanticipated severe droughts or extremely anomalous climatic events within the span of decades on the basis of the range of decadal climatic fluctuations, such as the Younger Dryas, are likely to have an impact on human societies.

Nevertheless, abrupt climatic events do not *determine* culture change, which ultimately depends on local ecological conditions, previous cultural modalities and norms, and unpredictable social dynamics. Such dynamics are shaped primarily by political organization, technological aptitude, economy, and religion/ideology, but are also subject to individual initiatives and decisions made under uncertainty.

No society can stop climate change, but efforts can be made to minimize vulnerability to the deleterious impact and enhance the resilience of the social system. Societal upheavals as a result of extreme climatic events may provide an incentive to relocate, intensify food production, or reorganize political institutions governing labor, exploitation of resources, distribution, consumption, and environmental impact. A case in point is the "collapse" of the Old Kingdom in Egypt. During the span of 200 years that began with chaos and disorder, the Egyptians reconsidered and revised the function and responsibility of centralized government, the organization of the administrative structure, and social ethics (Callender 2000; Seidlmayer 2000).

Any response to climate and environmental change has short-term and long-term consequences, and most individuals and interest groups are likely to pay more attention to their own short-term gains than those of society at large, future generations, or the planet. The current attempt at the greening of politics, which aims to remedy this shortsighted modus operandi of societies, is at least a start in remedying the potential catastrophic consequences of this approach, which cannot be sustained under current global conditions (see Barry 1999 on environment and social theory).

Climate change is only one of many variables that can lead to the collapse of a complex state society. There are more insidious internal forces that are likely to undermine the resilience of the state. What is important and potentially fruitful is to expose the structural weakness in a complex society that can hasten its breakdown and increase its vulnerability to climate change.

Responses to environmental stresses and opportunities caused by the transition from the last Ice Age to the postglacial warming have led in many areas to the invention and spread of a variety of food-producing economies. Incipient social complexity based on communal and intercommunal management of labor, storage, and sharing shortly followed, perhaps in response to the vulnerability of agrarian regimes to failures. This may be due to the nature of agrarian ecology and management, as well as decadal and centennial climatic fluctuations affecting soil moisture, seasonal distribution of rainfall, and the amount of heat available in the growing season. In certain ecological settings, where a large number of communities could be coordinated and where the agricultural potential is high, more advanced complex (state) organizations emerged providing both psychological comfort in the face of environmental adversities and a modicum of "relief" (or at least the promise of relief) in times of famine. One example is provided by one of Egypt's governers, Ankhtifi (Hassan 1997; Vandier 1950), at the time when

Egypt was suffering from the famines that led to the breakdown of the Old Kingdom. However, those beneficent lords were also top consumers who benefited from greater yields and hence from expanding the labor force (the "taxpayers"), the agrarian land, and agrarian productivity per unit area. It was thus to their benefit that societies became larger and larger, which meant greater complexity and a progressively higher cost of managing the collection of revenues, transport from the distant corners of states and empires, payment and rewards for a religious establishment to pacify the population, and payment and rewards to a police force and an army to prevent or suppress revolts and to conquer neighboring lands or defend their possessions (Hassan 2006).

The aggrandizement and maintenance of complex state societies have been sustained against climatic perturbations and repeated climatic upheavals. This was achieved through successive transformations of state organization, the adoption of technological innovations, colonization, and the reorganization of economic strategies. The latest and most remarkable shift as far as we are concerned has been the marked transition over the last two hundred years to industrial production; the use of mechanized and techno-farming; and the use of fossil fuel, electricity, and nuclear energy as main sources of energy (for the expansion and consolidation of European states from 1415 to 1980, see Abernethy 2000).

The consequences of the managerial and productive revolution that began two hundred years ago has now led to a spiraling world population, local depletion or pollution of critical resources (e.g., water), as well as concentration of population in urban hotspots, and progressively increasing consumption rates by ever-increasing segments of the world population.[11]

Today, the mesmerizing depiction of "climate change" as the culprit that will bring our civilization down should not lead us to think that climate change is the only threat to humanity. It should in fact make us realize that we have created a social system that appears to be vulnerable to climate change. There are already more than one and half billion human beings who are in a state of abject misery, with no access to clean drinking water in a world of rabid consumerism and ostentatious consumption. We already suffer from spiraling world population that with or without climate change will spell the end of human existence if continued unchecked. If the "scare" of climate change does not lead us to a deeper examination of the ills of our global, complex society with its serious mismanagement, inequities, regional and national rivalries, sectarian conflicts, and heavy toll on natural resources, then we are likely to continue to slide closer and closer to higher levels of inequities, violence, disorder, and repression.

Although it is feasible and potentially possible to redeploy existing capital, human resources, science, and technology on a global scale, we will not likely escape from the nationalist and sectarian agendas that belong to previous historical eras. The continuation of such policies within the context of the current potential for disseminating information and mobilizing masses outside the control of the nation states, as well as access to the means of armed confrontation outside the norms of the state and international conventions, make this world particularly vulnerable to cataclysmic internal disturbances and external factors, like climate change. The strategies of repression, exploitation, and conquest can no longer be entertained as a response to those who pose a threat to organized societies. Perhaps the threat of climate change with its transnational scale and the common global threat of water shortages and planetary ecological morbidity can foster a realignment of current national and international policies to provide world governance, at least in the domains of matters that threaten our collective survival (see Conca and Dabelko 2002; Orlie 1997).

As a technical note, Taylor predicted in 1999 that it would be another twenty years before the climate changes that are predicted to be associated with the greenhouse effect become large enough to be unambiguously differentiated from naturally occurring variations in climate. In less than a decade, we are already witnessing dramatic signals of a global warming. As Taylor recommended, we should act swiftly, since procrastination will prevent making timely and urgent informed decisions and will increase the social and economic costs.

Notes

1. See www.pages.unibe.ch.
2. This refers to mechanisms that "force" the climate to change. See www.ace.mmu.ac.uk/eae/Climate_Change/Older/Climate_Forcing.html, www.ncdc.noaa.gov/paleo/forcing.html.
3. Known as Dansgaard-Oeschger (D/O) events. These events started with an abrupt warming by around 5°C within a few decades or less, followed by gradual cooling over several hundred or thousand years. However, the cooling phase often ends with an abrupt final temperature drop back to cold conditions (stadial). Another major type of climatic event in glacial times is the Heinrich (H) event. This type of event consists of cold spells due to surging of the Laurentide Ice Sheet through the Hudson Strait, occurring in the cold stadial phase of some D/O cycles (Rahmstorf 2001).
4. There are several studies of global climatic changes for the last 100,000 years including data from ice cores as well as a high-resolution palaeoenvironmental records from other sites like Lago Grande di Monticchio, Italy by Allen et al. 1999.
5. A summary of the peopling of the earth from 100 kyr to 10 kyr is provided by Haywood 2001.
6. *Cal BP* refers to age determinations based on radiocarbon dating calibrated into equivalent solar years before present.
7. *Maquis* is a shrubland in Mediterranean countries, typically consisting of densely growing evergreen shrubs such as sage, juniper, and myrtle.
8. See also the work by Sandweiss and coworkers in Peru (Sandweiss 2003; Sandweiss et al. 2004).
9. The Intertropical Convergence Zone (ITCZ) is a belt of low pressure girdling Earth at the equator. It is formed by the vertical ascent of warm, moist air from above and below the equator.
10. Phytoliths are microscopic bodies, mostly of silica, that occur in many plants. They can be retrieved after the decay of the plants.

11. Compare with Messerli et al. 2000, http://www.geohive.com/showcase/atlas.html for world population and http://www.globalchange.umich.edu/globalchange2/current/lectures/human_pop/human_pop.html for world population growth in history, and Hassan 1983 on population and cultural evolution.

REFERENCES

Abernethy, D. 2000. *Global dominance: European overseas empires 1415–1980.* New Haven, CT: Yale University Press.

Adams, J., M. Maslin, and E. Thomas. 1999. Sudden climate transitions during the Quaternary. *Progress in Physical Geography* 23(1): 1–36.

Allen, B. D. and R.Y. Anderson 1993. Evidence from western North America for rapid shifts in climate during the last glacial maximum. *Science* 2.

Allen, M. R., P. A. Stott, J. F. B. Mitchell, R. Schnur, and T. L. Delworth. 2000. Quantifying the uncertainty in forecasts of anthropogenic climate change. *Nature* 407: 617–20.

Allen et al. 1999. Rapid environmental changes in southern Europe during the last glacial period. *Nature* 400: 740–43.

Alley, R. B., P. A. Mayewski, T. Sowers, M. Stuiver, K. C. Taylor, and P. U. Clark. 1997. Holocene climatic instability, a prominent widespread event 8200 yr ago. *Geology* 483–86.

Ambelard, S. 1996. Agricultural evidence and its interpretation on the Dhars Tichitt and Oulata, south-eastern Mauritania. In *Aspects of African archaeology*, eds. G. Pwiti and R. Soper, 421–27. Harare: University of Zimbabwe Publications.

Asouti, E. 2006. Beyond the Pre-pottery Neolithic B interaction sphere. *Journal of World Prehistory* 20: 87–126.

Barker, G. and D. Gilbertson, eds. 2000. *The archaeology of drylands: Living at the margin.* London: One World Books, Routledge.

Bar-Yosef, O. 1998. Agricultural origins: Caught between hypotheses and a lack of hard evidence. In *The transition to agriculture in the Old World*, special issue eds. O. Bar-Yosef, *The Review of Archaeology* 19: 58–64.

Barry, J. 1999. *Environment and social theory.* London: Routledge.

Bellwood, Peter S. 2005. *First farmers: The origins of agricultural societies.* London: Blackwell.

Binford, L. R. 1968. Post-Pleistocene adaptations. In *New perspectives in archaeology*, eds. S. R. Binford and L. R. Binford. Chicago: Aldine.

Bond, G. et al. 1997. A pervasive millennial-scale cycle in North Atlantic Holocene and glacial climates. *Science* 278: 1257–66.

Bradley, S., K. Briffa, T. J. Crowley, M. K. Hughes, P. D. Jones, and M. E. Mann. 2001. The scope of medieval warming. *Science* 292: 2011–12.

Bradley R. S., M. K. Hughes, H. F. Diaz. 2003. Climate change in medieval time. *Science* 17: 404–05.

Bryan, A. L. and R. Gruhn. 2003. Some difficulties in modeling the original peopling of the Americas. *Quaternary International* 109–110: 173–79.

Breunig, P. and K. Neumann. 2002. From hunters and gatherers to food production: New archaeological and archaeobotanical evidence from the West African Sahel. In *Droughts, food and culture: Ecological change and food security in Africa's later prehistory*, ed. F.A. Hassan, 123–156. New York: Kluwer Academic/Plenum Publishers.

Bridges, M. 1997. Soils. In *Applied climatology: Principles and practice*, eds. R. D. Thompson and A. Perry, 111–21. London: Routledge.

Burel, F. and J. Baudry. 2003. *Landscape ecology: Concepts, methods and applications.* Enfield, NH: Science Publishers.

Burroughs, W. J. 2005. *Climate change in prehistory: The end of the reign of chaos.* Cambridge: Cambridge University Press.

Butzer, K. W. 1996. Ecology in the long view: Settlement histories, agrosystemic strategies, and ecological performances. *Journal of Field Archaeology* 23: 141–50.
Callender, G. 2000. The middle kingdom renaissance (c. 2055–1650 BC). In *The Oxford history of ancient Egypt*, ed. I. Shaw, 148–83. Oxford: Oxford University Press.
Cappers, R. T. J. and S. Bottema, eds. 2002. *The dawn of farming in the Near East*, Studies in early Near Eastern production, subsistence, and environment 6. Berlin: Ex Oriente.
Childe, V. Gordon. 1928. *The most ancient East*. London: Kegan Paul.
———. 1934. *New light on the most ancient East*. London: Routledge.
———. 1936. *Man makes himself*. London: Qatts.
Cohen, D. J. 1998. The origins of domesticated cereals and the Pleistocene-Holocene transition in East Asia. *The Review of Archaeology* 19: 22–29.
Conca, K. and G. D. Dabelko. 2002. *Environmental peacekeeping*. Baltimore: John Hopkins University Press.
Cordova, Carlos E. 2007. *Millennial landscape change in Jordan*. Tucson: University of Arizona Press.
Costanza, G., L. Graumlich, and W. Steffen, eds. 2007. *Sustainability or collapse? An integrated history and future of people on earth*. Cambridge, MA: MIT Press in cooperation with Dahlem University Press.
Costanza, R, L. Graumlich, W. Steffen, C. Crumley, J. Dearing, K. Hibbard, R. Leemans, C. Redman, and D. Schimel. 2007. Sustainability or collapse: What can we learn from integrating the history of humans and the rest of nature? *Ambio: A Journal of the Human Environment* 36(7): 522–27.
Craik, K. H. 1972. An ecological perspective on environmental decision-making. *Human Ecology* 1: 69–80.
Curtis, J. H., D. A. Hodell, and M. Brenner. 1996. Climate variability on the Yucatan Peninsula (Mexico) during the past 3500 years, and implications for Maya cultural evolution. *Quaternary Research* 46: 37–47.
Dalfes, N. H., G. Kukla, and H. Weiss, eds. 1997. *Third millennium BC climate change and Old World collapse*. Berlin and Heidelberg: Springer-Verlag.
Davis, M. B. and R. Shaw. 2001. Range Shifts and Adaptive Responses to Quaternary Climate Change. *Science* 292: 673–78.
Dearing, J., L. Cromer, and T. Kiefer. 2007. Past human-climate ecosystem interactions. *Pages* 15(1): 1–32.
Derbyshire, E. 1997. Geomorphic processes and landforms. In *Applied climatology: Principles and practice*, eds. R. D. Thompson and A. Perry, 89–110. London: Routledge.
de Menocal, P. 2001. Cultural responses to climate change during the Late Holocene. *Science* 292: 667–73.
Drysdale, R. et al. 2006. Late Holocene drought responsible for the collapse of Old World civilization is recorded in an Italian cave flowstone. *Geology* 34: 101–04.
Edens, C. and T. Wilkinson. 1998. Southwest Arabia during the Holocene. *Journal of World Prehistory* 12: 55–119.
Garcea, E. A. 2004. An alternative way towards food production: The perspective from the Libyan Sahara. *Journal of World Prehistory* 18: 107–54.
Gifford-Gonzalez, D. 2000. Animal disease challenges to the emergence of pastoralism in Sub-Saharan Africa. *African Archaeological Review* 17(3): 95–139.
Goudie, A. 1993 (fourth edition). *The Human Impact on Natural Environment*. 4th. ed. Oxford: Blackwell.
Grosjean, M., L. Lautaro Núñez, I. Cartajena, and B. Messerli. 1997. Mid-Holocene climate and culture change in the Atacama Desert, Northern Chile. *Quaternary Research* 48: 239–46.

Halstead, P. 1996. The development of agriculture and pastoralism in Greece: When, how and what? In *The Origins and Spread of Agriculture and Pastoralism in Eurasia*, ed. D. Harris, 296–310. London: University College London.

Harris, D. 1996. The origin and spread of agriculture and pastoralism in Eurasia: An overview. In *The origins and spread of agriculture and pastoralism in Eurasia*, ed. D. Harris, 552–73. London: University College London.

———. 1998. The origins of agriculture in Southwest Asia. *The Review of Archaeology* 19(2): 5–11.

Hassan, F. A. 1977. The dynamics of agricultural origins in Palestine: A theoretical model. In *Origins of agriculture*, ed. C. Reed, 589–607. The Hague: Mouton.

———. 1981. *Demographic archaeology*. New York: Academic Press.

———. 1983. Earth resources and population: An archaeological perspective. In *How Humans Adapt: A Biocultural Odyssey*, ed. D. J. Ortner, 191–226. Washington, D. C.: Smithsonian International Symposia Series.

———. 1988. The predynastic of Egypt, ed F. Wendorf and A. E. Close. *Journal of World Prehistory* 2(2): 135–85.

———. 1992. The ecological consequences of evolutionary cultural transformations. In *Nature and humankind in the age of environmental crisis*. Kyoto: International Research Center for Japanese Studies.

———. 1993. Population ecology and civilization in Ancient Egypt. In *Historical ecology*, ed. C. L. Crumley, 155–81. Santa Fe, NM: School of American Research.

———. 1997. Climate, famine and chaos: Nile floods and political disorder in early Egypt, in *Third millennium BC climate change and Old World collapse*, NATO ASI Series, vol. I 49, eds. H. Nüzhet Dalfes, G. Kukla, and H. Weiss, 1–23. Berlin and Heidelberg: Springer-Verlag.

———. 2000a. Holocene environmental change and the origins of food production in the Middle East. *Adamatu* 1: 7–28.

———. 2000b. Environmental perception and human responses in history and prehistory. In *The way the wind blows: Climate, history and human action*, eds. R. J. McIntosh, J. A. Tainter, and S. K. McIntosh, 121–140. New York: Columbia University Press.

———, ed. 2002a. *Droughts, food and culture: Ecological change and food security in Africa's later prehistory*. New York: Kluwer Academic/Plenum Publishers.

———. 2002b. Holocene environmental change and the transition to agriculture in south-west Asia and North-east Africa. In *The origins of pottery and agriculture*, ed. Y. Yasuda. Delhi: Lustre Press, Roli Books.

———. 2003. Climatic changes and cultural transformations in Farafra Oasis, Egypt. *Archaeology International* 2003/4: 35–39.

———. 2006. The lie of history: Nation-states and the contradictions of complex societies. In *Integrated history and future of people on earth (IHOPE)*, eds. R. Costanza, L. J. Graumlich, and W. Steffen, 169–96. Cambridge, MA: MIT Press in cooperation with Dahlem University Press.

———, ed. 2007a. *Traditional water techniques for a sustainable future*. Cairo and Alexandria: Bibliotheca Alexandrina.

———. 2007b. Extreme Nile floods and famines in medieval Egypt (AD 930–1500) and their climatic implications. *Quaternary International* 173–74: 101–12.

———. 2007c. Droughts, famine and the collapse of the Old Kingdom: Re-reading Ipuwer. In *The archaeology and art of ancient Egypt: Essays in honor of David B. O'Connor*, eds. Z. Hawass and J. Richards, 357–77. Cairo: SCA.

Hassan, F., B. Barich, M. Mahmoud, and M. A. Hemdan. 2001. Holocene playa deposits of Farafra Oasis, Egypt, and their palaeoclimatic and geoarchaeological significance. *Geoarchaeology* 16(1): 29–46.

Haywood, J. 2001. *Cassell's atlas of world history*. London: Cassell.

Hodell, D. A., J. H. Curtis, and M. Brenner. 1995. Possible role of climate in the collapse of Classic Maya civilization. *Nature* 375: 391–94.
Hoffecker, J. F., R. Powers and T. Goebel. 1993. The colonization of Beringia and the peopling of the New World. *Science* 259(5091): 46–53.
Huang, F. and M. Zhang. 2000. Pollen and phytolith evidence for rice cultivation during the Neolithic at Longqiuzhuang, eastern Jianghuai, China. *Vegetation History and Archaeobotany* 9: 161–68.
Hughes M. K. and H. F. Diaz. 1994. Was there a 'medieval warm period,' and if so, where and when? *Climatic Change* 26: 109–42.
Hulme, M. 2001. Climatic perspectives on Sahelian desiccation: 1973–1998. *Global Environmental Change* 11(1): 19–29.
Kohler-Rollefson, I. 1988. The aftermath of the Levantine Neolithic in the light of ecological and ethnographic evidence. *Paléorient* 14: 87–93.
Issar, A. S. and M. Zohar. 2004. *Climate change: Environment and civilization in the Middle East.* New York: Springer.
Kuijit, I. and N. Goring-Morris. 2002. Foraging, farming and social complexity in the pre-pottery Neolithic of the southern Levant, a review and synthesis. *Journal of World Prehistory* 16(4): 361–440.
Lamb, H. H. 1982. *Climate, history and the modern world.* London: Routledge Kegan & Paul.
Lawlor, D. W. 1997. Agriculture and fisheries. In *Applied climatology: Principles and practice*, eds. R. D. Thompson and A. Perry, 215–27. London: Routledge.
Lees, S. H. and D. G. Bates. 1974. The origins of specialized nomadic pastoralism: A systemic model. *American Anthropologist* 39: 187–93.
Liu, Li. 2004. *The Chinese Neolithic: Trajectories to early states.* Cambridge: Cambridge University Press.
Lutz, W., W. C. Sanderson, and S. Scherbov, eds. 2004. *The end of world population growth in the 21st century: New challenges for human capital formation and sustainable development.* London: Earthscan.
Mace, R. 1993. Transition between cultivation and pastoralism in sub-Saharan Africa. *Current Anthropology* 34: 363–82.
Mannion, A. M. 1997. Vegetation. In *Applied climatology: Principles and practice*, eds. R. D. Thompson and A. Perry, 123–39. London: Routledge.
Marshall, F. 1994. Archaeological perspectives on East African pastoralism. In *African Pastoralist Systems*, eds. E. Fratkin, K. Galvin, and E. Roth, 17–43. Boulder: Lynne Rienner.
Mayewski, P. A. et al. 2004. Holocene climatic variability. *Quaternary Research* 62: 243–55.
Mellars, P. 2002. The *Homo sapiens* peopling of Europe. In *The peopling of Britain: The shaping of a human landscape*, eds. P. Slack and R. Ward, 39–68. Oxford: Oxford University Press.
Messerli, B., M. Grosjean, T. Hofer, L. Nunez, and C. Pfister. 2000. From nature-dominated to human-dominated environmental changes. *Quaternary Science Review* 19: 459–79.
McClure, H.A. 1976. Radiocarbon chronology of the Quaternary lakes in the Arabian desert. *Nature* 363: 755.
Núñez. L., M. Grosjean, and I. Cartajena. 2001. Human dimensions of Late Pleistocene/Holocene arid events in southern South America. In *Interhemispheric climatic linkages*, ed. V. Markgraf, 105–17. New York: Academic Press.
Orlie, M. A. 1997. *Living ethically, acting politically.* Ithaca: Cornell University Press.
O'Connor, D. 1974. Political systems and archaeological data in Egypt: 2600–1780 B.C. *World Archaeology* 6(1): 15–38.

Osborn T. J. and K. R. Briffa. 2006. The spatial extent of 20th century warmth in the context of the past 1200 years. *Nature* 311: 841–44.
Rahmstorf, Stefan 2001. *Climate Abrupt change*. DOI:10.1006/rwos.0269.
Redman, C. 2004. Archaeology of global change: The impact of humans on their environment. Washington, DC: Smithsonian.
Reed, Charles 1977. *Origins of agriculture*. The Hague: Mouton.
Reynolds, J. F. and M. Stafford Smith. 2001. *Global desertification: Do human cause deserts?* Berlin: Dahlem University Press.
Rollefson, G. 1992. Neolithic settlement patterns of the Levant: A synthesis. *Paléorient* 15: 168–73.
Rosen, A. M. 2006. *Civilizing climate: The social impact of climate change in the ancient Near East*. London: AltaMira.
Sandweiss, D. H. 2003. Terminal Pleistocene through Mid-Holocene archaeological sites as paleoclimatic archives for the Peruvian coast. *Palaeogeography, Palaeoclimatology, Palaeoecology* 194: 23–40
Sandweiss, D.H. et al. 2004. Geoarchaeological evidence for multidecadal natural climatic variability and ancient Peruvian fisheries. *Quaternary Research* 61: 330–34.
Scarborough, V. 2003. *The flow of power: Ancient water systems and landscapes*. Santa Fe, NM: School of American Research Press.
Scheffer, M., S. Carpenter, J. Foley, C. Folke, and B. Walker 2001. Catastrophic shifts in ecosystems. *Nature* 413: 591–96.
Schurr, T.G. 2004. The peopling of the New World: Perspectives from molecular anthropology. *Annual Review of Anthropology* 33(3): 551–83.
Seidlmayer, S. 2000. The first intermediate period (c. 2160–2055 BC). In *The Oxford History of Ancient Egypt*, ed. I. Shaw, 118–147. Oxford: Oxford University Press.
Shimada, I., C. B. Schaaf, L. G. Thompson, and E. Mosley-Thompson. 1991. Cultural impacts of severe droughts in the prehistoric Andes: Application of a 1,500-year ice core precipitation record. *World Archaeology* 22(3): 247–70.
Shulmeister J., D. T. Rodbell, M. K. Gagan, and G. O. Seltzer. 2006. Inter-hemispheric linkages in climate change: Paleo-perspectives for future climate change. *Climate of the Past* 167–85.
Simmons, A. H. 2007. *The Neolithic revolution in the Near East*. Tuscon: University of Arizona Press.
Smithers, J. and B. Smit. 1997. Human adaptability to climatic variability and change. *Global Environmental Change* 7: 129–46.
Stanley, D. J. and Z. Chen. 1996. Neolithic settlement distributions as a function of sea level-controlled topography in the Yangtze delta, China. *Geology* 24: 1083–86.
Stine, S. 1994. Extreme and persistent drought in California and Patagonia during mediaeval time. *Nature* 369: 546–49.
Stonehouse, B. 1997. Animal responses to climate. In *Applied climatology: Principles and practice*, eds. R. D. Thompson, R. D. and A. Perry, 141–151. London: Routledge.
Street-Perrot, F. A. 1994. Drowned trees record dry spells. *Nature* 369: 518.
Street-Perrott, F. A. and R. A. Perrott. 1990. Abrupt climate fluctuations in the tropics: The influence of Atlantic Ocean circulation. *Nature* 343: 607–12.
Tainter, J. A. 1988. *The collapse of complex societies*. Cambridge: Cambridge University Press.
Taylor, K. 1999. Rapid climate change. *American Scientist* 87: 320–27.
Turner, C. S. et al. 2001. Early human occupation at Devil's Lair, southwestern Australia 50,000 years ago. *Quaternary Research* 55: 3–13.
Vandier, J. 1950. Mo'alla: La tombe d'Ankhtifi et la tombe de Sébekhotep. Cairo: Institut français d'archéologie orientale.

Walther, G-R et al. 2002. Ecological responses to recent climate change. *Nature* 416: 389–95.
Wang, Y. et al. 2005. The Holocene Asian monsoon: Links to solar changes and North Atlantic Climate. *Science* 308: 857.
Weiss, H. 2000. Beyond the Younger Dryas: Collapse as adaptation to abrupt climatic change in ancient west Asia and the eastern Mediterranean. In *Environmental disaster and the archeology of human response*, eds. G. Bawden and R. M. Reycraft, 75–98. Maxwell Museum of Anthropology, Anthropological paper No. 7.
Weiss, H. and R. S. Bradley. 2001. What derives societal collapse? *Science* 291: 609–10.
Weiss, H., M. A. Courty, W. Wetterstrom, F. Guichard, L. Senior, R. Meadow, and A. Curnow. 1993. The genesis and collapse of third millennium north Mesopotamia. *Science* 261: 995–1004.
Wenxiang, W. and L. Tungsheng. 2004. Possible role of the "Holocene Event 3" on the collapse of Neolithic cultures around the central plain of China. *Quaternary International* 117: 153–66.
Wetterstrom, W. 1998. The origins of agriculture in Africa: With particular reference to sorghum and pearl millet. In *The transition to agriculture in the Old World*, ed. O. Bar-Yosef, special issue of *The Review of Archaeology* 19: 30–46.
Whitehead, P. 1997. Hydrological processes and water resources. In *Applied climatology: Principles and practice*, eds. R. D. Thompson, R. D. and A. Perry, 65–73. London: Routledge.
Yasuda, Y. et al. 2004. Environmental archaeology at the Chengtoushan site, Hunan Province, China and implications for environmental change and the rise and fall of the Yangtze civilization. *Quaternary International* 123–125: 149–58.
Zeng, N. 2003. Drought in the Sahel. *Science* 302: 999–1000.
Zárate, M., G. Neme, and A. Gil. 2004. Mid-Holocene palaeoenvironments and human occupation in southern South America. *Quaternary International* 132: 1–3.

Chapter 2

Climate and Weather Discourse in Anthropology: From Determinism to Uncertain Futures

Nicole Peterson and Kenneth Broad

Introduction

Global climate change has become an increasingly visible topic in public culture over the past few decades, and will likely dominate environmental, political, and social agendas for some time to come. Only in the last few years has a critical mass of anthropologists begun to focus on the social practices and cultural implications surrounding the production of climate change models and scenarios, the communication and interpretation of climate information, climate change causes and solutions, and the implications of its impacts for people worldwide.

The topic of climate has a long history in the social sciences. As far back as Hippocrates, scholars theorized about how climate shapes society, assessing how climate differences, extremes, and seasonal patterns affected human activity (Harris 1968, 41–42). Focus on these topics often led to ethically dubious and racist theories like climatic (or geographic) determinism, in which climate led to certain cultural or social behaviors.

A focus on climate constraints and human responses, or "adaptation," is central to today's multidisciplinary academic discourse, and three major simplifications of the past are being acknowledged: climate is only one of several drivers affecting human behavior, climate on most timescales is not static, and we are capable of influencing and changing global climate. Acknowledging these deviations from geographic determinism opens territory for anthropological exploration.

With a broad acceptance that, to differing degrees, climate organizes and shapes central aspects of our lives, we are now organizing our understanding of climate in epistemologically complex ways: from temporally and spatially explicit predictions of seasonal climate characteristics (e.g., rain, temperature, fires, freezes, pest outbreaks, etc.) to the anthropogenic influence of postindustrial greenhouse gases, with global projections reaching decades into the future. Subsequently, in contrast to the subject of "the weather" that was long considered in some Western cultures to be a politically and emotionally neutral (if not boring and unsophisticated) topic for the "insufficiently prepared conversationalist" (Strauss and Orlove 2003, 29), climate as an issue is playing an increasing role in reinforcing and shifting ideological positions (Leiserowitz 2006).

The main ethical motivation and financial justification for this globalized research effort in climate is to allow better understanding and prediction of our climate in order to manage it in better ways for "societal benefit" (Pielke Jr. and Glantz 1995). Herein lies the social science community's raison d'etre to turn its focus to the issue of climate. Assuming that the anthropological community fully recognizes that climate is intrinsic to both nature *and* society,[1] this chapter highlights some of the directions that contemporary anthropological approaches have gone and can go in critically engaging with climate discourses. In the process of consciously (re)constructing some of the narratives surrounding climate in anthropology, this review aims to provide a context for current climate anthropology (Nelson and Finan 2000), both represented by the contents of this volume and other sources.

Anthropologists often use the term *narrative* to refer to stories and myths shared by field consultants but also to describe how academic research is itself a form of storytelling or mythmaking. Natural and social science research ideas and agendas, including those in anthropology, are directed in part by past studies and narratives about what we do and how we do it (Kuhn 1962). These narratives can suggest and preclude ideas and norms, and thus actions or perspectives. In recognizing and acknowledging the power of these imagined trajectories, we can more easily see alternatives that might aid us in our work of understanding how the world works.

Narrating anthropology's past produces a series of themes and perspectives that appear, shift, and sometimes disappear. Anthropologists have sometimes taken stock of where the field as a whole has been and have shown how it has reflected the biases, interests, and directions of previous anthropological studies (i.e. Harris 1968; Marcus and Fischer 1986; Ortner 1993, 2006). Anthropological research on climate change similarly reflects the biases, interests, and directions of previous climate-related studies (i.e., weather and seasonal climate variability), as well as how social scientists have addressed general environment-human and environment-society interactions (as reviewed by Orlove 1980; Vayda and McCay 1975). While we touch upon some of the same trends and issues as these authors, our review focuses specifically on climate and weather.

We have organized this chapter to reflect a shift in theoretical and practical interests for anthropologists concerned with climate. We suggest that current discussions about climate and weather differ in two ways from earlier interests in local weather, seasonal variations, extreme events, and cosmologies. Attention is now focused on both the global nature of anthropogenic climate change and the production and distribution of forms of scientific knowledge. Accordingly, this current focus deals explicitly with group perceptions and behavior under conditions of uncertainty.

Climate as Local Context and Determinant

It is common for sociocultural anthropologists to include in their case studies details about the climate of their study area, including average rainfall, seasonal variation, and weather anomalies such as droughts, hurricanes, or floods. Such descriptions often treated climate and weather as a static quality that framed the study. Anthropologists rarely integrated the physical characteristics of the location, including climate, into the explicit analysis of other cultural elements such as political organization, perceptions, or identity issues.

Early anthropologists were not just content with cataloguing the variety of human lives around the world; they were also interested in the drivers of societal and cultural variation, one of which was the environment (Orlove 1980). However, geographical (or environmental) determinism had dominated pre-Enlightenment intellectual thought, beginning with Hippocrates and Aristotle and lasting well into the twentieth century (Harris 1968; Moran 1982). In fact, writers drawing on the early Greek philosophers, most prominently Ibn Khaldun, Montesquieu, and Compte, considered climate an important factor for health, physical and personality characteristics, and sociopolitical organization (Boia 2005; Harrison 1996; Huntington 1912). Not surprisingly, the home location of the scholar was claimed to have the optimal climate for supporting "normal" civilization; as one moves from north and south, climate becomes harsher (i.e., less hospitable to European clothing, housing, and agricultural customs of the time), and (in modern parlance) "otherness" increases. This division of the world into north and south still conceptually symbolizes relations of power, domination, and control by some states/individuals over others, as analyzed in many of the social sciences.

Nineteenth-century anthropologists readily explained cultural or physiological differences based on climate variation: why did one group have darker skin, longer noses, low population density, co-sleeping, or matrilocal residence, while other groups didn't? Average temperatures, seasonal variation, and other climate-related variables were used as explanatory factors for both physical and cultural variation (Brookfield 1964; Whiting 1964). For instance, Wissler (1926) argued that the overlap of climatic and cultural zones in North American native communities was based in an ecological relationship between the two. Unfortunately, environmental determinism arguments were often used to justify racism and imperialism (Frenkel 1992). This, and the rise of Boasian cultural anthropology, led to its gradual disappearance from the discipline.

With the late-nineteenth-century introduction of Boas's historical possibilism (Boas 1896), mainstream anthropology rejected environment and climate as the sole determiners of societal and cultural tendencies and either posited other influencing factors (see Ember and Ember 2007 for a recent example), or juxtaposed deterministic orientations by illuminating the role of culture in shaping human responses to climate (Krogman 1943; Laughlin 1974; McCullough 1973; Sahlins 1964). Contemporary analyses continue in this vein. For instance, Reyna's (1975) study of Barma in Chad suggests that cultural devices like bride wealth are used to adapt to rainfall cycles.

Anthropologists began to identify the importance of other explanatory factors in addition to or in place of climate—like political, economic, or social issues—that drove such cultural phenomena as migration (Siegel 1971). In tandem, researchers continued to consider how variation might be due to climate differences, or exist despite similarities, such as comparing production strategies in similar alpine environments around the world (Rhoades and Thompson 1975; Wolf 1972). Cultural ecologists, including Steward and White, focused on how societies adapted to their environment and available technologies. Other research critiqued and built on this work, providing the foundation for ecological or environmental anthropology (Moran 1982; Orlove 1980).

British anthropologists similarly rejected determinism but turned to an emphasis on structural-functionalism, focused almost entirely on the social structure of a society, and rejecting deterministic and cultural ecological explanations. Radcliffe-Brown and others focused on how human-environment relations are dependent upon, the result of, and the means of maintaining a certain social structure (Harris 1968).

Rayner (2003) argues that the "chauvinistic approach" of determinism ultimately resulted in an eighty-year backlash, in which anthropologists purposefully avoided climate as a research topic. The impetus to reengage with climate as a topic can arguably be retraced to the rise of political economy, and the anthropological interests in disaster research and cognition. Political economy, the analysis of the relations between the political, economic, and social spheres of a society, along with structural Marxism in the 1970s, would have a continuing influence on anthropological engagements with many topics, including environment and climate (Ortner 1993).[2] While in structural Marxism ecological relations remain secondary to social relations and ideologies, political economy has had a more lasting impact on environmental and climate anthropology, particularly through its interest in interactions and inequity within a world system. Political ecology, combining this interest with cultural ecology, focused on how relations between humans and their environments are mediated by wealth and power (Netting 1996).

Political economy also sparked debates about how inequalities in access to resources arise and are maintained, leading some to question development policies and practices around the world. For example, studies of so-called natural disasters began to explore the social relationships that increase the risks and dangers of certain populations (Oliver-Smith 1996). In addition, later anthropological studies of climate forecast use drew upon development critiques emphasizing the ways that wealth or power direct technologies, like those associated with the Green Revolution (Agrawala and Broad 2002; Finan 2003; Roncoli, Ingram, and Kirshen 2001).

Disaster anthropology emerged in the 1960s, and worked to bring together previous anthropological accounts of floods, volcanoes, earthquakes, fires, and droughts with an explicit focus on disasters as a topic of research (Hoffman and Oliver-Smith 2002; Oliver-Smith 1996; Torry 1979). Anthropologists like Tony Oliver-Smith proposed that anthropology had a unique role to play in disaster research, based of the discipline's strengths in attending to all the areas of life that disasters touch (Oliver-Smith 2002; see also his contribution to this volume). Anthropology was also considered a "good fit" since disasters are the outcomes of the interaction between nature and society, a central focus of the discipline. Cultural ecologists urged anthropologists to approach disasters or hazards (defined as the "natural" part of disasters) from an ecological and social organizational perspective—focusing on the ability of individuals to adapt to changes in the environment (Vayda and McCay 1975). Anthropology's focus on the social elements of disasters increased interest in further defining the social experiences and situations of populations prone to be victims. Since the 1980s, researchers have focused on concepts of vulnerability, resilience, and adaptation as ways to understand the social bases of disasters (Oliver-Smith 1996). More recent work with vulnerability emphasizes unpacking the concept to see the underpinnings, including the problems of seeing nature separate from society (Oliver-Smith 2002). In addition, some anthropologists are assessing the propensity for technological tools used to monitor disasters to overshadow other interpretations of the event, as in the 1997–98 El Niño exacerbated fires in Indonesia (Harwell 2000).

In addition to this increased interest in disasters, the 1980s saw anthropologists sharpening their focus on ideas and symbolic systems, in part generated by emerging psychological and cognitive studies in other disciplines (D'Andrade 1995). Cognitive anthropology emerged at this time from a combination of earlier Boasian interests in ethnoscience and ideas and methods from linguistics (D'Andrade 1995). Ethnobiologists constructed folk taxonomies from the rich native knowledge of plant and animal species, which led to greater understanding about human cognitive talents (Berlin 1992). Studies of ethnoscience or ethnoecology asked how human ideas and knowledge about meteorology affected adaptation to climatic conditions (Brookfield 1964; Grivetti 1981; Waddell 1975). For example, Sillitoe (1993) examined the ethnometeorology of a group in Papua New Guinea, suggesting that their understanding of climate and weather affects their ritual life. Specifically, local people ask a white woman's spirit to act morally and so ensure a return to normal weather. More recent work also focuses on ethnometeorology, but with a particular interest in climate change (Huber and Pedersen 1997; Ingold and Kurttila 2000; Strauss and Orlove 2003; Vedwan 2006).

Current Anthroclimatology (or Climate Anthropology)

Several changes mark the transition between earlier work with climate and newer engagements. Anthropologists in the 1990s began to engage global movements of information, people, and objects in their work, most noticeably in research dealing with debates about the hegemony of Western scientific knowledge, recognition of El Niño's global reach, greater concerns about inequities and vulnerabilities, increased application of climate forecasts and other technologies, the effects of participation on understanding uncertainties, and the role of power and inequalities in the effects of and responses to climate change. While prior work with climate and weather emphasized the local nature of meteorological experience,[3] recent projects represent a shift in how human-climate interactions are conceptualized, particularly in the context of understanding the global scale of these interactions.

First of all, interest in "folk" climate models inevitably created a contrast between traditional and scientific knowledge, leading many to ask how certain kinds of knowledge (i.e., Western scientific discourses) become privileged. An interest in culturally specific knowledge and ethnoscience led naturally to an interest in the intersection of traditional ecological knowledge (TEK) and Western "scientific" forecasts (Grivetti 1981). This line of research asked how locally generated knowledge has been used to predict weather or climatic events, and was a move towards recognizing a plurality of scientific knowledge about the world (Cruikshank 2001).

Similarly, studies of local knowledge focused on cultural and decision-making models, in which climate ideas and information are one influence on behavior (Durrenberger and Pálsson 1986; Paolisso 2002). In anthropological studies of agricultural decision-making, models of information and its application have become one way to understand how different individuals and groups frame their environment. On a more general, aggregated social scale (vs. cognitive mapping), this research has also included an explicit interest in American environmental values and cultural models, including how these relate to attitudes about climate change (Kempton 1997; Kempton, Boster, and Hartley 1996). Much of the recent focus, in step with broader trends of environmental activism and media coverage, has been a focus on environmental groups and energy use (Henning 2005; Kitchell, Hannan, and Kempton 2000). Halvorsen and her colleagues (2007), for example, emphasize the problem of relying on older cultural models, such as for ozone depletion, for understanding current climate problems, even among experts.

A second change influencing recent climate research and understanding its impacts is the recognition of the global influence of El Niño and La Niña on climate variability. The idea of a global climate has specific implications for how anthropologists have undertaken place-based research. In attempting to understand this mix of local and global events, climate anthropologists have drawn on globalization theories from anthropology and beyond, often bringing new theoretical insights into the debates that highlight the role of identity, imagery, and the nation-state in mediating these informational assemblages (Broad and Orlove 2007; Sturken 2001). In considering the relationships between places and information, anthropologists have rethought relationships between media, governance, and society to critique assumptions about control and power. Strauss (2003) illustrates this in her fascinating analysis of the "synopticon" in which local information becomes part of a global weather discourse through outlets like the Weather Channel.

Thirdly, drawing from the work on vulnerability coming out of disaster research and earlier work on occupational multiplicity (Comitas 1973), anthropologists have started to focus on the central role of institutional flexibility for successful adaptation. For example, inasmuch as climate is a constraint on decision-making, it can also provide an opportunity for creative agency (Bennett 1982; Jennings 2002). These constraints might be evinced by market changes (Smit et al. 2000), social network extensions (Adger 2003), or other mechanisms. This interest in adaptive capacity is evident in a variety of research topics, from natural resource and livestock management (Galvin et al. 2001; Jennings 2002) to Arctic responses to climate change (Duerden 2004). At least one issue of the interdisciplinary journal *Climatic Change* has focused on the relationships among climate variability and vulnerability, incorporating risk, uncertainty, learning, and resource management into various case studies (de Loë and Kreutzwiser 2000; Eakin 2000; Reilly and Schimmelpfennig 2000). Other researchers also explicitly apply vulnerability concepts to climate change (Adger et al. 2001; Magistro and Roncoli 2001; Vásquez-León 2002). In drawing upon and extending research regarding vulnerability, these social scientists have incorporated local experiences with global events to focus on who is at risk, why they are at risk, and what might be done about it.

Importantly, the earlier work on ethnoecology has led to important studies of local knowledge of climate change and social adaptations to change, particularly in the Arctic. Researchers have started to think about perceived differences between TEK and scientific information, both in how TEK can add to scientific observations of climate change and how local knowledge can be valued and included in the coproduction of both mitigation and adaptation. Specifically, emphases on co-research or community-based research with indigenous groups point towards a constructive integration of different sources of knowledge, particularly when an emphasis on complementarities replaces struggles over authority (Berkes 2002). The Krupnik and Jolly volume (2002) includes many examples of such partnerships forged through innovative techniques, including daily diaries, youth-elder camps, and expert-to-expert interviews. Sometimes, however, the dichotomization of TEK and scientific knowledge reproduces a potentially artificial division of scientific and nonscientific information rather than recognizing the social construction of all knowledge, and the underlying similarities among various information sources.

The fourth major shift is reflected in anthropology's interest in the role of technology in vulnerability and adaptation, including GIS and climate forecasts (Breuer et al. in press; Finan and Nelson 2001; Harwell 2000; Ziervogel and Calder 2003). Seasonal climate forecast use has sparked an interest in anthropology at the intersection of environment, development, and agriculture. Consequently, anthropologists have forged collaborations with climate forecasters, agricultural development agencies, and others in an attempt to grasp the significance and utility of climate forecasts and El Niño events for people around the world. Recent ethnographies have reflected these interests, turning their gaze on the subculture of meteorology (Fine 2007), the forecast production and dissemination process (Agrawala, Broad, and Guston 2001; Finan 2003), and working to understand the relationships between forecasts and climate-based decisions (Taddei 2005). These studies have generated their own discourse, arguing that climate information—including its content, format, timing, dissemination approach, etc.—must be tailored to very local contexts and activities to be of use to the targeted populations. In addition, the targeted populations may have difficulty understanding or implementing technically complex forecasts (Lemos et al. 2002). All these issues must be considered in research on climate change as groups and individuals find themselves struggling with uncertainties about when and to what extent changes will occur, their participation in identifying and enacting adaptation and mitigation strategies, and the role of technology in this process. To some extent these discussions have already started, in the Arctic and elsewhere (Krupnik and Jolly 2002).

Similarly, anthropologists have also begun to study the underlying tension in forecast dissemination between scientific forecasts and local climate prediction practices (Ajibade and Shokemi 2003; Orlove et al. under review; Pennesi 2007; Taddei 2005). While most focus on the differences between TEK and scientific knowledge, some research shows important points of congruence (Orlove, Chiang, and Cane 2002). Furthermore, by comparing the various means of understanding weather and climate it is clear that social, political, and ideological positioning also influences the acceptance of information (Leiserowitz 2006; Taddei 2005).

The fifth change addresses how uncertainty is understood. Climate forecasts are characterized by a great deal of uncertainty, often expressed probabilistically, drawing on a history of work in disasters and risk in anthropology and elsewhere (Boholm 2003; Davis 1998; Hackenberg 1988; Hoffman and Oliver-Smith 2002). More recent research suggests that uncertainty can be difficult (Cash, Borck, and Patt 2006), but not impossible, to communicate (Phillips and Orlove 2004; Suarez and Patt 2004). Participatory processes can help communicate this information (Patt, Suarez, and Gwata 2005), but may not be sufficient on their own (Broad et al. forthcoming). At the same time, climate research on uncertainty connects to a history of research on risk, both in terms of its social construction (Douglas and Wildavsky 1983; Slovic, Fischhoff, and Lichtenstein 1986) and its perception (Leiserowitz 2006).

Lastly, anthropologists have also started to work on the human rights and social justice issues of climate change. For example, Crate (2008) considers the interactions of climate change, culture change, and human rights, and Adger (2003) investigates the role of social capital in adaptation. Political ecologists have been particularly interested in connecting risk to various social conditions and relationships, and have begun to think about vulnerability to climate in the same way (e.g., Vásquez-León, West, and Finan 2003; Ziervogel, Bharwani, and Downing 2006).

The Future of Climate in Anthropology

Studying human behavior linked to climate change poses challenges that differ from earlier studies of weather or seasonal climate. Our mental models of the world's natural processes are shaped by experience, evolutionary processes, and our daily experiences. As events become spatially and temporally distant—either forward or backward in time—our ability to tease out relative objectivity vanishes. In this way, weather versus climate becomes an important distinction in understanding human responses to climate happenings. The statistical average of weather events is what we know intellectually to be climate. How recent storms or droughts, floods or famines, have affected us personally—physically or emotionally—and how they are framed by key intermediary groups such as the media, are more likely to account for our perception of the climate. Reconstructions of past climate from proxies such as corals, tree rings, and gases trapped in ancient ice reveal the dramatic changes in temperature and precipitation that our planet has undergone over the millennia; sometimes in just days (e.g., volcanic eruptions) or decades ("The Little Ice Age"), and on a global scale. On timescales that we can embrace full cycle are El Niño and La Niña; a recurring phenomenon collectively known as ENSO—the El Niño Southern Oscillation. "Strange weather" and seasonal disruptions in distant continents are linked to ENSO, and through a combination of an earth observatory system and computer models, we can predict these events months in advance, and with variable skill we can anticipate the ensuing impacts in large regions of North America, Asia, Africa, and South America. As the quest for order continues, climatologists are perpetually in search of statistical patterns in the climate data. They make (often highly disputed) claims of identifying phenomenon that recur on decadal or longer timescales, and add more acronyms to their unwieldy lexicon[4]. Beyond these patterns of natural variability, we humans have been fingered as the culprits in affecting the "natural" climate, resulting in global warming.

One approach to organizing the material and symbolic implications of climate is to conceptualize them in terms of their sociopolitical scales. For example, the debate surrounding the (non)signing of the Kyoto Protocol brings into relief charged linkages of climate with concerns of development, equity, environmentalism, and globalization critics (see Litfin 1994). Most generally, the increasing inseparability of climate from discourse of universal values, when framed as an atmospheric common property dilemma, brings us to the "luxury" versus "survival" emissions debate, symbolized respectively by the "North" versus "South" labels (Jasanoff 1993). How these conflicts play out in performances on the global stage such as United Nations meetings, in national settings with acts such as *An Inconvenient Truth*, or through ethically framed social movements (e.g., *Christians against Climate Change*) has not been studied with the ethnographic methods that anthropology has directed toward other multisited themes. As international agencies adopt climate change into their proactive social agendas, questions of the sort posed by Ferguson (1990) regarding the imposition of others' priorities on local actors should arise. As in past projects, anthropologists will play a critical checks and balances role, as both defenders and critics of such prioritization and approaches of implementation.

Continuing on the global scale, a subject that deserves increased focus is the real (e.g., Intergovernmental Panel on Climate Change) and imagined (liberal elite university academics) communities of climate scientists that shape the form of global political debates. On the information supply side, these distinct communities, including their methods, models, and worldviews—in the vein of Bruno Latour and other science and technology studies scholars—are worthy subjects of study in situ (e.g., Lahsen 2005). How the craft skills of climate analysis and prediction are transformed (or not) with increasing reliance on computational power should be of interest beyond climate, as other labor spaces become technocratically syncretized and automated and where the honorific of scientist is threatened by both machine and ideological positioning, depending on the audience. Echoing work done on weather-related disaster studies (e.g., Peacock, Morrow, and Gladwin 1997), attention is turning to issues of gender-based vulnerabilities to climate change (Masika 2002) and other issues of justice in climate change (Page 2006). Yet still untouched (to our knowledge) are many potential topics, including an evaluation of the gendered roles of climate science, the social process of developing climate models, and the ways these models become active objects in understanding climate change.

How these distinct groups are viewed and trusted by the public clearly affects willingness to act on information, but this is only one part of the cognition affecting behavior. Only recently have the cognitive aspects, including linguistic and visual analysis of multiple types of information, been approached (Marx et al. in press; Taddei 2005). Questions, some more or less context dependent, remain wide open: What are the roles of memory, framing of uncertainty, and cultural models of environmental resilience? How do these intangibles interact with the more traditionally studied socioeconomic constraints influencing proactive and reactive adaptation?

We have mostly discussed, reflecting the anthropological focus, human adaptation to weather and climate impacts. Climate change discourse, however, necessitates equal consideration of mitigation issues—i.e., issues surrounding the prevention of the sources of greenhouse gases. Doing so will quickly draw anthropologists into the debate surrounding alternative energy and policy choices, and most profoundly, into revisiting our longstanding fascination with consumption. What choices will individuals, groups, or governments need to make with regard to consumption choices? At the heart of theories of consumption is the often-implicit role of natural resources, and the transformation of these into products. Yet in climate change, consumption (such as buying a car or using electricity) continues to deplete resources, beyond the initial production and through an entire life of an object (Appadurai 1986). Climate change thus brings about growing recognition of the real value of such goods, obscured by time scales, hidden costs, and even inequalities constructed through markets. To some extent, climate change is unavoidably about our global thirst for goods, and in making these links more visible, anthropologists have the potential to shift the discussion about both topics (Wilk 2002).

CONCLUSION

Climate or weather—to varying degrees—link all scales of human activity, objects, and ideas. The anthropological study of climate has evolved from early work drawing on climate to explain civilizations' cultural characters and racial diversity, or "anthropogeography" (Geertz 1963), to specific studies of local adaptations to weather and climate, motivated by diverse theoretical and applied projects. Anthropological attention is now moving back to the global scale in its still-nascent study of climate change, a twenty-first-century phenomenon addressed by multisited assemblages of activists and scientists (e.g., IPCC) getting widespread media coverage and having unprecedented global impacts.

Ethnographic research into the distinct subcultures that functionally link the climate information supply and demand linkages is in its infancy, and has been focused primarily on seasonal timescale (i.e., ENSO) predictions and adaptations. Organizational aspects of the local, regional, and supranational groups that handle information strongly influence the interpretation and representation of uncertain information (Fine 2007). Beyond the cultural influence on organizational interaction there exist broader issues of political economy linked to the privatization of weather and climate data collection, sharing, and forecasts (e.g., The Weather Channel, AccuWeather, etc.). How these reduce or exacerbate the societal inequity that is evident in climate impacts is yet another topic of analysis. Anthropologists may likely find themselves arguing against the importance of global warming as a major risk factor versus more immediate (and longstanding) drivers of vulnerability including property rights, education, and access to water and health care. Climate change discourse has the potential to obfuscate unequal power relations, letting governments off the hook for poor environmental and social policies and practices.

Up to the present, anthropology has been for the most part reflexively concerned with ethnographies categorizing impacts and adaptations to weather and seasonal climate impacts. The uncertainty in our knowledge of how the very public science of climate change will be perceived and acted upon, and the intended and unintended consequences of action, is daunting. How much time before environmental and social impacts become unacceptable is a question that is unanswerable. How, as individual scholars and citizens, we chose to balance the study of this global phenomenon versus trying to more directly affect the political order, is a question only the readers of this chapter can answer.

Notes

1. Note that historical ecologists among others have long been aware of the interconnectedness of nature/culture, but generally applied the ideas only to landscapes and animals.
2. Marx had a significant influence on anthropology prior to this, including cultural ecology (Orlove 1980), but Marxism in the 1970s dominated the field.
3. Studies of worldviews and cosmologies still present a local conception of interactions between humans and their environment, even if the environment is global in scope.
4. For example, the NAO (North Atlantic Oscillation), alleged to influence Europe's climate; the PDO (Pacific Decadal Oscillation) that drives storm patterns and influences coastal ecosystems in North America's Pacific Northwest; etc.

References

Adger, W. N. 2003. Social capital, collective action and adaptation to climate change. *Economic Geography* 79: 387–404.

Adger, W. N., T. A. Benjaminsen, K. Brown, and H. Svarstad. 2001. Advancing a political ecology of global environmental discourses. *Development and Change* 32: 681–715.

Agrawala, S. and K. Broad. 2002. Technology transfer perspective on climate forecast applications. In *Research in science and technology studies: Knowledge and technology transfer*, ed. M. D. Laet, 45–69.

Agrawala, S., K. Broad, and D. H. Guston. 2001. Integrating climate forecasts and societal decision making: Challenges to an emergent boundary organization. *Science, Technology, and Human Values* 26: 454–77.

Ajibade, L. and O. Shokemi. 2003. Indigenous approaches to weather forecasting in Asa L.G.A., Kwara State, Nigeria. *Indilinga: African Journal of Indigenous Knowledge Systems* 2: 37–44.

Appadurai, A., ed. 1986. *The social life of things: Commodities in cultural perspective.* New York: Cambridge University Press.

Bennett, J. W. 1982. *Of time and the enterprise: North American family farm management in a context of resource marginality*. Minneapolis: University of Minnesota Press.

Berkes, F. 2002. Epilogue: Making sense of Arctic environmental change? In *The earth is faster now: Indigenous observations of Arctic environmental change*, eds. I. Krupnik and D. Jolly, 335–49. Fairbanks, AK: ARCUS.

Berlin, B. 1992. On the making of a comparative ethnobiology. In *Ethnobiological classification: Principles of categorization of plants and animals in traditional societies*. Princeton, NJ: Princeton University Press.

Boas, F. 1896. *Race, language, and culture*. New York: Macmillan.

Boholm, Å. 2003. The cultural nature of risk: Can there be an anthropology of uncertainty? *Ethnos* 68: 159–78.

Boia, L. 2005. *The weather in the imagination*. London: Reaktion Books.

Breuer, N. E., V. Cabrera, K. Ingram, K. Broad, and P. Hildebrandt. In press. AgClimate: A case study in participatory decision support system development. *Climatic Change*.

Broad, K. and B. Orlove. 2007. Channeling globality: The 1997–98 El Niño climate event in Peru. *American Ethnologist* 34: 285–302.

Broad, K., N. Peterson, B. Orlove, A. Pfaff, and C. Roncoli, et al. Forthcoming. Participation, social interaction, and the use of climate information. *Science and Public Policy*.

Brookfield, H. C. 1964. The ecology of highland settlement: Some suggestions. *American Anthropologist* 66: 20–38.

Cash, D. W., J. C. Borck, and A. G. Patt. 2006. Countering the loading-dock approach to linking science and decision making. *Science, Technology, and Human Values* 31: 1–30.

Comitas, L. 1973. Occupational multiplicity in rural Jamaica. In *Work and family life—West Indian perspectives*, eds. L. Comitas and D. Lowenthal. New York: Anchor Books.

Crate, S. A. 2008. Gone the bull of winter: Grappling with the cultural implications of and anthropology's role(s) in global climate change. *Current Anthropology* 49(4): 569–95.

Cruikshank, J. 2001. Glaciers and climate change: Perspectives from oral tradition. *Arctic* 54: 377–93.

D'Andrade, R. G. 1995. *The development of cognitive anthropology*. Cambridge, UK: Cambridge University Press.

Davis, M. 1998. *Ecology of fear: Los Angeles and the imagination of disaster*. New York: Metropolitan Books.

de Loë, R. C., and R. D. Kreutzwiser. 2000. Climate Variability, Climate Change and Water Resource Management in the Great Lakes. *Climatic Change* 45: 163–79.

Douglas, M. and A. Wildavsky. 1983. *Risk and culture: An essay on the selection of technological and environmental dangers*. Berkeley and Los Angeles: University of California Press.

Duerden, F. 2004. Translating climate change impacts at the community level. *Arctic* 57: 204.

Durrenberger, E. P. and G. Pálsson. 1986. Finding fish: The tactics of Icelandic skippers. *American Ethnologist* 13: 213–29.

Eakin, H. 2000. Smallholder maize production and climatic risk: A case study from Mexico. *Climatic Change* 45: 19–36.

Ember, C. R. and M. Ember. 2007. Climate, econiche, and sexuality: Influences on sonority in language. *American Anthropologist* 109: 180–85.

Ferguson, J. 1990. *The anti-politics machine: "Development," depoliticization, and bureaucratic power in Lesotho*. Cambridge: Cambridge University Press.

Finan, T. J. and D. R. Nelson. 2001. Making rain, making roads, making do: Public and private responses to drought in Ceará, Brazil. *Climate Research* 19: 97–108.

Finan, T. J. 2003. Climate science and the policy of drought mitigation in Ceará, Northeast Brazil. In *Weather, climate, culture*, eds. S. Strauss and B. Orlove, 203–16. New York: Berg.

Fine, G. A. 2007. *Authors of the storm: Meteorologists and the culture of prediction*. Chicago: University of Chicago Press.

Frenkel, S. 1992. Geography, empire, and environmental determinism. *Geographical Review* 82: 143–53.

Galvin, K. A, R. B. Boone, N. M. Smith, and S. J. Lynn. 2001. Impacts of climate variability on East African pastoralists: Linking social science and remote sensing. *Climate Research* 19: 161–72.

Geertz, C. 1963. *Agricultural involution: The process of ecological change in Indonesia.* Berkeley and Los Angeles: Univ. of California Press.

Grivetti, L. E. 1981. Geographical location, climate and weather, and magic: Aspects of agricultural success in the eastern Kalahari, Botswana. *Social Science Information* 20: 509–36.

Hackenberg, R. A. 1988. Scientists or survivors? The future of applied anthropology under maximum uncertainty. In *Anthropology for tomorrow: Creating practitioner oriented programs in applied anthropology*, ed. R. Trotter, 170–85. Washington, D.C.: American Anthropological Association.

Halvorsen, K. E., S. M. Kramer, S. Dahal, and B. S. Solmon. 2007. Cultural models, climate change, and biofuels. In *XVth international conferences of the society for human ecology*. Rio de Janeiro, Brazil.

Harris, M. 1968. *Rise of anthropological theory: A history of theories of culture.* New York: Harper and Row.

Harrison, M. 1996. "The tender frame of man": Disease, climate, and racial difference in India and the West Indies, 1760–1860. *Bulletin of the History of Medicine* 70: 68–93.

Harwell, E. E. 2000. Remote sensibilities: Discourses of technology and the making of Indonesia's natural disaster. *Development and Change* 31: 307–40.

Henning, A. 2005. Climate change and energy use: The role for anthropological research. *Anthropology Today* 21: 8–12.

Hoffman, S. M and A. Oliver-Smith, eds. 2002. *Catastrophe and culture: The anthropology of disaster*. Santa Fe: School of American Research Press.

Huber, T. and P. Pedersen. 1997. Metereological knowledge and environmental ideas in traditional and modern societies: The case of Tibet. *Journal of the Royal Anthropological Institute* 3: 577–97.

Huntington, E. 1912. The fluctuating climate of North America. *Geographical Journal* 40: 264–79.

Ingold, T. and T. Kurttila. 2000. Perceiving the environment in Finnish Lapland. *Body & Society* 6: 183–96.

Jasanoff, S. 1993. India at the crossroads in global environmental policy. *Global Environmental Change* 3: 32–52.

Jennings, T. L. 2002. Farm family adaptability and climate variability in the northern Great Plains: Contemplating the role of meaning in climate change research. *Culture & Agriculture* 24: 52–63.

Kempton, W. 1997. How the public views climate change. *Environment* 39: 12–21.

Kempton, W., J. S. Boster and J. Hartley. 1996. *Environmental values in American culture.* Cambridge, MA: MIT Press.

Kitchell, A, E. Hannan and W. Kempton. 2000. Identity through stories: Story structure and function in two environmental groups. *Human Organization* 59: 96–105.

Krogman, W. M. 1943. Climate makes the man: Clarence A. Mills. *American Anthropologist* 45: 290–91.

Krupnik, I. and D. Jolly, eds. 2002. *The earth is faster now: Indigenous observations of Arctic environmental change*. Fairbanks, AK: Arctic Research Consortium of the United States.

Kuhn, T. S. 1962. *The structure of scientific revolutions*. Chicago: University of Chicago Press.

Lahsen, M. 2005. Seductive simulations? Uncertainty distribution around climate models. *Social Studies of Science* 35: 895–922.
Laughlin, C. D. 1974. Maximization, marriage and residence among the So. *American Ethnologist* 1: 129–41.
Leiserowitz, A. 2006. Climate change risk perception and policy preferences: The role of affect, imagery, and values. *Climatic Change* 77: 45–72.
Lemos, M. C., T. J. Finan, R. W. Fox, D. R. Nelson, and J. Tucker. 2002. The use of seasonal climate forecasting in policymaking: Lessons from northeast Brazil. *Climatic Change* 55: 479–507.
Litfin, K. 1994. *Ozone discourse: Science and politics in global environmental cooperation.* New York: Columbia University Press.
Magistro, J. and C. Roncoli. 2001. Anthropological perspectives and policy implications of climate change research. *Climate Research* 19: 91–96.
Marcus, G. and M. Fischer. 1986. *Anthropology as cultural critique: An experimental moment in the human sciences*. Chicago: University of Chicago Press.
Marx, S. M., E. U. Weber, B. S. Orlove, A. Leiserowitz, D. H. Krantz, et al.In press. Communication and mental processes: Experiential and analytic processing of uncertain climate information. *Global Environmental Change.*
Masika, R, ed. 2002. *Gender, development and climate change.* Oxford UK: Oxfam.
McCullough, J. M. 1973. Human Ecology, Heat adaptation, and belief systems: The hot-cold syndrome of Yucatan. *Journal of Anthropological Research* 29: 32–36.
Moran, E. 1982. *Human adaptability: An introduction to ecological anthropology.* Boulder, CO: Westview Press.
Nelson, D. R. and T. J. Finan. 2000. The emergence of a climate anthropology in northeast Brazil. *Practicing Anthropology* 22: 6–10.
Netting, R. M. 1996. Cultural ecology. In *Encyclopedia of cultural anthropology*, eds. D. Levinson, and M. Ember, 267–71. New York: Henry Holt.
Oliver-Smith, A. 1996. Anthropological research on hazards and disasters. *Annual Review of Anthropology* 25: 303–28.
———. 2002. Theorizing disasters: Nature, power and culture. In *Catastrophe and culture: The anthropology of disaster*, eds. S. M. Hoffman and A. Oliver-Smith, 23–48. Santa Fe: School of American Research Press.
Orlove, B. 1980. Ecological anthropology. *Annual Review of Anthropology* 9: 235–73.
Orlove, B., J. Chiang, and M. Cane. 2002. Ethnoclimatology in the Andes: A cross-disciplinary study uncovers a scientific basis for the scheme Andean potato farmers traditionally use to predict the coming rains. *American Scientist* 90: 428–35.
Orlove, B., C. Roncoli, M. Kabugo, and A. Majugu. Under review. Indigenous knowledge of climate variability in southern Uganda: The multiple components of a dynamic regional system. *Climatic Change.*
Ortner, S. 1993. Theory in anthropology since the sixties. In *Culture/power/history: A reader in contemporary social theory*, eds. N. B. Dirks, G. Eley, and S. B. Ortner. Princeton, NJ: Princeton University Press.
———. 2006. *Anthropology and social theory: Culture, power, and the acting subject.* Durham, NC: Duke University Press.
Page, E. 2006. *Climate change, justice and future generations*. Cheltenham, UK: Edward Elgar Publishing.
Paolisso, M. 2002. Blue crabs and controversy on the Chesapeake Bay: A cultural model for understanding watermen's reasoning about blue crab management. *Human Organization* 61: 226–39.
Patt, A., P. Suarez, and C. Gwata. 2005. Effects of seasonal climate forecasts and participatory workshops among subsistence farmers in Zimbabwe. *PNAS* 102: 12623–28.
Peacock, W. G., B. H. Morrow, and H. Gladwin, eds. 1997. *Hurricane Andrew: Ethnicity, gender, and the sociology of disasters*. London: Routledge.

Pennesi, K. 2007. *The predicament of prediction: Rain prophets and meteorologists in northeast Brazil*. University of Arizona, Tucson.

Phillips, J. G. and B. S. Orlove. 2004. *Improving climate forecast communications for farm management in Uganda*. Final Report to the NOAA Office of Global Programs, Silver Spring, Maryland.

Pielke Jr., R. A. and M. H. Glantz. 1995. Serving science and society: Lessons from large-scale atmospheric science programs. *Bull. Amer. Meteorol. Soc.* 76: 2445–58.

Rayner, S. 2003. Domesticating nature: Commentary on the anthropological study of weather and climate discourse. In *Weather, climate, culture*, eds. S. Strauss and B. Orlove, 277–90. New York: Berg.

Reilly, J. and D. Schimmelpfennig. 2000. Irreversibility, uncertainty, and learning: Portraits of adaptation to long-term climate change. *Climatic Change* 45: 253–78.

Reyna, S. P. 1975. Making do when the rains stop: Adjustment of domestic structure to climatic variation among the Barma. *Ethnology* 14: 405–17.

Rhoades, R. E. and S. I. Thompson. 1975. Adaptive strategies in alpine environments: Beyond ecological particularism. *American Ethnologist* 2: 535–51.

Roncoli, C., K. Ingram and P. Kirshen. 2001. The costs and risks of coping with drought: Livelihood impacts and farmers' responses in Burkina Faso. *Climate Research* 19: 119–32.

Sahlins, M. 1964. Culture and environment: The study of cultural ecology. In *Horizons of anthropology*, ed. S. Tax, 132–47. Chicago: Aldine.

Siegel, B. J. 1971. Migration dynamics in the interior of Ceara, Brazil. *Southwestern Journal of Anthropology* 27: 234–58.

Sillitoe, P. 1993. A ritual response to climatic perturbations in the highlands of Papua New Guinea. *Ethnology* 32: 169–87.

Slovic, P., B. Fischhoff and S. Lichtenstein. 1986. The psychometric study of risk perception. In *Risk evaluation and management*, eds. V. R. Covello, J. Menkes, and J. Mumpower, 3–24. New York: Plenum.

Smit, B., I. Burton, R. J. T. Klein and J. Wandel. 2000. An anatomy of adaptation to climate change and variability. *Climatic Change* 45: 223–51.

Strauss, S. 2003. Weather wise: Speaking folklore to science in Leukerbad. In *Weather, climate, culture*, eds. S. Strauss and B. Orlove, 39–60. New York: Berg.

Strauss, S. and B. Orlove. 2003. Up in the air: The anthropology of weather and climate. In *Weather, climate, culture*, eds. S. Strauss and B. Orlove. New York: Berg.

Sturken, M. 2001. Desiring the weather: El Niño, the media, and California identity. *Public Culture* 13: 161–89.

Suarez, P. and A. Patt. 2004. Cognition, caution and credibility: The risks of climate forecast application. *Risk, Decision and Policy* 9: 75–89.

Taddei, R. 2005. *Of clouds and streams, prophets and profits: The political semiotics of climate and water in the Brazilian northeast*. PhD Diss., Department of Anthropology, Columbia University, New York.

Torry, W. I. 1979. Anthropological studies in hazardous environments: Past trends and new horizons. *Current Anthropology* 20: 517–40.

Vásquez-León, M. 2002. Assessing vulnerability to climate risk: The case of small-scale fishing in the Gulf of California, Mexico. *Investigaciones Marinas* 30: 204–05.

Vásquez-León, M., C. T. West and T. J. Finan. 2003. A comparative assessment of climate vulnerability: Agriculture and ranching on both sides of the US-Mexico border. *Global Environmental Change* 13: 159–73.

Vayda, A. P. and B. J. McCay. 1975. New directions in ecology and ecological anthropology. *Annual Review of Anthropology* 4: 293–306.

Vedwan, N. 2006. Culture, climate and the environment: Local knowledge and perception of climate change among apple growers in northwestern India. *Journal of Ecological Anthropology* 10: 4–18.

Waddell, E. 1975. How the Enga cope with frost: Responses to climatic perturbations in the central highlands of New Guinea. *Human Ecology* 3: 249–73.
Whiting, J. 1964. The effects of climate on certain cultural practices. In *Explorations in cultural anthropology*, ed. Goodenough.
Wilk, R. 2002. Consumption, human needs, and global environmental change. *Global Environmental Change* 12: 5–13.
Wissler, C. 1926. *The relation of nature to man in aboriginal America*. New York: Oxford University Press.
Wolf, E. 1972. Ownership and political ecology. *Anthropological Quarterly* 45: 201–05.
Ziervogel, G., S. Bharwani and T. E. Downing. 2006. Adapting to climate variability: Pumpkins, people and policy. *Natural Resources Forum* 30: 294–305.
Ziervogel, G. and R. Calder. 2003. Climate variability and rural livelihoods: Assessing the impact of seasonal climate forecasts in Lesotho. *Area* 35: 403–17.

Chapter 3

Fielding Climate Change in Cultural Anthropology

Carla Roncoli, Todd Crane, and Ben Orlove

Introduction

Within the last ten years, anthropologists have become involved in climate research to an unprecedented degree (Batterbury 2008; Brown 1999; Magistro and Roncoli 2001; Rayner 2003; Strauss and Orlove 2003). Three conditions are responsible for this development: the irrevocable transformations that climate change is bringing to the people and places traditionally studied by anthropologists (Boko et al. 2007), the general recognition of the importance of research on the human dimensions of climate change (Vogel et al. 2007), and the growing opportunities for anthropologists to participate in interdisciplinary climate application and adaptation research (Roncoli 2006).

To this challenge anthropology brings its core theoretical tenet: that culture frames the way people perceive, understand, experience, and respond to key elements of the worlds which they live in. This framing is grounded in systems of meanings and relationships that mediate human engagements with natural phenomena and processes. This framing is particularly relevant to the study of climate change, which entails movement away from a known past, though an altered present, and toward an uncertain future, since what is recalled, recognized, or envisaged rests on cultural models and values. Individual and collective adaptations are shaped by common ideas about what is believable, desirable, feasible, and acceptable (Nazarea-Sandoval 1995; Rappaport 1979). Anthropology's potential contributions to climate research are the description and analysis of these mediating layers of cultural meaning and social practice, which cannot be easily captured by methods of other disciplines, such as structured surveys and quantitative parameters.

This chapter examines a number of studies that exemplify the way anthropologists have engaged with various aspects of climate change. We do not intend to present a comprehensive review, but we seek to identify the epistemological and methodological approaches that have led to particularly valuable insights. We recognize that a great deal of research on climate change and its effects on cultural systems and social organization has been carried out in archaeology (Balter 2007; Kuper and Kropelin 2006; Migowski et al. 2006; Richerson, Boyd, and Bettinger 2001), historical ecology (Crumley 1994; McIntosh, Tainter, and McIntosh 2000; Oldfield 1993), and cultural ecology (Bogin 1982; de Menocal 2001; Peterson and Haug 2005). In this chapter, however, we focus on the ways that cultural anthropologists address present-day issues related to global climate change, issues that are confronting both local communities and global scientific and policy communities with unparalleled urgency and severity (Batterbury 2008).

Our discussion begins by highlighting the distinctiveness of ethnographic fieldwork as a way to gain insights into the relationship between climate and culture. For the remainder of the chapter, we focus on four overlapping axioms that elucidate the different ways cultures engage their world through the prism of climate change: how people perceive climate change through cultural lenses ("perception"); how people comprehend what they see based on their mental models and social locations ("knowledge"); how they give value to what they know in terms of shared meanings ("valuation"); and how they respond, individually and collectively, on the basis of these meanings and values ("response"). In the conclusions, we argue that, since climate change is about global fluctuations and interconnections, cultural anthropologists are challenged to broaden their field horizons and venture out on uncharted epistemological terrains. At the same time, given the ideological and politicized nature of climate science and its influential role in policy decisions that affect the lives of indigenous communities, marginalized groups, and the poor, anthropologists should stand firm in their tradition of committed localism and ethnographic reflexivity (Marcus 1995).

Being There

Ethnographic fieldwork, based on extended periods of residence and research at a community level, has been anthropology's dominant approach to capture the elusive domains of cultural meaning and practice. Anthropology's emphasis on fieldwork and participant observation stems from the recognition that engaging in daily life and social relationships provides a contextual understanding of cultural realities that cannot be captured by structured survey methods alone (DeWalt and DeWalt 2002; Jorgesen 1989; Schensul et al. 1999). Fieldwork allows for a slower accumulation of evidence and understanding and for key insights to arise unexpectedly, during experiences that allow glimpses of how the world is perceived and experienced by local peoples—for example, while participating in ancestral rituals in Tanzania (Sanders 2003), witnessing impacts of El Niño drought on daily life in Papua New Guinea (Ellis 2003), or drinking early morning coffee with Maryland crab fishermen (Paolisso 2003). While ethnographic interviewing and participatory research techniques are also deployed to elicit information on various issues, it is this full immersion in fieldwork that constitutes anthropology's trademark tool.

"Being there" is increasingly being embraced beyond anthropology by other social scientists, such as sociologists and cultural geographers who work on vulnerability and adaptation to global environmental change. In describing methods used to gather data for her ethnography on climate and economic change in rural Mexico, geographer Hallie Eakin (2006, 213) acknowledges that "some of my greatest insights into the livelihoods of farmers in the Puebla-Tlaxcala Valley came from simply being there: helping with the harvest, chatting with mothers outside the primary school, attending a wedding celebration or school graduation. None of these methods and data sources would have been sufficient on their own to understand the full complexity of the farmers' vulnerability." Accordingly, most chapters in her book open with accounts of fieldwork events or conversations that provided clues to crucial aspects of climate vulnerability, risk management, and adaptive capacity. In particular, field interactions provided her with insights into the way livelihoods are infused with cultural meanings and adaptations reflect agency in the way people endeavor to make the best out of their circumstances. Also trained as a geographer, Petra Tschakert (2004a, 2004b) has conducted fieldwork in rural communities in Senegal, illustrating the value of participatory methodologies in eliciting farmers' views of climate change, its causes, and its impacts (Tschakert 2007b). By comparing conceptual models of farmers and "experts," she highlights where they diverge and where they may complement each other through collective learning. She also applies ethnographic methods to identify soil fertility strategies with carbon sequestration potential, and then uses soil measures to assess their potential for carbon sequestration and economic analyses and agent-based modeling to assess their feasibility and profitability for different wealth categories of farmers (Tschakert 2007a). Her approach is noteworthy for highlighting how African farmers can participate in mitigation efforts, rather than merely adapting to climatic changes.

Two studies of sub-Arctic regions, Susan Crate's (2006a, 2008) political ecology of Viliui Sakha, agro-pastoralists of northeastern Siberia, Russia, and Julie Cruikshank's (2005) ethnohistory of Yukon Territory First Nations peoples, exemplify the epistemological value of long-term involvement with and deep personal commitments to particular communities. Even as they conduct multisited ethnographies, the researchers' studies build on applied efforts to produce educational and advocacy resources for local people, such as educational and documentary materials on local history, language, culture, and environment (Crate 2006a, xiii–xiv; Cruikshank 2005, x–xi) and generate knowledge that is brought to bear in land claims negotiations between an indigenous people and the central government (Cruikshank 2005, 287). These intimate ties with research communities allow the coproduction of ethnographic narratives that involve both local experts and research partners. These cases suggest how long-term participant observation can develop into engaged ethnography.

Anthropologist Sarah Strauss (2003) takes the significance of "being there" into the realm of weather observation and forecasting. Writing about the development of a national weather service in Switzerland, she points out how the work of volunteers who record local weather data supports the generation of scientific knowledge about regional and global weather phenomena by professional meteorologists. Yet, in order to make sense of scientific forecasts, lay users must recontextualize them based on intimate knowledge of their particular localities. Generalizing from her Swiss material to other cases, Strauss notes that "farmers, sailors, mountain guides, and others, who make their living by their skills at navigating nature's complex rhythms and random disturbances, know that to trust the weatherman's forecast *alone* is to cast one's lot to the wind—there is no substitute, no matter how sophisticated, for being there" (Strauss 2003, 55). Along these same lines, Roncoli et al. (2003) show how the collective experience and sustained observation of nature helps African farmers to interpret scientific seasonal climate forecasts in ways that are adaptive in their immediate context. When anthropologists head to the field they seek to grasp this level of experiential competence found among the members of the communities in which they work. "Being there" enables them to gain a much greater depth of insight into those systems of practice and meaning that define such communities, be they subsistence farmers or research scientists.

PERCEPTION

By emphasizing collective experience and cultural framing, anthropology gives voice to folk narratives of climate change, expanding the discussion beyond the broader spheres of earth sciences, policy debates, and media headlines. Visual and sensory perceptions are key elements of the folk epistemology of climate (Strauss and Orlove 2003). The human body's senses are important avenues through which people get to know their local weather in its particular manifestations, such as rain, hail, snow, wind, and temperature. For example, rain may be experienced corporeally and emotionally by seeing, hearing, feeling, and even smelling. Biologist and ethnoscientist Gary Nabhan (2002) describes how indigenous inhabitants of the arid American Southwest and of northwestern Mexico note with delight that after rare rainstorms, "the desert smells like rain." In Uganda, farmers might see clouds in the sky and, based on their color and shape, know whether it will rain; they may feel the wind and, based on its direction or strength, recognize whether it will bring rain or chase it away; they may hear thunder and see lightning flashes on the horizon and, based on their orientation, predict whether the storm will head their way; they may feel heat at night, and, based on its intensity and the time of year, discern whether planting time is approaching (Orlove et al. under review).

Anthropological research on the modes and shifts of collective attention highlights how public understandings of climate events and of climate science incorporate local agency and meanings (Broad and Orlove 2007; Vedwan 2006). In communities that rely on natural resources for their livelihood, people are keenly watchful of the landscape and quickly discern climatic anomalies and their effects. Farmers in the Sahelian region of West Africa point to shrinking water bodies, disappearing plants and crops, and changing settlement patterns as evidence of reduced rainfall over the last three decades of the twentieth century (Tschakert 2007b; West, Roncoli, and Ouattara 2008). In many parts of the world, trees, wind, and birds have been subjects of attentive scrutiny by local farmers who rely on them to predict seasonal rainfall. But where climate change is eroding the reliability of such indicators, public attention may shift from them to different, external sources of information, such as radio and television weather forecasts (Roncoli, Ingram, and Kirshen 2002). On the other hand, in areas where rainfall variability is becoming an increasingly severe problem, such as the humid tropics of southern Uganda, farmers are becoming more attuned to environmental clues to orient their planting choices (Orlove et al., under review). For example, while discussing their expectations for the upcoming rainy season, Ugandan farmers stressed the need to be alert as not to be "tricked" by the climate, meaning not to misread signs and be caught unprepared. This element of concentrated attention is sometimes translated in the Luganda language as *amaanyi*, the same term that denotes mental and physical effort, including energy, resolve, and confidence, traits that are upheld as essential to coping with uncertainty.

Ethnographic interviews and participant observation provide important entry points into ways which reveal the phenomena that people use as evidence that climate is changing. Some expressions of climatic change are especially salient due to their striking visibility, such as diminished snowfall or glacial retreat (Orlove, Wiegandt, and Luckman 2008; Vedwan and Rhoades 2001). For example, in a study in the western Himalayas, Vedwan and Rhoades (2001) interviewed apple farmers about the reasons for declining production. If farmers spontaneously mentioned climate change, researchers followed up by asking what specific aspects had changed and what caused such changes. Farmers identify changing snowfall events, specifically shifts in its timing and intensity, as the main evidence that climate is changing. Such changes were framed largely in terms of variation in snowfall: for example, they point to increasing occurrence of late snowfall as a sign of variations in rainfall or temperature. The salience of snowfall highlights the significance of visual indicators in farmers' understanding of apple-weather interactions, particularly regarding how weather variability affects fruit color and appearance. Such sensitivity to subtle changes in the environment allows farmers to contribute understanding of local manifestations of global climate change.

The visual and narrative representations that people evoke when describing their environments provide unique insights for an anthropological study of the human dimensions of climate change. In their investigation of cultural perceptions of environmental change surrounding Mount Shasta in northern California, Wolf and Orlove (2008) allowed interviewees to express at length their views and feelings about the mountain. Analysis of those responses shows that perceptions varied according to the respondents' birthplace, residence, experience, and worldview. For example, locally born people emphasized utilitarian functions (e.g., the mountain as provider of water), while those who moved in were more attuned to aesthetic and spiritual meanings and more likely to mention the snow and glaciers that give the summit its characteristic "whiteness." Orlove et al. (2008) highlight the role of glacier-covered peaks as visual icons of both nature and culture. Powerful symbols of unspoiled, unconquered nature, glaciers attract tourists and mountaineers from different parts of the world. At the same time, they are emblematic of cultural identities, featured in official imagery of cities and countries that claim a particular relationship to mountains and glaciers. The massive glacierized peak of Mt Ararat, for example, is depicted in Armenia's coat of arms and serves as a symbol of the Armenian people and nation, of their greatness and of their indomitable will to survive. The representations, sense of attachment, and economic importance of mountains and glaciers contribute to shaping the way local people respond to accounts of climate change from scientists, government agencies, and other organizations. Yet, while a great deal has been written on glacier retreat, very little empirical research has been conducted on human responses to its varied impacts (Orlove et al. 2008).

Anthropologists have used interactive visual methods to elicit local understandings of impacts of climate change and to stimulate reflections on adaptive responses at the community level. Working in the highlands of Ecuador, Rhoades, Zapata, and Aragundy (2006, 2008) examine historical paintings, photographs, and other documents, including accounts of early explorers and climbers. These illustrate the gradual disappearance of glaciers and snow fields covering the Cotacachi volcano, a place of great cultural significance to surrounding communities. These images were then used to elicit local peoples' stories and commentaries about the climatic changes that have transformed the face of the mountain and the surrounding landscape. The researchers also constructed a three-dimensional physical model of the watershed, including the volcano, and used it to stimulate direct discussions of environmental change. At the end of the project, when the model was transferred to the community, local people proceeded to paint the volcano's peak white, an action that affirmed the cultural significance of the mountain in face of the loss of identity and control over its natural resources. The whiteness of the snow was a constant reference point in people's collective memories and oral histories, which related to the mountain as an animated and feminine presence ("Mama Cotacachi"). In popular imagination, the lost snowcap on her summit was seen as a sign of her fading beauty and youth. As she declines, so do people's sense of well being and social harmony (Rhoades, Zapata, and Aragundy 2006, 2008).

The personification of landscape features reflects a view that nature *includes* humanity and culture, rather than being juxtaposed to them. In this perspective, which has historically been central to the worldview of some indigenous people of Arctic and sub-Arctic regions, natural elements such as glaciers, mountains, seas, and animals are seen as sentient beings, having agency, emotions, and interest in human affairs (Cruikshank 2005; Krupnik and Jolly 2002; Laidler 2006; Nadasdy 2005). Though the environment can be observed, explored, exploited, and managed by humans and altered by climatic change, the same environment can also respond in favorable or punishing ways. Drawing on travelers' reports, old illustrations, folk narratives, clan histories of First Nations peoples, and, in particular, the life stories of indigenous women elders in the Yukon Territory, Cruikshank (2001, 2005) braids a multistranded account of climate change as seen from a myriad of vantage points. The cast of characters include nineteenth-century travelers, scientists, environmentalists, and explorers whose itineraries trace global connections of colonial expansion ranging from Alaska to Africa. Early ethnographers are also featured among human actors, as one voice among many rather than as authoritative sources. Among the main protagonists and the chief narrators, as both the center and the source of the stories, are the glaciers, whose emptiness and whiteness denote an imaginative space where recollections of the past and projections to the future meet. This textuality of landscape and seascape embodying collective experiences of climatic change is highlighted by other anthropologists of Arctic regions. By building on Nuttall's concept of "memoryscapes" (Nuttall 1991, 1992) and by combining paleoecology, archaeology, and oral history, Henshaw (2003) documents how Inuit place-names encode information on fluctuating ice conditions, wildlife behavior, and other natural phenomena that help orient people's movements over the territory and transfer cultural knowledge across generations.

Knowledge

In indigenous epistemologies, seeing and knowing are understood as closely related. Youth often learn technical practices by actively watching and practicing with adults (Barnhardt and Kawagley 2005; Laidler 2006). Elders are vested with authoritative knowledge, because those who have lived many years have seen things, including climate events, changes, and impacts, like the Sahelian droughts and famines of the 1970s and 1980s (Roncoli et al. 2002; West et al. 2008), or the surging glaciers and perilous migrations of the "Little Ice Age" (Cruikshank 2005). Thus, open-ended interviews with local elders and recording of life histories are often used in anthropological research to elicit local knowledge and cultural memory (Crate 2002, 2006a, 2006b; Cruickshank 2005).

Research on indigenous environmental knowledge has a long-standing tradition in anthropology and ranges from ethnoscience, to applied anthropology, to the more recent political ecology and environmental movements

(Nazarea 2006)[1]. A wealth of studies document the importance of indigenous knowledge in agricultural development and environmental management (Brokensha, Warren, and Werner 1980; DeWalt 1994; McCorkle 1989; Rhoades and Bebbington 1995; Richards 1985; Sillitoe 1998; Stevenson 1996; Thompson and Scoones 1994; Warren, Slikkerveer, and Brokensha 1995) and its resilience in the face of commercialization and government control (McDaniel, Kennard, and Fuentes 2005). Farmers' knowledge and experience are being increasingly recognized as valuable assets for building the resilience of rural livelihoods to climate variability and change (Nyong, Adesina, and Osman Elasha 2007; Stigter et al. 2005). The Fourth Technical Assessment of the Intergovernmental Panel on Climate Change (IPCC) emphasizes the value of indigenous knowledge systems for climate predictions, adaptive management, and policy making, and calls for more studies in this area (Boko et al. 2007).

Until recently, relatively few studies in the indigenous knowledge literature directly focused on climate (Ingold and Kurttila 2000; Katz, Lammel, and Goloubinoff 2002; Roncoli et al. 2002; Sillitoe 1996).[2] This may be due to the fact that, while recent research on local ecological knowledge is propelled by concerns about environmental conservation and intellectual property rights, knowledge about climate cannot be managed, transferred, appropriated, or consumed the same as cultural or natural resources. Cultural anthropologists have used different approaches to explore local knowledge of climate. Ethnoscientific research sought to understand local knowledge as a system of taxonomies and classifications (Atran 1985; Berlin, Breedlove, and Raven 1974; Hunn 1982; Posey 1984, 1986). Using ethnographic techniques such as free listing, sorting, ranking, and triads, anthropologists have documented how farmers distinguish many different types of clouds, rains, winds, and other phenomena (Roncoli et al. 2002; Sillitoe 1996). In the process of eliciting these typologies, basic principles underlying cultural notions of climate are revealed. For example, in categorizing rain events, Sahelian farmers look at the duration, distribution, and timing of precipitation, suggesting that the latter is understood in terms of process rather than amount of rainfall. Farmers appreciate rains that occur during the night and last several hours, allowing for rainwater to infiltrate and for soil to remain moist for several days (Roncoli et al. 2002). Climate variation is perceived in relation to salient categories, such as the decreased frequency of "big rains" that fall in July and August over the Sahel (West et al. 2008) and the early snowfalls that favor apple production in the western Himalayas (Vedwan and Rhoades 2001).

The detection of anomalous patterns of wind, rain, hail, snow, frost, and temperature hinges on local understandings of time. By revealing the ways that people organize cyclical and linear time into meaningful segments, linguistic anthropology elucidates the kinds of variations people are able to discern and adapt to (Puri 2007). Seasonality is the most basic scaffolding of people's sense of time, not only structuring perceptions of fluctuations in resource availability but also deployment of adaptive responses. Comparing data from twenty-eight language groups, Orlove (2003) examines the names and attributes of seasons and their subcomponents. All the language systems in this sample have names for seasons, which are defined by atmospheric and environmental indicators, though in some areas the notion of the calendar year was a colonial introduction. Even in equatorial regions that register minimal seasonal variation, ethnometeorological knowledge includes a rich terminology for the cyclical manifestation of climatic events during the year (Sillitoe 1996). In agricultural, pastoral, and fishing communities, where seasonality shapes livelihoods, climate change is often understood in terms of deviations from a cognized normative calendar. Seasonal calendars are often used in ethnographic research as a way of eliciting and systematizing local knowledge of climate, although Vedwan and Rhoades (2001) caution that they need to be treated as conceptual models rather than as factual representations of climate-related events and activities.

Researchers have studied the traditional practices that indigenous peoples use to forecast seasonal variability in order to increase the reception, understanding, and use of scientific forecasts (Roncoli 2006). Local predictive systems are relevant to climate change research as they point to the salient parameters and normative frameworks of seasonal variation. Methods such as open-ended interviews and focus groups with farmers, elders, and local experts have been deployed in order to elicit rich repertoires of shared and specialized knowledge based on environmental observations and ritual practices (Eakin 1999; Finan 1998; Huber and Pedersen 1998; Luseno et al. 2003; Roncoli et al. 2002), while surveys have been used to assess the distribution of such knowledge in the population. By showing that knowledge is embedded in systems of social relations and cultural meanings, anthropologists counteract the common tendency to reduce it to decontextualized inventories of signs and beliefs. For example, surveys point to the quality and quantity of fruits from certain wild trees as among the forecasting indicators most commonly mentioned by African farmers (Kihupi et al. 2003; Phillips, Makandze, and Unganai 2001). But ethnographic research clarifies that farmers do not rely on random observations of generic specimens, but rather on the sustained observation of particular trees that stand near their homes and farms as constant witnesses to the unfolding of social life and seasonal time (Roncoli et al. 2002).

Anthropologists have begun to explore the empirical relation between local knowledge and climate phenomena. Efforts to link ethnographic and meteorological data remain hindered by several challenges, including a) the dearth of long-term data series for local indicators, such as fruiting of wild tree species or behavior of birds or insects; b) the difficulty of operationalizing local experiences of climate variability and change in ways that permit a correlation with scientific records; and c) the discrepancies in the spatial scale and time-frame of local experiences and decisions on the one hand and of regional and global processes on the other. Some researchers have sought to address these challenges through interdisciplinary research and innovative research design. A study on the ethnoclimatology of Andean communities in Peru and Bolivia combines analyses of ethnographic, historical, agricultural, atmospheric, and astronomic data to show that traditional forecasts, which use observations of the Pleiades star cluster in June to forecast the onset of the rains in October and November, are reasonably accurate and based on natural phenomena (Orlove, Chiang, and Cane 2000, 2002). Conversely, research among the Twareg pastoralists of Niger found a lack of correspondence between climate data and local perceptions of climate impacts (Sollod 1990).[3] In this study, pastoralists' qualitative assessments of rainfall and pastoral life were coded according to a 1–9 point Likert scale, ranging between catastrophe and celebration, and then compared with the rainfall record for the previous forty years (Sollod 1990, 272). Findings suggest that rainfall alone is not a reliable indicator of stress but needs to be contextualized by parameters relevant to pastoral habitats.

Anthropologists Marcela Vásquez-León, Colin West, and Timothy Finan pioneered comparative analysis of farmers' perceptions and meteorological records in two contexts, representative of dryland environments—the American Southwest (West and Vásquez-León 2003) and the Sudan-Sahel region of West Africa (West et al. 2008). In both cases, researchers used open-ended interviews and fieldwork interactions to elicit farmers' views of climate variability. Drawing on their results, the researchers constructed culturally specific indices of climatic variability that they then compared to meteorological data from local weather stations. The analysis shows concordance between the two cases, suggesting that people are indeed able to discern climatic changes, beyond the limited timeframe of weather fluctuations (West and Vásquez-Léon 2003; also Puri 2007). People's assessment of climate variation, however, are grounded in localized contexts and processes of livelihood adaptation and can, therefore, diverge from regional trends inferred from scientific technologies, such as remote sensing imagery and global circulation models (Rautman 1994; West et al. 2008). Multiple sources of information—drawing on qualitative and quantitative methods, spanning local and regional scales, and covering both folk and expert knowledge—are needed to develop a holistic understanding of climate change.

Valuation

People's perceptions and knowledge systems are framed by cultural contexts with which they ascribe meaning and value to what they see and know. This understanding has led some anthropologists to question efforts to systematize indigenous knowledge into decontextualized and discrete data sets and subordinate it to validity standards established by Western scientific institutions (Agrawal 1995; Nadasdy 1999; Nuttall 1998; Purcell 1998, 1999; Purcell and Onjoro 2002). While the integration of indigenous and scientific knowledge may be useful in developing adaptive systems, such projects are sometimes animated by a view of knowledge as a transferable package of skills and technologies, devoid of the cultural values that make it meaningful to those who depend on it for their livelihood (Cruikshank 2005; Nadasdy 2005). This issue is especially irksome in areas where local sovereignty and state bureaucracies clash over environmental management, contrasting very different ways of thinking about and valuing landscape and wildlife. Anthropologists working in Arctic and sub-Arctic or alpine regions, for example, illustrate how animals, mountains, glaciers, and other landscape features are conceived by local people as more than assets to be managed or measured. They are rather to be embraced as part of a moral universe that includes both humans and nature, and their decline, due to unsustainable use or to climatic change, is mourned as a loss of cultural identity and meaning (Crate 2006a, 2008; Cruikshank 2005; Nadasdy 2005; Nuttall et al. 2005; Rhoades et al. 2008; Salick and Byg 2007).

In many communities, weather and climate are understood as part of a universe infused with spiritual significance. Perturbations are often interpreted in terms of violation of religious, moral, and social norms (Ellis 2003; Roncoli et al. 2002; Salick and Byg 2007; Sanders 2003; Vedwan 2006). At the same time, global environmental change is seen as threatening the integrity of the spiritual world and its benevolent relationship to humanity (Crate 2008). Accounts of environmental and climatic change are often embedded in moral and mythological discourse and should not, therefore, be taken at face value (Ellis 2003). Cultural and spiritual values shape people's attitudes towards remembrance of the past and predictions of the future. Weather patterns of the past may be idealized in nostalgic recollections of one's childhood and serve as a cognitive framework for remembering significant events (Harley 2003). Religious convictions induce Muslim farmers in Burkina Faso to see efforts to predict rainfall as a lack of humility and trust in God (Roncoli et al. 2002). But it is not only aboriginal communities in marginal environments who attribute spiritual and moral values to nature. Among crab fishermen of the East Coast of the United States, for example, the unpredictable nature of weather and weather-dependent resources is valued as a way to ensure that God remains in control of creation and to restrain humans from plundering the latter because of greed (Paolisso 2003). In discussing perceptions of climate change and glacial processes around Mount Shasta, the mountain that contains the largest glaciers in California, Wolf and Orlove (2008) note that many people comment on the spiritual value of the mountain, and speak of the mountain itself as acting consciously, for example, purposefully sending avalanches to destroy ski areas built in pristine areas.

It is as important to understand environmental values in industrialized countries, which produce most of world's greenhouse gases and have a major political role in addressing climate change, as in the developing world.

In one of the first anthropological studies exploring such issues, Kempton, Boster, and Hartley (1995) show that Americans' attitudes towards the environment are framed in terms of widely shared religious beliefs, ethical principles, and social obligations (e.g., to one's descendants) as well as by utilitarian considerations. The researchers found this framing to be not only for environmentalists but also for a wide range of social groups with different relationships to environmental resources, including people employed in sawmills, coal industry, and dry cleaning, and the general public in California. Similarly, Henning (2005) used phone interviews with consumers and experts, such as municipal energy advisers and employees of utility companies, to study the underlying cultural values and social goals shaping attitudes toward energy use in Scandinavia. Her research demonstrates that desire for autonomy and flexibility and idealized notions of private and public space define the way that users expect and evaluate comfort and convenience, a logic that diverges from the strictly economic reasoning embraced by the professionals.

Such studies are noteworthy in that they demonstrate the value of ethnographic methods, such as open-ended interviews and discourse analysis, in uncovering unanticipated issues and underlying cultural models.[4] These studies also venture beyond the habitual domain of anthropology by turning the analytical gaze on the researchers' own culture and, in some cases (Henning 2005; Kempton et al. 1995), on communities that are defined in terms of common interests, collective action, or consumer behavior rather than physical localities and kinship bonds. By including industry professionals and policy makers among their informants, they also chart important new territory for the ethnography of climate change.

Scientific communities themselves are increasingly an important site of anthropological research. Drawing theoretical inspiration from science and technology studies (Jasanoff and Wynne 1998; Shackley and Wynne 1995, 1996), researchers are beginning to interrogate scientific debates and practices and to demonstrate how they are no less shaped by cultural perceptions and social context than indigenous knowledge systems. They note that the production and circulation of scientific research is as shaped by cultural and political factors as other human activities. This strand of research does not aim to undermine the validity of scientific analysis, but it denies the nature of scientific analysis as objective, culture-free statements about external reality, and it shows that this analysis takes place in a specific cultural milieu with its own set of values, assumptions, and power dynamics. Building on several years' experience as an anthropologist employed in climate research institutions and on hundreds of interviews with climate scientists and decision-makers, Myanna Lahsen illuminates the social construction of scientific authority in climate research and policy (Lahsen 2005a, 2005b, 2007a, 2007b). Her work, like other work in recent decades (Latour, Woolgar, and Salk 1986; Shapin and Schaffer 1989), challenges the notion of science as a value-free process in pursuit of an objective truth, and shows that ideology and power configure particular understandings and uses of science. Even among climate scientists, different ideas about the relationship of science and technology and about the role of science in society underlie assessments of what constitutes good science and what knowledge is worth making public (Lahsen 1998; Shackley et al. 1998). In the case of Global Circulation Models, which are widely used to generate projections of climate change, Lahsen (2005a) finds that the intense personal investment modelers make in developing models, the fragmented nature of the modeling process, and the highly competitive funding environment in the United States all hinder modelers' ability to realistically represent, and even perceive, the models' accuracy (see also Shackley 2001).

The uncertainties inherent in climate predictions allow some latitude for spinning or contesting the meaning of scientific knowledge, a fact that underscores the importance of understanding the social and cultural processes of its production and circulation (Broad and Orlove 2007; Broad, Pfaff, and Glantz 2002; Broad et al. 2007a; Pfaff, Broad, and Glantz 1999). In the context of modern geopolitics, ambiguities and contradictions surrounding climate science become political weapons in negotiations between developed and less-developed countries or among key actors within countries. Examples include the struggle between conservation-oriented NGOs and a central government favoring environmentally destructive development in Brazil (Lahsen 2007b), or between an embattled president and political constituencies in Peru (Broad and Orlove 2007). Basic societal values, such as the perceived objectivity and fairness of scientific practice, influence the reception and use of scientific information by policymakers and the public at large. This can be seen in ethnographic research that elucidates how an interplay of subjective judgments, cultural meanings, and political agendas can shape representations of and responses to climate science, be they climate change projections (Lahsen 2007b), El Niño-based seasonal forecasts (Broad and Orlove 2007; Roncoli et al. 2003) or extreme weather advisories (Broad et al. 2007a, Sherman-Morris 2005).

Response

Understanding the interactions of culture and climate, and in particular the role of perceptions, knowledge, and values as elements of these interactions, brings us to focus on adaptive responses. Several anthropologists emphasize that, while information on cultural meanings and attitudes can be elicited through various methodological techniques, it is primarily through ethnographic research of climate-centered practices in localized contexts that we can really understand their livelihood significance (Puri 2007; Vedwan 2006; Vedwan and Rhoades 2001). As for other kinds of indigenous technologies and environmental knowledge, climate adaptations enacted by rural producers are often based on what Richards (1993) terms "performative knowledge," a competence that is ingrained in farmers' time-honored and place-based experience rather than encoded in abstract principles. Among African dryland farmers, for example, responses to climate variability consist of iterative sequences of improvised strategic adjustments rather than of the implementation of consciously established plans that can be articulated a priori (Batterbury 1996). In other words, farmers do not formulate a strategy for the growing season, based on their expectations of what the climate might be, and then proceed to carry out. Rather, they get their bearings by scrutinizing the environment during the weeks that lead up to the onset of the rains, and continue to do so until the viable planting time is over, a process that relies on a mix of sensory intuition, cumulative experience, and learned skills (Orlove et al. under review; Roncoli et al. 2001).

While it cannot be assumed that current practices for coping with climate variability will easily translate into long-term adaptations to climate change, valuable insights can be gleaned from understanding the contexts and processes that contribute to adaptation to climate variability (Eakin 2006). Because these strategies are complex and culturally embedded, they are not easily captured by snapshot assessments and structured surveys. Participant observation and in-depth interviewing are more suited to elucidating the intricate decision-making processes and the influences and negotiations that shape them (Roncoli 2006). Anthropology has a long-standing tradition of research on agricultural decision-making (Barlett 1980; Gladwin 1989; Nazarea-Sandoval 1995). Decision tree modeling has been productively used to describe and analyze adaptive strategies by African farmers and pastoralists (Little, Mahmoud, and Coppock 2001; Roncoli et al. 2000). This approach has enabled researchers to identify critical junctures where climate affects livelihood decisions and where management practices might be modified to adapt to different climate scenarios, showing how future options may progressively narrow as choices are made. Ethnographic research on decision-making can help researchers recognize the material and institutional constraints that hinder adaptation, the trade-offs inherent in different options, and the criteria and considerations that influence choices among them. For example, adaptive decisions may be shaped not only by climate conditions and economic constraints, but also by livelihood needs and goals (e.g., a health crisis or desire for education) and by cultural values (e.g., preferences for certain staple foods, cultural identity invested in a pastoralist lifestyle).

Fieldwork among farmers and pastoralists reveals that adapting to climate variability often involves balancing risk and uncertainties in one area with those in other areas (Roncoli et al. 2001). In making livelihood decisions, people constantly juggle different kinds of risk, not only related to climate variation but also to livestock disease, price fluctuations, violent attacks, legal prosecution, and social marginalization. Efforts to mitigate one type of risk may expose households to a set of different threats, as in the case of pastoralists who respond to drought by migrating to areas where they may face harassment (Crane 2006; Little, Mahmoud, and Coppock 2001). In this context, diversification of options, flexibility of responses, and tactical decision-making are recognized as key determinants of household resilience (Eakin 2006; Mishra, Prins, and Van Wieren 2003; Thornton et al. 2007; Tschakert 2004b, 2007a). This capacity hinges on the availability of resources that allow for swift adjustments to fit changing environmental and economic conditions. In Sahelian farming communities, for example, this flexibility is afforded by access to landholdings that have different soil conditions and water retention capacities; the ability to mobilize experienced and disciplined labor at short notice; and the timely availability of viable seed, liquid capital, and productive technology (Batterbury 1996, 2001; Ingram, Roncoli, and Kirshen 2002; Roncoli et al. 2001; Tschakert 2007b). Among African pastoralists, mobility is key to survival, allowing access to grazing areas, water sources, favorable markets, and safe routes, but timely information and support networks are essential for exercising these options (Galvin et al. 2001; Little et al. 2001; Thornton et al. 2007). These adaptive capacities are grounded in cultural identities and social relations that are mediated by kinship and community.

Field research in a wide range of field settings has shown that adaptation to climate variability and change is not only a function of technical solution. Rather it stems from a web of social reciprocities and obligations, which may be intentionally pursued or manipulated to secure access to resources and assistance at critical times (Crate 2006a; Eakin 2006; Finan and Nelson 2001; Little et al. 2001; Nelson 2007; Nuttall 1992; Puri 2007; Waddel 1975). A hallmark of economic anthropology, this understanding was developed and popularized during the 1990s as the "livelihood approach" (Scoones 1998). This approach builds on the core tenet that livelihoods involve more than the satisfaction of basic needs through direct production of material goods. Rather, livelihood draws on ties beyond the household unit and rests on social networks and institutions, human health and capabilities, knowledge and competences, as well as environmental resources and services. Different groups and individuals have varied combinations of rights, claims, privileges, and liberties that largely determine whether and how they may access and use these assets for their own benefit. But these configurations of opportunities and constraints are shaped beyond the farm and the household, by the policy and institutional arrangements fashioned by supralocal players, such as international development agencies, urban markets, and the state. The livelihood approach has often been applied to sustainable development and environmental management efforts, but rarely to climate adaptation (Eakin 2006; Keil et al. in press; Ziervogel and Calder 2003).

Understanding the decision processes by which households select and enact adaptive responses, and the institutional context that shapes those decisions, is important because even successful adaptations entail alternative risks and costs that may be borne by less powerful groups and sectors.

Within households, women, children, and the elderly may see their needs curtailed, their work burden increased, and their assets diminished or appropriated by others (Denton 2002; Roncoli et al. 2001). Ethnic or religious minorities, immigrants, lower castes, and the poor may face greater exclusions and hostilities as climate change impacts intensify competition over natural resources. For these marginal groups, resource access often rests on ambiguous rights and informal agreements that can be easily revoked at times of heightened demand (Peters 2004). Thus, the experience of climate change, the exposure to its negative impacts, and the efforts to ensure household survival are functions of one's social location and of the ability to negotiate positive terms of engagement from such a location (Eakin 2006; Finan and Nelson 2001). Fieldwork can play a key role in detecting these dynamics of power, conflict, agency, and resistance, while participatory methods such as livelihood assessments, vulnerability mapping, and free listing and ranking can uncover divergent perceptions and definitions of risk held by differentially positioned social groups (Finan 2007; Finan and Nelson 2001; Little et al. 2001; Roncoli et al. 2008; Tschakert 2007b).

Heterogeneities in risk exposure and response capacity are constituted spatially as well as socially. Vertical and regional movements are critical attributes of resilience to climate impacts in mountain ecosystems (Orlove et al. 2002; Rhoades et al. 2006; Vedwan 2006) and in pastoral habitats (Crate 2006a; Galvin et al. 2001; Thornton et al. 2007; Tyler et al. 2007) respectively. The spatial nature of adaptation in these environments means that boundaries, territories, and passages also delineate a landscape of selective vulnerabilities and entitlements. The extent to which observations from field studies can be scaled up to regional levels is one of the greatest challenges faced by anthropologists working in climate adaptation research. Geographic information systems (GIS), remote sensing, and modeling tools have demonstrated their utility in generalizing localized information (Cliggett et al. 2007; Galvin et al. 2001; Moran and Liverman 1998; Moran et al. 2007; Tschakert and Tappan 2004). These methods can help document the intra- and interseasonal variation in resource availability and access and determine the landscape impacts of local adaptations, such as the expansion of cultivation into wetlands, rangelands, or marginal areas. This may be of particular value in high mountain ecosystems and extreme latitudes where climate change is having dramatic impacts (Meehl et al. 2007), and in the African drylands, which are characterized by a high degree of variability and well suited to pastoralist livelihoods (Trench et al. 2007).

Adaptation to climate variability and change is not only a matter of individual and household decisions. It also requires institutional and policy measures that support agricultural production, food security, water resource management, and infrastructural development (Broad et al. 2007b; Eakin et al. 2007). Among anthropologists, Tim Finan and his team have spearheaded interdisciplinary research on the articulation of private and public responses to drought in northeast Brazil (Finan 2003; Finan and Nelson 2001; Lemos et al. 2002). The study combines multiple methods, integrating ethnographic fieldwork and a survey of about five hundred households; analysis of public documents and media reports; and an institutional analysis based on interviews with about fifty stakeholders including political leaders, bank managers, rural extension agents, labor unions, NGOs, and the media. The researchers analyze the process from the issuing of a drought forecast to the mobilization of drought relief, showing how scientific information is used in official discourses and decisions in ways that reinforce established patron-client relations. At the same time, constituencies can appeal to the alleged "neutrality" of science to counteract political favoritism and demand greater transparency of decision-making (Lemos 2003; Lemos et al. 2002). The articulation of climate change and collective action is an increasingly relevant domain that has received relatively little attention by anthropologists, beyond research on environmental attitudes that may motivate and mobilize activism (Kempton, Boster, and Hartley 1995; Pendergraft 1998). In-depth ethnographic research in social movements can push this line of analysis further by elucidating the ways groups decide on strategic responses and how those responses help define new communities of practice in the politicized field of global climate change (Adger 2003).

Conclusion

By entering public discourse and affecting interconnected decision-making systems at multiple scales, from local to global, climate change is becoming an increasingly salient issue for cultural anthropology. Cultural anthropologists can draw on a diversity of intellectual traditions that provide resources for understanding perceptions, knowledge, values, and practices relative to global climate change. Ethnoscience pioneered the study of traditional environmental knowledge, documenting its significance in terms of sustainability and adaptation (Conklin 1954; Moock and Rhoades 1992; Warren, Slikkerveer, and Brokensha 1995; Winkler Prins 1999). Political ecology has broadened the scope by situating environmental change and natural resource management in relation to dynamic power relations from the local to the global level (Bebbington and Batterbury 2001; Blaikie et al. 1994; Bryant 1998). Science and technology studies provide an analytical foothold for examining the production of scientific knowledge and its use in policy as social processes configured by a context of political and ideological struggles (Darier et al. 1999; Demeritt 2001, 2006; Gusterson 1996; Miller 2004; Sarewitz 2000).

Anthropological research on climate reflects the multidimensional nature of the impacts of climate change and the adaptive responses of humans to these impacts. As the studies reviewed in this chapter show, anthropological research can illuminate cognitive, symbolic, and even linguistic aspects of climate change, as well as behavioral responses and power dynamics at both micro- and macro-scales. The use of ethnographic methods continues to be essential to capturing the full gamut of lived experiences and cultural meanings associated with climate. These approaches reveal how climate impinges on human life, not only through its impacts on wellbeing and livelihoods, but also as a dimension of collective narratives, structuring memories of the past and aspirations and anxieties about the future.

At the same time, the multiscale and long-range nature of climate change is leading anthropologists to field settings that do not always lend themselves to approaches familiar to anthropologists, particularly those that hinge on personal interactions and sustained observation of everyday life. This shift is not new to anthropology: for the last twenty years, the practice of "ethnography" has been expanding from physical localities, in which people have long-lasting social ties built on kinship and proximity, to multisited networks composed of people whose lives are connected and who share meanings and practices through media, institutions, and technology (Amit 2000; Marcus 1995, 1998). But an interest in climate change is beckoning anthropologists to explore new frontiers at the intersection of global science, governance, markets, and culture, where the tools of any one discipline are no longer sufficient to achieve a systemic understanding. In addition, anthropologists are increasingly practicing their discipline not only as independent academic researchers, but as professionals embedded in those institutions and supported by government or private sector funding.

Global climate change, therefore, confronts anthropologists with a host of challenges, although some are not entirely new to the discipline. To make effective contributions to interdisciplinary research, anthropologists must learn to collect and manage data in ways that are consistent with those of other scientific traditions. To make anthropological insights relevant to policy, anthropologists must translate them into programmatic prescriptions for decision-makers. This shift will require immersion into systems of knowledge relative to climate science and attainment of new competencies and proficiency in the idioms of science and policy. It may also necessitate tactful negotiations on how research is conducted and how results are interpreted and disseminated. Anthropologists are well equipped to address these challenges, which are inherent to ethnographic fieldwork and to the discipline's epistemological grounding.

Yet it is also important that anthropologists do not compromise their intellectual identities and disciplinary traditions, and that they keep doing what their training, experience, and theoretical inclinations prepare them to do best. Anthropologists should continue to focus on local-level processes and on the consequences of policy and institutional decisions on individuals, households, and communities (Galvin 2007). By highlighting the complexities, ambiguities, uncertainties, and conflicts that characterize vulnerability and adaptation to climate change, anthropologists can offer a necessary counterweight to the tendencies to reduce and simplify reality that sometimes characterizes economic analyses and environmental assessments. This emphasis requires them to educate the scientific community on the value of in-depth research based on nonrandom, small-sized samples, open-ended interviews, and face-to-face engagement of researchers and research participants that characterize ethnographic approaches.

As fieldwork is reconfigured in terms of multisited ethnography, institutional embeddedness, and advocacy-oriented research, anthropologists also come to face new ethical dilemmas that arise from potentially conflicting commitments and accountabilities to research participants, scientific peers, funding agencies, and employers (Marcus 1995). Anthropologists should maintain their long-standing critical discernment, and neither give into pressures to conform to value systems that prevail in other disciplinary or institutional contexts nor compromise on core ideals of cultural sensitivity and social equity. Given the ideological and political polarization surrounding global climate change, it is imperative that anthropologists participate in scientific and policy debates with critical reflexivity. Doing fieldwork in the interstices of climate science, policy, and politics is as challenging as it is crucial, due to the difficulties that arise in studying culture and power within one's own communities and institutions.

Above all, if anthropology is to assert itself as an authoritative voice in climate change debates, engagement must go beyond the individual involvement of a handful of frontliners, many of whom, significantly, are working outside anthropology departments or even academia. Rather, the anthropological community should endorse climate change research as an urgent research priority for the discipline (Finan 2007; Lahsen 2007a). As the climate research and policy communities increasingly acknowledge the value of anthropological research, anthropology as a discipline also needs to recognize that participation in climate policy debates and efforts to build capacity for adaptation at all levels, from local communities to global institutions, are central to anthropology's intellectual mandate and field-grounded epistemology. In other words, engaged ethnography (Batterbury 2008) must be embraced as a vital way of "being there."

Acknowledgments

The authors acknowledge contributions and comments by Simon Batterbury, Kate Dunbar, Preston Hardison, Myanna Lahsen, Heather Lazrus, Peat Leith, Shiloh Moates, Robert Rhoades, and Colin West, as well as constructive feedback by the editors. Research and writing were partly supported by grants from the National Oceanic and Atmospheric Administration to the Southeast Climate Consortium (SECC) and by the National Science Foundation to the Center for Research on Environmental Decisions (CRED).

Notes

1. In this chapter we use the qualifiers "local", "indigenous", "traditional", and "farmers" in association with knowledge as contextually appropriate. We not intend to analyze the plethora of terminologies or acronyms that have been used in various contexts (Antweiler 2004: 3–5). We also do seek to define such knowledge in itself or as opposed to science, although some general attributes may be recognized: for example, the fact that it tends to be holistic-integrative, place-based, orally-transferred, functional, habitual, dynamic, and, in various degrees, shared or specialized (Ellen and Harris 2000).
2. A World Bank database on indigenous knowledge has no keyword for climate: http://econ.worldbank.org/external/default/main?menuPK=633473&pagePK=64165395&piPK=64165418&theSitePK=469372
3. The Niger study used what the author (a veterinarian) refers as a "methodology . . . unusual for pastoral systems research" (Sollod 1990: 271). The interviews were carried out by an educated Twareg who could speak the language, dressed as a nomadic herder, traveled by camel, and spend the night in the pastoral camps, thus functioning as "an indigenous anthropological research agent."
4. The Kempton, Boster, and Hartley (2005) study used quotes from open-ended interviews to construct statements that were later proposed for agreement or disagreement in the course of a survey. This inductive approach contrasts with that used by Pendergraft (1998) who draws on public discourse on environmental issues to formulate similar statements for a survey on individual and collective attitudes toward climate change (but fails to cite the previous study by Kempton et al.).

References

Adger, N. 2003. Social capital, collective action, and adaptation to climate change. *Economic Geography* 79(4): 387–404.

Agrawal, A. 1995. Dismantling the divide between indigenous and scientific knowledge. *Development and Change* 26: 413–39.

Agrawala, S. and K. Broad. 2002. Technology transfer perspectives on climate forecast applications. *Science & Technology Studies* 13: 45–69.

Amit, V. 2000. *Constructing the field: Ethnographic fieldwork in the contemporary world.* London, New York: Routledge.

Antweiler, C. 1998. Local knowledge and local knowing: An anthropological analysis of contested cultural products in the context of development. *Anthropos* 93: 469–94.

Atran, S. 1985. The nature of folk-biological life forms. *American Anthropologist* 87: 89: 315.

Balter, M. 2007. Mild climate, lack of moderns let last Neanderthals linger in Gibraltar. *Science* 313: 1557.

Barlett, P., ed. 1980. *Agricultural decision-making: Anthropological contributions to rural development.* San Diego: Academic Press.

Barnhardt, R. and A. O. Kawagley. 2005. Indigenous knowledge systems and Alaska native ways of knowing. *Anthropology and Education Quarterly* 36: 8–23.

Batterbury, S. 1996. Planners or performers? Reflections on indigenous dryland farming in northern Burkina Faso. *Agriculture and Human Values* 13: 12–22.

———. 2001. Landscapes of diversity: A local political ecology of livelihood diversification in South-western Niger. *Ecumene* 8: 437–64.

———. 2008. Anthropology and global warming: The need for environmental engagement. *The Australian Journal of Anthropology* 19(1): 62–68.

Bebbington A. and S. Batterbury. 2001. Transnational livelihoods and landscapes: Political ecologies of globalization. *Ecumene* 8: 369–80.

Berlin B., D. Breedlove and P. Raven. 1974. *Principles of Tzeltal plant classification.* New York: Academic Press.

Blaikie, P., T. Cannon, I. Davis, and B. Wisner. 1994. *At risk: Natural hazards, people's vulnerability, and disasters*. London: Routledge.

Bogin, B. 1982. Climage change and human behavior on the Southwest coast of Ecuador. *Central issues in Anthropology* 4(1): 21–31.

Boko, M., I. Niang, A. Nyong, C. Vogel, A. Githeko, M. Medany, B. Osman-Elasha, R. Tabo, and P. Yanda. 2007. Africa: Climate change 2007: Impacts, adaptation and vulnerability. In *Contribution of working group II to the fourth assessment report of the intergovernmental panel on climate change*, eds. M. Parry, O. Canziani, J. Palutikof, P. van der Linden and C. Hanson, 433–67. Cambridge: Cambridge University Press.

Broad, K. and B. Orlove. 2007. Channeling globality: The 1997–98 El Niño climate event in Peru. *American Ethnologist* 34: 285–302.

Broad, K., A. Pfaff and M. Glantz. 2002. Effective and equitable dissemination of seasonal-to-interannual climate forecasts: Policy implications from the Peruvian fishery during El Niño 1997–98. *Climatic Change* 54: 415–38.

Broad, K., A. Leiserowitz, J. Weinkle, and M. Steketee. 2007a. Misinterpretations of the cone of uncertainty in Florida during the 2004 hurricane season. *Bulletin of the American Meteorological Society* May: 651–67.

Broad, K., A. Pfaff, R. Taddei, A. Sankarasubramanian, U. Lall, and F. de A. Souza Filho. 2007b. Climate, stream flow prediction and water management in northeast Brazil: Societal trends and forecast value. *Climatic Change* 84: 217–39.

Brokensha D., D. Warren, and O. Werner, eds. 1980. *Indigenous knowledge systems and development*. Waltham, MD: University Press America.

Brown, K. 1999. Climate anthropology: Taking global warming to the people. *Science* 283: 1440–41.

Bryant, R. 1998. Power, knowledge, and political ecology in the third world: A review. *Progress in Physical Geography* 22: 79–94.

Cliggett, L., E. Colson, R. Hay, T. Scudder, and J. Unruh. 2007. Chronic uncertainty and momentary opportunity: A half century of adaptation among Zambia's Gwembe Tonga. *Human Ecology* 35: 19–31.

Conklin, H. 1954. An ethnoecological approach to shifting agriculture. *Transactions of the New York Academy of Sciences* 17: 133–42.

Crane, T. 2006. Changing times, changing ways: Local knowledge, political ecology and development in central Mali. PhD diss., University of Georgia.

Crate, S. 2002. Viliui Sakha oral history: The key to contemporary household survival. *Arctic Anthropology* 39(1): 134–54.

———. 2006a. *Cows, kin, and globalization: An ethnography of sustainability*. Walnut Creek, CA: AltaMira Press.

———. 2006b. Elder knowledge and sustainable livelihoods in post-Soviet Russia: Finding dialogue across generations. *Arctic Anthropology* 43(1): 40–51.

———. 2008. Gone the bull of winter: Grappling with the cultural implications of and anthropology's role(s) in global climate change. *Current Anthropology* 49(4): 569–95.

Cruikshank, J. 2001. Glaciers and climate change: Perspectives from oral tradition. *Arctic* 54: 372–93.

———. 2005. *Do glaciers listen? Local knowledge, colonial encounters, and social imagination*. Vancouver, Toronto: UBC Press.

Crumley, C. 1994. The ecology of conquest: Contrasting agropastoral and agricultural societies' adaptation to climatic change. In *Historical ecology*, ed. C. Crumley, 183–20. Santa Fe, NM: SAR Press.

Darier, E., S. Shackley, S. Wynne, and B. Wynne. 1999. Towards a 'folk integrated assessment' of climate change? *International Journal of Environment and Pollution* 11(3): 351–72.

Demeritt, D. 2001. The construction of global warming and the politics of science. *Annals of the Association of American Geographers* 91: 307–37.
———. 2006. Science studies, climate change and the prospects for constructivist critique. *Economy and Society* 35: 453–79.
Denton, F. 2002. Climate change vulnerability, impacts, and adaptation: Why does gender matter? *Gender and Development* 10(2): 10–20.
de Menocal, P. 2001. Cultural responses to climate change during the Late Holocene. *Science* 292: 667.
DeWalt, B. 1994. Using indigenous knowledge to improve agriculture and natural resource management. *Human Organization* 53: 540–52.
DeWalt, K. and B. DeWalt. 2002. *Participant observation: A guide for fieldworkers.* Walnut Creek, CA: AltaMira Press.
Eakin, H. 1999. Seasonal climate forecasting and the relevance of local knowledge. *Physical Geography* 20: 447–60.
———. 2005. Institutional change, climate risk, and rural vulnerability: Cases from Central Mexico. *World Development* 33(11): 1923–38.
———. 2006. *Weathering risk in rural Mexico: Climatic, institutional, and economic change.* Tucson: University of Arizona Press.
Eakin, H. V. Magan, J. Smith, J. L. Moreno, J. M. Martínez, and O. Landavazo. 2007. A stakeholder driven process to reduce vulnerability to climate change in Hermosillo, Sonora, Mexico. *Mitigation and Adaptation Strategies for Global Change* 12: 935–55.
Ellen, R. and H. Harris. 2000. Introduction. In *Indigenous environmental knowledge and its transformations*, eds. R. Ellen, P. Parkes, and A. Bicker, 1–34. Amsterdam: Harwood.
Ellis, D. 2003. Changing earth and sky: Movement, environmental variability, and responses to El Niño in the Pio-Tura region of Papua New Guinea. In *Weather, climate and culture,* eds. S. Strauss and B. Orlove, 161–80. Oxford: Berg.
Finan, T. 1998. Of bird nests, donkey balls, and El Niño: The psychology of drought in Northeast Brazil. Paper presented at the annual meetings of the American Anthropological Association Annual Meeting, December 2–5, Philadelphia.
———. 2003. Climate science and the policy of drought mitigation in Ceará, northeast Brazil. In *Weather, climate and culture,* eds. S. Strauss and B. Orlove, 203–16. Oxford: Berg.
———. 2007. Is "official" anthropology ready for climate change? *Anthropology News*, December, 10–11.
Finan, T. and D. Nelson. 2001. Making rain, making roads, making do: Public and private adaptations to drought in Ceará, Northeast Brazil. *Climatic Research* 19: 97–108.
Galvin, K. 2007. Adding the human component to global environmental change research. *Anthropology News*, December. 11–12.
Galvin, K., R. Boone, N. Smith, and S. Lynn. 2001. Impacts of climate variability on East African pastoralists: Linking social science and remote sensing. *Climate Research* 19(2): 161–72.
Gladwin, C. 1989. *Ethnographic decision tree modeling.* Newbury Park, CA: Sage.
Gusterson, H. 1996. *Nuclear rites: A weapons laboratory at the end of the Cold War.* Berkeley: University of California Press.
Harley, T. 2003. "Nice weather for the time of the year: The British obsession with the weather." In *Weather, climate and culture,* edited by S. Strauss and B. Orlove, 103–18. Oxford: Berg.
Henning, A. 2005. Climate change and energy use. *Anthropology Today* 21(2005): 8–12.
Henshaw, A. 2003. Climate and culture in the North: The interface of archaeology, paleoenvironmental science, and oral history. In *Weather, climate and culture,* eds. S. Strauss and B. Orlove, 217–31. Oxford: Berg.

Huber, T. and P. Pedersen. 1998. Meteorological knowledge and environmental ideas in traditional and modern societies: The case of Tibet. *Journal of the Royal Anthropological Institute* 3: 577–98.

Hunn, E. 1982. The utilitarian in folk biological classification. *American Anthropologist* 84: 830–47.

Ingold, T., ed. 2000. *The perception of the environment: Essays in livelihood, dwelling, and skill.* London: Routledge.

Ingold, T. and T. Kurtilla. 2000. Perceiving the environment in Finnish Lapland. *Body and Society* 6: 183–96.

Ingram, K., C. Roncoli, and P. Kirshen. 2002. Opportunities and constraints for farmers of West Africa to use seasonal precipitation forecasts with Burkina Faso as a case study. *Agricultural Systems* 74: 331–49.

Jasanoff, S. and B. Wynne. 1998. "Science and decision-making," In *Human choice and climate change*, vol. 1, eds. S. Rayner and E. Malone, 1–87. Columbus, OH: Batelle Press.

Jorgensen, D. 1989. *Participant observation: A methodology for human studies.* Newbury Park, CA: Sage Publications.

Katz, E., A. Lammel, and M. Goloubinoff, eds. 2002. *Entre ciel et terre; climat et sociétés.* Paris, IRD: Ibis Press.

Keil, A., M. Zeller, A. Wida, B. Sanim, and R. Birner, in press. What determines farmers' resilience towards ENSO-related drought? An empirical assessment in Central Sulawesi, Indonesia. *Climatic Change.*

Kempton, W., J. Boster, and J. Hartley. 1995. *Environmental values in American culture.* Cambridge, MA: MIT Press.

Kihupi N., R. Kingamkono, H. Dihenga, M. Kingamkono, and W. Rwamugira. 2003. Integrating indigenous knowledge and climate forecasts in Tanzania. In *Coping with climate variability: The use of seasonal climate forecasts in Southern Africa*, eds. C. Vogel and K. O'Brien, 155–69. Burlington, VT: Ashgate.

King, V.T. 1996. Environmental change in Malaysian Borneo: Fire, drought and rain. In *Environmental change in Southeast Asia: People politics and sustainable development,* eds. M. Parnwell and R. Bryant, 165–89. New York: Routledge.

Krupnik, I. and D. Jolly, eds. 2002. The earth is faster now: Indigenous observations of arctic environmental change. Fairbanks, AK: Arctic Consortium of the United States.

Kuper, R. and S. Kropelin. 2006. Climate-controlled Holocene occupation in the Sahara: Motor of Africa's evolution. *Science* 313: 803–07.

Lahsen, M. 1998. The detection and attribution of conspiracies: The controversy over Chapter 8. In *Paranoia within reason: A casebook on conspiracy as explanation*, ed. G. Marcus, 111–26. Chicago: University of Chicago Press.

———. 2005a. Seductive simulations: Uncertainty distribution around climate models. *Social Studies of Science* 35: 895–922.

———. 2005b. Technocracy, democracy and U.S. climate science politics: The need for demarcations. *Science, Technology, & Human Values* 30: 137–69.

———. 2007a. Anthropology and the trouble of risk society. *Anthropology News.* December, 9–10.

———. 2007b. "Trust through participation? Problems of knowledge in climate decision making." In *The social construction of climate change,* ed. M. Pettinger, 173–96. Aldershot: Ashgate.

Laidler, G. 2006. Inuit and scientific perspectives on the relationship between sea ice and climate change: The ideal complement? *Climatic Change* 78: 407–44.

Latour, S., S. Woolgar, and J. Salk. 1986. *Laboratory life: The social construction of scientific facts*, Princeton, NJ: Princeton University Press.

Lemos, M.C. 2003. A tale of two policies: The politics of climate forecasting and drought relief in Ceará, Brazil. *Policy Science* 36: 101–24.

Lemos, M.C., T. Finan, R. Fox, D. Nelson, and J.Tucker. 2002. The use of seasonal climate forecasting in policymaking: Lessons from Northeast Brazil. *Climatic Change* 55: 171–96.

Little, P., H. Mahmoud, and D. Coppock. 2001. When deserts flood: Risk management and climatic processes among East African pastoralists. *Climate Research* 19(2): 149–59.

Luseno, W., J. McPeak, C. Barrett, P. Little, and G. Gebru. 2003. Assessing the value of climate forecast information for pastoralists: Evidence from Southern Ethiopia and Northern Kenya. *World Development* 31: 1477–94.

Magistro J. and C. Roncoli. 2001. Anthropological perspectives and policy implications of climate change research. *Climate Research* 19: 91–96.

Malone, E. and S. Rayner. 2001. Role of the research standpoint in integrating global-scale and local-scale research. *Climate Research* 19(2): 173–78.

Marcus, G. 1995. Ethnography in/of the world system: The emergence of multi-sited ethnography. *Annual Review of Anthropology* 24: 95–117.

———. 1998. *Ethnography through thick and thin*. Princeton, NJ: Princeton University Press.

McCorkle, C. 1989. Towards a knowledge of local knowledge and its importance for agricultural RD&E. *Agriculture & Human Values* 4: 4–13.

McDaniel, J., D. Kennard, and A. Fuentes. 2005. Smokey the tapir: Traditional fire knowledge and fire prevention campaigns in lowland Bolivia. *Society & Natural Resources* 18: 921–31.

McIntosh, R., J. Tainter, and S. Keech McIntosh, eds.. 2000. *The way the wind blows: Climate, history, and human action*. In *Series in historical ecology*, eds. W. Balée and C. Crumley. New York: Columbia University Press.

Meehl, G. A., T. F. Stocker, W. D. Collins, P. Friedlingstein, A. T. Gaye, J. M. Gregory, A. Kitoh, R. Knutti, J. M. Murphy, A. Noda, S. C. B. Raper, I. G. Watterson, A. J. Weaver, and Z.-C. Zhao. 2007. Global climate projections. In *Climate change 2007: The physical science basis*. Contribution of Working Group I to the Fourth Assessment Report of the Intergovernmental Panel on Climate Change, eds. S. Solomon, D. Qin, M. Manning, Z. Chen, M. Marquis, K. B. Averyt, M. Tignor, and H. L. Miller, 747–845, Cambridge and New York: Cambridge University Press.

Migowski, C., M. Stein, S. Prasad, J. Negendank, and A. Agnon. 2006. Holocene climate variability and cultural evolution in the Near East from the Dead Sea sedimentary record. *Quaternary Research* 66: 421–31.

Miller, C. 2004. Climate science and the making of a global political order. In *States of knowledge: The co-production of science and social order*, ed. S. Jasanoff, 46–66. London and New York: Routledge.

Minnegal, M. and P. Dwyer. 2000. Responses to a drought in the interior lowlands of Papua New Guinea: A comparison of Bedamuni and Kubo-Konai. *Human Ecology* 28: 493–526.

Mishra, C., H. Prins, and S. Van Wieren. 2003. Diversity, risk mediation, and change in a Trans-Himalayan agropastoral system. *Human Ecology* 31: 595–609.

Moock, J. L. and R. Rhoades, eds. 1992. *Diversity, farmer knowledge, and sustainability*. Ithaca, NY: Cornell University Press.

Moran, E. and D. Liverman, eds. 1998. *People and pixel: Remote sensing and social science*. Washington, DC: National Academy Press.

Moran, E., R. Adams, B. Bakoy'Ema, S. Fiorini, and B. Boucek. 2007. Human strategies for coping with El Niño related drought in Amazônia. *Climatic Change* 77: 343–61.

Mutiso, S. 1996. Indigenous knowledge in drought and famine forecasting in Machakos District, Kenya. In *Indigenous knowledge and change in African agriculture*, eds. W. Adams, L. Slikkerveer and I. Ames, 67–86. Center for Indigenous Knowledge for Agriculture and Rural Development. Ames: Iowa State University.

Nabhan, G. 2002. *The desert smells like rain: A naturalist in O'Odham country*. Tucson: University of Arizona Press.
Nadasdy, P. 1999. The politics of TEK: Power and the "integration" of knowledge. *Arctic Anthropology* 36(1999): 1–18.
———. 2005. *Hunters and bureaucrats: Power, knowledge, and aboriginal-state relations in the southwest Yukon*. Vancouver: University of British Columbia Press.
Nazarea, V. 2006. Local knowledge and memory in biodiversity conservation. *Annual Review of Anthropology* 35: 317–35.
Nazarea-Sandoval, V. 1995. *Local knowledge and agricultural decision-making in the Philippines: Class, gender and resistance*. Ithaca, NY: Cornell University Press.
Nelson, D. 2007. Expanding the climate change research agenda. *Anthropology News*. December, 12–13.
Nelson, D. and T. Finan. 2000. The emergence of a climate anthropology in Northeast Brazil. *Practicing Anthropology* 22: 6–10.
Nuttall, M. 1991. Memoryscape: A sense of locality in Northwest Greenland. *North Atlantic Studies* 1(2): 39–51.
———. 1992. *Arctic homeland: Kinship, community and development in Northwest Greenland*. Toronto: University of Toronto Press.
———. 1998. *Protecting the Arctic: Indigenous peoples and cultural survival*. Amsterdam: Harwood.
Nuttall, M., F. Berkes, B. Forbes, G. Kofinas, T. Vlassova, and G. Wenzel. 2005. Hunting, herding, fishing and gathering: Indigenous peoples and renewable resource use in the Arctic. In *ACIA (Arctic Climate Impact Assessment) Scientific report*. Cambridge: Cambridge University Press.
Nyong, A., F. Adesina, and B. Osman Elasha. 2007. The value of indigenous knowledge in climate change mitigation and adaptation strategies in the African Sahel. *Mitigation and Adaptation Strategies for Global Change* 12: 787–97.
Oldfield, F. 1993. Forward to the past: Changing approaches to quaternary palaeoecology. In *Climate change and human impact on the landscape*, ed. F. Chambers, 14–21. London: Chapman Hall.
Orlove, B. 2003. How people name seasons. In *Weather, climate and culture*, eds. S. Strauss and B. Orlove, 121–40. Berg: Oxford.
———. 2005. Human adaptation to climate change: A review of three historical cases and some general perspectives. *Environmental Science and Policy* 8: 589–600.
Orlove, B., J. Chiang, and M.Cane. 2000. Forecasting Andean rainfall and crop yield from the influence of El Niño on Pleiades visibility. *Nature* 403: 68–71.
———. 2002. Ethnoclimatology in the Andes: A cross-disciplinary study uncovers the scientific basis for the scheme Andean potato farmers traditionally use to predict the coming rains. *American Scientist* 90: 428–35.
Orlove B., K. Broad, and A. Petty. 2004. Factors that influence the use of climate forecasts. *Bulletin of the American Meteorological Society* 85: 1–9.
Orlove, B. and M. Kabugo. 2005. Signs and sight in Southern Uganda: Representing perception in ordinary conversation. *Etnofoor* 1: 124–41.
Orlove, B., E. Wiegandt, and B. Luckman. 2008. The place of glaciers in cultural and natural landscapes: Environment, history, and culture as influences on perceptions of glacier dynamics. In *Darkening peaks: Mountain glacier retreat in social and biological contexts*, eds. B. Orlove, E. Wiegandt, and B. Luckman, 3–19. Berkeley: University of California Press.
Orlove, B., C. Roncoli, M. Kabugo, and A. Majugu. Under review. Indigenous knowledge of climate variability in southern Uganda: The multiple components of a dynamic regional system. *Climatic Change*.
Ovuka, M. and S. Lindqvist. 2000. Rainfall variability in Murang'a district, Kenya: Meteorological data and farmers' perceptions. *Geografiska Annaler* 82: 107–19.

Paolisso, M. 2003. Chesapeake Bay watermen, weather, and blue crabs: Cultural models and fishery policies. In *Weather, climate and culture*, eds. S. Strauss and B. Orlove, 61–83. Berg: Oxford.
Pendergraft, C. 1998. Human dimensions of climate change: Cultural theory and collective action. *Climatic Change* 39: 643–66.
Peters, P. 2004. Inequality and social conflict over land in Africa. *Journal of Agrarian Change* 4(3): 269–314.
Peterson, L. and G. Haug. 2005. Climate and the collapse of Maya civilization. *American Scientist* 93: 322–29.
Pettinger, M., ed. 2007. *The social construction of climate change*. London: Ashgate.
Pfaff A., K. Broad, and M. Glantz. 1999. Who benefits from climate forecasts? *Nature* 397: 645–46.
Phillips, J., E. Makaudze, and L. Unganai. 2001. Current and potential use of climate forecasts for resource-poor farmers in Zimbabwe. In *Impacts of El Niño and climate variability in agriculture*, special publication no. 63, ed. C. Rosenzweig, 87–100. Madison, WI: American Society of Agronomy.
Posey, D. 1984. Ethnoecology as applied anthropology in Amazonian development. *Human Organization* 43: 95–107.
———. 1986. Topics and issues in ethnoentomology, with some suggestions for the development of hypothesis generation and testing in ethnobiology. *Journal of Ethnobiology* 6: 99–120.
———. 2000. Ethnobiology and ethnoecology in the context of national laws and international agreements affecting indigenous and local knowledge, traditional resources and intellectual property rights. In *Indigenous environmental knowledge and its transformations*, eds. R. Ellen, P. Parkes, and A. Bicker, 35–54. Amsterdam: Harwood.
Purcell, T. 1998. Indigenous knowledge and applied anthropology: Questions of definition and direction. *Human Organization* 57(3): 258–72.
Purcell, T. and E. A. Onjoro. 2002. Indigenous knowledge, power and parity: Models of knowledge integration. In *Participating in development: Approaches to indigenous knowledge*, eds. P. Sillitoe, A. Bicker, and J. Pottier, 162–88. London: Routledge.
Puri, R. 2007. Responses to medium-term stability in climate: El Niño, droughts, and coping mechanisms in foragers and farmers in Borneo. In *Modern crises and traditional strategies: Local ecological knowledge in island southeast Asia*, ed. R. Ellen, 46–83. New York: Berghahn Books.
Rappaport, R. 1979. *Ecology, meaning and religion*. Berkeley, CA: North Atlantic Books.
Rautman, A. 1994. Regional climate records and local experience: 'Drought' and the decline of dryfarming in Central New Mexico. *Culture and Agriculture* 49: 12–15.
Rayner, S. 2003. Domesticating nature: Commentary on the anthropological study of weather and climate discourse. In *Weather, climate and culture*, eds. S. Strauss and B. Orlove, 277–90. Oxford: Berg.
Rayner, S., H. Ingram, and D. Lach. 2005. Weather forecasts are for wimps: Why water resource managers do not use climate forecasts. *Climatic Change* 69: 197–277.
Rhoades, R. and A. Bebbington. 1995. Farmers who experiment: An untapped resource for agricultural research and development. In *Indigenous knowledge systems: The cultural dimension of development*, eds. M. Warren, L. J. Slikkerveer, and D. Brokensha, 296–307. London: Intermediate Technology Publications.
Rhoades, R., X. Zapata, and J. Aragundy. 2006. Climate change in Cotacachi. In *Development with identity: Community, culture, and sustainability in the Andes*, ed. R. Rhoades, 64–74. Cambridge, MA: CABI Publishing.
———. 2008. Mama Cotacachi: Local perceptions and societal implications of climate change, glacier retreat, and water availability. In *Darkening peaks: Mountain glacier retreat in social and biological contexts*, eds. B. Orlove, E. Wiegandt, and B. Luckman, 218–27. Berkeley: University of California Press.

Richards, P. 1985. *Indigenous agricultural revolution.* London: Huchinson & Co.
———. 1993. "Cultivation: Knowledge or performance?" In *An anthropological critique of development: The growth of ignorance*, ed. M. Hobart, 61–78. New York: Routledge.
Richerson, P., R. Boyd, and R. Bettinger. 2001. Was agriculture impossible during the Pleistocene but mandatory during the Holocene? A climate change hypothesis. *American Antiquity* 6: 387–411.
Roncoli, C. 2006. Ethnographic and participatory approaches to research on farmers' responses to climate predictions. *Climate Research* 33: 81–99.
Roncoli, C., K. Ingram, P. Kirshen, and I. Flitcroft. 2000. Opportunities and constraints to using seasonal precipitation forecasting to improve agricultural production systems and livelihood security in the Sahel-Sudan region: A case study of Burkina Faso, CFAR-Phase 1. Proceedings of the International Forum on Climate Prediction, Agriculture, and Development, International Research Institute for Climate Predictions, April 26–28, Palisades, New York.
Roncoli, C., K. Ingram, and P. Kirshen. 2001. The costs and risks of coping with drought: Livelihood impacts and farmers' responses in Burkina Faso. *Climate Research* 19: 119–32.
———. 2002. Reading the rains: Local knowledge and rainfall forecasting among farmers of Burkina Faso. *Society and Natural Resources* 15: 411–30.
Roncoli, C., K. Ingram, P. Kirshen, and C. Jost. 2003. Meteorological meanings: Understandings of seasonal rainfall forecasts by farmers of Burkina Faso. In *Weather, climate and culture*, eds. S. Strauss and B. Orlove, 181–202. Oxford: Berg.
Roncoli, C., C. Jost, P. Kirshen, M. Sanon, K. Ingram, M. Woodin, L. Somé, F. Ouattara, J. Sanfo, C. Sia, P.Yaka, and G. Hoogenboom. 2008. From accessing to assessing forecasts: An end-to-end study of participatory forecast dissemination in Burkina Faso (West Africa). *Climatic Change*, DOI: 10.1007/s10584-008-9445-6.
Salick, J. and Anja Byg, eds. 2007. *Indigenous people and climate change.* Oxford: Tyndall Centre for Climate Change Research.
Sanders, Todd. 2003. (En)Gendering the weather: Rainmaking and reproduction in Tanzania. In *Weather, climate and culture*, eds. S. Strauss and B. Orlove, 181–202. Oxford: Berg.
Sarewitz, D. 2000. Science and environmental policy: An excess of objectivity In *Earth matters: The earth sciences, philosophy, and the claims of community*, ed. R. Frodeman, 255–75. Upper Saddle River, NJ: Prentice Hall.
Schensul, S., J. Schensul, and M. LeCompte. 1999. *Essential ethnographic methods: Observations, interviews, and questionnaires.* Walnut Creek, CA: AltaMira.
Scoones, I. 1998. Sustainable rural livelihoods: A framework for analysis. IDS Working paper no. 72. Brighton, UK: Institute of Development Studies.
Shackley, S. 2001. Epistemic lifestyles in climate change modeling. In *Changing the atmosphere: Expert knowledge and environmental governance*, eds. C. Miller and P. Edwards, 107–34. Cambridge, MA: MIT Press.
Shackley, S. and B. Wynne. 1995. Integrating knowledges for climate change: Pyramids, nets and uncertainties. *Global Environmental Change* 52: 113–26.
———. 1996. Representing uncertainty in global climate science and policy: Boundary ordering devices and authority. *Science, Technology and Human Values* 213: 275–302.
Shakeley, S., P. Young, S. Parkinson, and B. Wynne. 1998. Uncertainty, complexity, and concepts of good science in climate change modeling: Are GCM the best tools? *Climatic Change* 8: 159–205.
Shapin, S. and S. Schaffer. 1989. *Leviathan and the air-pump.* Princeton, NJ: Princeton University Press.
Sherman-Morris, K. 2005. Tornadoes, television, and trust: A closer look at the influence of the local weathercaster during severe weather. *Environmental Hazards* 6: 201–10.

Sillitoe, P. 1996. *A place against time: Land and environment in the Papua New Guinea highlands*. London: Routledge.

———. 1998. The development of indigenous knowledge: A new applied anthropology. *Current Anthropology* 39: 223–52.

———. 2007. Local science vs. global science: An overview. In *Local science vs. global science: Approaches to indigenous knowledge in international development*, ed. P. Sillitoe, 1–22. New York: Berghahn Books.

Sollod, A. 1990. Rainfall variability and Twareg perceptions of climate impacts in Niger. *Human Ecology* 18: 267–81.

Stern, P. 1999. Learning to be smart: An exploration of the culture of intelligence in a Canadian Inuit community. *American Anthropologist* 101: 502–14.

Stevenson, M. 1996. Indigenous knowledge in environmental assessment. *Arctic* 49: 278–91.

Stigter, C., Z. Dawei, L. Onyewotu, and M. Xurong. 2005. Using traditional methods and indigenous technologies for coping with climate variability. *Climatic Change* 70: 255–71.

Strauss, S. 2003. Weather wise: Speaking folklore to science in Leukerbad. In *Weather, climate and culture*, eds. S. Strauss and B. Orlove, 39–61. Oxford: Berg.

Strauss, S. and B. Orlove. 2003. Up in the air: The anthropology of weather and climate. In *Weather, climate and culture*, eds. S. Strauss and B. Orlove, 3–14. Oxford: Berg.

Thompson, J. and I. Scoones. 1994. Challenging the populist perspective: Rural people's knowledge, agricultural research and extension practice. *Agriculture & Human Values* 11: 58–76.

Thornton, P., R. Boone, K. Galvin, S. BurnSilver, M. Waithaka, J. Kuyiah, S. Karanja, E. González-Estrada, and M. Herrero. 2007. Coping strategies in livestock-dependent households in East and Southern Africa: A synthesis of four case studies. *Human Ecology* 35: 461–76.

Torry, W. 1983. Anthropological perspectives on climate change. In *Social science research and climate change*, ed. R. Chen, E. Boulding, and S. Schneider, 207–88. Dordrecht: Springer.

Tschakert, P. 2004a. Carbon for farmers: Assessing the potential for carbon sequestration in the Old Peanut Basin of Senegal. *Climatic Change* 67: 273–90.

———. 2004b. The costs of soil carbon sequestration: An economic analysis for small-scale farming systems in Senegal. *Agricultural Systems* 81: 227–53.

———. 2007a. Environmental services and poverty reduction: Options for smallholders in the Sahel. *Agricultural Systems* 94 (1): 75–86.

———. 2007b. Views from the vulnerable: Understanding climatic and other stressors in the Sahel. *Global Environmental Change* 17(3): 381–96.

Tschakert, P. and G. Tappan. 2004. The social context of carbon sequestration: considerations from a multi-scale environmental history of the Old Peanut Basin of Senegal. *Journal of Arid Environments* 59(3): 535–64.

Trench, P., J. Rowley, M. Diarra, F. Sano, and B. Keita. 2007. *Beyond any drought: Root causes of chronic vulnerability in the Sahel.* The Sahel Working Group, June. http://www.iied.org/mediaroom/docs/Beyond%20Any%20Drought.pdf.

Tyler, N., J. Turi, M. Sundset, K. Strom Bull, M. Sara, E. Reinert, N. Oskal, C. Nellemann, J. McCarthy, S. Mathiesen, M. Martello, O. Magga, G. Hovelsrud, I. Hanssen-Bauer, N. Eira, I. Eira, and R. Corell. 2007. Saami reindeer pastoralism under climate change: Applying a generalized framework for vulnerability studies to a sub-arctic social-ecological system. *Global Environmental Change* 17: 191–206.

Vasquez-León, M., C. West, and T. Finan. 2003. A comparative assessment of climate vulnerability: Agriculture and ranching on both sides of the US-Mexico border. *Global Environmental Change* 13: 159–73.

Vedwan, N. 2006. Culture, climate and the environment: Local knowledge and perception of climate change among apple growers in Northwestern India. *Journal of Ecological Anthropology* 10: 4–18.
Vedwan, N. and R. Rhoades. 2001. Climate change in the western Himalayas of India: A study of local perception and response. *Climate Research* 9: 109–17.
Vogel, C., S. Moser, R. Kasperson, and G. Dabelko. 2007. Linking vulnerability, adaptation, and resilience science to practice: Pathways, players, and partnerships. *Global Environmental Change* 17: 349–64.
Waddell, E. 1975. How the Enga cope with frost: Responses to climatic perturbations in the Central Highlands of New Guinea. *Human Ecology* 3: 249–73.
Warren, M., L. Slikkerveer, and D. Brokensha, eds. 1995. *The cultural dimension of development: Indigenous knowledge systems.* London: Intermediate Technology Publications.
West, C. and M. Vásquez-León. 2003. Testing farmers' perceptions of climate variability: A case study from the Sulphur Sping Valley, Arizona. In *Weather, climate and culture*, eds. S. Strauss and B. Orlove, 233–50. Oxford: Berg.
West, C., C. Roncoli, and F. Ouattara. 2008. Local perceptions and regional rainfall trends in the Central Plateau, Burkina Faso. *Land Degradation and Development.* DOI: 10.1002/ldr.842
Wiegandt, E. and R. Lugon. 2008. Challenges of living with glaciers in the Swiss Alps, past and present. In *Darkening peaks: Mountain glacier retreat in social and biological contexts*, eds. B. Orlove, E. Wiegandt, and B. Luckman, 35–48. Berkeley: University of California Press.
Winkler Prins, A. M. G. A. 1999. Local soil knowledge: A tool for sustainable land management. *Society and Natural Resources* 11(7): 151–61.
Wolf, B. and B. Orlove. 2008. Environment, history, and culture as influences on perceptions of glacier dynamics. In *Darkening peaks: Mountain glacier retreat in social and biological contexts*, eds. B. Orlove, E. Wiegandt, and B. Luckman, 49–67. Berkeley: University of California Press.
Wynne, B. 1996. May the sheep safely graze? A reflexive view of the expert-lay knowledge divide. In *Risk, environment and modernity: Towards a new ecology*, eds. S. Lash, B. Szerszynski, and B. Wynne, 27–44. London: Sage.
Ziervogel, G. and R. Calder. 2003. Climate variability and rural livelihoods: Assessing the impact of seasonal climate forecasts in Lesotho. *Area* 35: 403–18.

Chapter 4

Climate Change and Population Displacement: Disasters and Diasporas in the Twenty-first Century

Anthony Oliver-Smith

Introduction: Environmental Change and Migration

Throughout history humans have had to adapt to both short- and long-term environmental change. Adaptations have taken many forms, but migration, whether permanent or temporary, has always been a central response and survival strategy (Hugo 1996; Paul 2005). Traditional patterns of human adaptation have involved temporary and seasonal migration, particularly in Africa, Asia, and Latin America, or more long-term migration responses, for example, of people facing cyclical droughts in places like the Sahel (Merryman 1982) and Northeast Brazil (Kenny 2002). Drought was a climatic feature of those environments and people developed adaptive strategies to cope with it, largely by temporarily migrating and then returning once the drought broke. Anthropologists have long been among the leading researchers of these adaptive patterns, particularly those related to issues of famine and famine relief (Shipton 1990). Now, however, we are seeing those temporary migrations becoming permanent, some because of climate change and others because of inadequate government response to either the threat or impact of drought.

Based on their work with peoples engaging in seasonal or stress-related migration and the process of involuntary migration and resettlement, anthropologists are well positioned to address the potentials of mass uprooting, migration, and resettlement that global climate change presents. Over the last thirty years anthropologists working with uprooted people around the world have addressed an extremely wide array of issues in their work, spanning activities as diverse yet related as applied research, policy formation, theory building, evaluation, planning, implementation, and resistance. Anthropologists were among the first to recognize and report on the impoverishment, social disorganization, and violation of human rights that occurred among uprooted populations. They have probed the similarities and differences among the different kinds of displacement for insights to improve theory, policy, and practice (Cernea and McDowell 2000;

Hansen and Oliver-Smith 1982). Anthropological knowledge and practice in areas associated with displacement and resettlement will take on increasing importance as more and more people and communities confront the impacts of global climate change.

Today, global environmental change is more extreme than at any other time in recorded history. Its local realities increasingly uproot large numbers of people. The complex interplay of social and economic factors in this backdrop of environmental change increases the vulnerability of both people and environments, intensifying the impacts of such changes when they occur. Moreover, greater numbers of people are more vulnerable to the impacts of such changes than ever before, due both to increases in population and density and to environmental degradation and residence in dangerous areas. One of the main environmental forces humans face in the twenty-first century is global climate change.

Global Climate Change and Migration

Global climate change is increasingly accepted by both the scientific community and the general public as a reality that must be addressed in both policy and practice. The recent reports from the Intergovernmental Panel on Climate Change (IPCC 2007a, 2007b) affirm that human-induced factors are responsible for generating significant increases in temperatures around the world. Among the consequences of this rise in temperature are increases in the rate of sea level rise; increases in glacial, permafrost, Arctic, and Antarctic ice melt; more rainfall in specific regions of the world and worldwide; more severe droughts in tropical and subtropical zones; increases in heat waves; changing ranges and incidences of diseases; and more intense hurricane and cyclone activity. Additionally, many of these changes feedback into others to accelerate overall effects. These changes increasingly affect natural systems globally, altering hydrological, terrestrial, biological, and aquatic subsystems. Accordingly, they have great potential to uproot large numbers of people, forcing them to migrate as individuals and families or permanently displacing them and/or relocating them as communities. Local effects of global climate change will also combine with other negative factors, such as environmental contamination, to force migration and displacement of people. Overall there are three major effects of global climate change that will contribute most to uprooting people: loss of ecosystem services, loss of land, and increased intensity and frequency of climate-based natural disasters. Each of these three also interacts with the others to compound and intensify effects. The next sections explore each of these effects in-depth.

Loss of Ecosystem Services

First, global climate change is expected to seriously alter the availability of and access to ecosystem services. Human life in specific environments is maintained by the provision of a number of ecosystem services, including food, water, fuel, and nutrition, as well as those cultural elements, largely spiritual and/or aesthetic, that sustain communities through expressive links to natural features (Renaud et al. 2007; UN Millennium Project 2005). However, in combination with other factors such as nutrient pollution, overexploitation, invasive species, and disease, and demographic, economic, sociopolitical, and cultural factors, global climate change will further strain the resilience of local socioecological systems beyond their capability to provide necessary services. The environmental degradation of local ecosystems compounded by the local effects of global climate change can result in the loss of sufficient resources and other ecosystem services, uprooting people and communities and forcing them to migrate. Drylands, which cover approximately 41 percent of terrestrial land surfaces and are home to roughly two billion people, are prone to desertification, a process that leaves environments depleted of ecosystem resources and services and no longer able to sustain human life. Increased temperatures and diminished rainfall in specific areas will exacerbate existing conditions and contribute to the deterioration of ecosystem services, already scant in many dryland regions (Renaud et al. 2007). Rising sea levels, as well as increased rainfall and flooding, particularly in low-lying river delta regions, may also result in loss of ecosystem services, obligating populations to relocate.

Loss of Land

Loss of land entails both the loss of ecosystem services and the actual disappearance of terrain. Current projections of temperature and sea level rise and increased intensity of droughts and storms suggest that population displacement due to loss of terrain will take place at significant scales within the next thirty to fifty years, particularly for populations in coastal zones. People who occupy global regions in the low elevation coastal zone (between one and ten meters above sea level) are vulnerable to the permanent inundation of their homes and livelihoods caused by global climate change–induced sea level rise. Although these areas constitute only 2 percent of total earth land surface, they contain 10 percent of the world's population and 13 percent of the urban population, including almost two-thirds of all cities larger than 5 million people (McGranahan, Balk, and Anderson 2007). Recent research estimates that "in all, 634 million people live within such areas—defined as less than 10m above sea level—and that number is growing. Of the more than 180 countries with populations in the low-elevation coastal zone, about 70 percent have urban areas of more than five million people that extend into it, including Tokyo; New York; Mumbai, India; Shanghai, China; Jakarta, Indonesia; and Dhaka, Bangladesh." Indeed, about 75 percent of all the people residing in low-lying areas are in Asia, and the most vulnerable are the poor (McGranahan et al. 2007). Over the last five years, some communities in coastal areas in Alaska, the South Pacific, and the Gulf Coast of the United States have been facing the threat of community-wide displacement and resettlement. The physical and social processes recently triggered by Hurricane Katrina on the Gulf Coast of the United States underscore the threat of this emerging reality. While some displacement is likely to be gradual as coastal land is increasingly inundated over the coming years, elevated sea levels will also increase the impacts of tropical storms, creating sudden, devastating disasters.

Increasing Intensity and Frequency of Climate-Driven Disasters

The third major effect of global climate change that will force migrations and displacements of peoples is the increasing intensity and frequency of climate-driven disasters, perhaps most profoundly understood by Hurricane Katrina. With the increase in ocean surface temperatures over the last fifty years, levels of damage from extreme weather events have also increased. On a global scale, losses from natural disasters have increased dramatically over the last half century, particularly so since the middle of the 1980s (Munich Re 1999, 16). As a purely economic index, annual dollar losses from environmental disasters increased from 3.9 billion to 40 billion a year by the 1990s (Renaud et al. 2007, 26). Dollar-based indices, while useful, do not capture the suffering endured by disaster victims, which generally extends far past the event, and is more and more frequently compounded by permanent displacement and resettlement. Global climate change will increase the risk of stronger tropical storms with higher storm surges, which, when combined with rising sea levels, will extend the onshore impacts of coastal flooding much further inland, particularly in the low coastal elevation zone. While damage from high winds in tropical cyclones and hurricanes can be devastating, the risk of increased flooding has the highest potential for the displacement and resettlement of communities. With increasing evidence that "what used to be 'a once in a 100 year event' is becoming more common"(Huq et al. 2007, 4), resulting displacements could become permanent as more and more coastal land is lost to the sea or eliminated as habitable zones because of tidal storm surges.

To some degree, the first two effects of global climate change that we have discussed—the loss of ecosystem services and the loss of land—can be glossed very often in terms of the third, that is, as disasters. Will the impacts of global climate change qualify as disasters? Indeed, the term *disaster* has been applied to the outcomes of all three of the climate change–based processes that will lead to displacement and resettlement. The choice of that term is often operationally driven. That is, in order to activate aid and assistance to affected populations, the drought/famine or the inundated coastline are defined as humanitarian emergencies or disasters. Disasters thus play a metonymic role for encompassing many processes and events currently unfolding in the contemporary world. While it is not my intention to digress into the deeply complex definitional debate around the concept of disaster

(Oliver-Smith 1999; Quarantelli 1998), linking the concept of disaster to these important issues correctly relocates the focus of analysis from an event to a process. In focusing on an environmental or social problem and glossing it as "a disaster," analysts are often actually dealing with part of the processual aspect of disaster, that is, the social and technological construction of conditions of vulnerability, often without being entirely aware of it. Moreover, regardless of how analysts view these processes, it is fairly clear that their outcomes will be seen and felt as disasters by the affected populations.

Vulnerability and Environmental Refugees

Although the problem seems to be an enduring one, until relatively recently most of the literature has characterized environmentally induced displacement as temporary, suggesting that people eventually return (Oliver-Smith 1991). However, most of this research has been event focused, both temporally and spatially, and has given far less attention to the longer term, more geographically dispersed aspects of postevent recovery or reconstruction. Today, with more precise spatial and temporal scales of analysis, natural disasters, rather than unanticipated and unique events caused by a natural agent, are seen to be much more explainable in terms of the "normal" order of things—that is, the conditions of inequality and subordination in the society rather than the accidental geophysical features of a place. This perspective has shifted the focus away from the disaster event and towards the vulnerability of peoples embedded in the "on-going societal and man-environment relations that prefigure [disaster]" (Hewitt 1983, 24–27).

The concept of vulnerability as it interplays with the dynamics of some sort of environmental change involves the totality of relationships in a given social situation producing a set of conditions that render a society unable to absorb the impacts of a natural or social agent without significant disruption of its capacity to fulfill the basic needs of its members. Risks and outcomes are thus largely socially produced. This more complex understanding of vulnerability and disasters enables researchers to analyze how social systems generate the conditions that place people, differentiated along axes of class, race, ethnic, gender, or age, at different levels of risk from the same hazard and subject to different forms of suffering from the same event. Vulnerability to these dynamic processes is particularly accentuated in the developing world, where people have fewer resources either to manage threats or to recover from impacts. Therefore, the impacts of global climate change, like any disaster, will be socially, politically, and economically mediated, distributed, and interpreted, with measures to mitigate and respond similarly structured.

Since the 1980s researchers have linked the issue of environmental change with human migration, explicitly designating people who are forced to leave their homes, temporarily or permanently, due to the threat, impact, or effects of a hazard or environmental change as "environmental refugees" (El-Hinnawi 1985). Although environmental studies have traditionally focused on the natural world, the impacts of pollution, deforestation, soil erosion, degradation, desertification, and other environmental processes on human beings have also been a source of both interest and concern to ecological and social scientists. Indeed, as mentioned, the impacts of many of these processes have often been framed as "disastrous" because they create stress, disrupt normal social processes, and force people to adapt by making temporary adjustments or permanent changes in how and where they live.

The debate over this issue, with claims of millions of environmental refugees being produced versus counterclaims that the evidence is uneven, unconvincing, and counterproductive, has been active since the 1980s. Norman Myers has asserted that recent human-induced environmental change, such as desertification, deforestation, or soil erosion, compounded by natural and man-made disasters, could force as many as 50 million people to migrate from their homes by 2010 (1997). Other researchers dispute the accuracy of the term *environmental refugee*, finding it misleading. They attribute the displacement of people to a complex pattern of factors including political, social, economic, and environmental forces (Black 2001; Castles 2002; Wood 2001). Environmental disruptions, including natural disasters, are seen to cause temporary displacement, but not some idea of authentic (i.e., permanent) migration. Indeed, if permanent migration does occur as the result of a disaster, it is seen as more the result of deficient responses of weak or corrupt states rather than the environment as expressed in the form of a natural hazard impact.

Black's critique, that focusing on environmental factors as causes of migration may obscure the role of political and economic factors, is well taken and echoes the position held by most disaster researchers today. Focusing solely on agents reveals little about the political or economic forces that also contribute to disasters or, for that matter, any forced migration that might ensue. But these objections in turn elide the fact that the environment, its resources, and its hazards are socially constructed and framed for people through social, economic, and political factors, even in the best of times (Oliver-Smith 2002). The environment cannot be separated from society to isolate it as a single cause. The displacement from the South Asian tsunami is probably the closest we come to migration as a result of a purely natural phenomenon. In most cases, however, nature and society produce socially constructed environments that are mutually constitutive. In other words, the environment cannot be a single cause because, in most cases, it is interwoven with society. A disaster is no longer defined in terms of its event aspect only, but also in terms of the processes that set it in motion and the postevent processes of adaptation and adjustment in recovery and reconstruction.

There is little question that some environmental disasters force people to migrate. Both socionatural and technological disasters, often in combination, may uproot communities by sudden destruction, as in the gas explosion of Guadalajara, Mexico (Macias and Calderon Aragon 1994), or by saturation of the environment with toxic substances, making it uninhabitable or unusable, as in Chernobyl; Valdez, Alaska; or parts of New Orleans (Button in press; Dyer, Gill, and Picou 1992; Petryna 2002). In some cases, depending on the agents, the scale of the event, and the physics of dispersal, saturation can be local, as in the Murphy oil spill in parts of New Orleans, whereas the contamination from Chernobyl affected large sections of a continent. The South Asian tsunami clearly displaced millions, and Hurricane Katrina uprooted more than one million people and left many hundreds of thousands permanently displaced. Their displacement, however, is not due to environmental reasons alone, but to the political economy of reconstruction as well.

The Great Flood of 1927 in the lower Mississippi Valley displaced nearly 700,000 people, approximately 330,000 of whom were African Americans who were subsequently interned in 154 relief "concentration camps," where they were forced to work (www.mvd.usace.army.mil/MRC-History-Center/gallery/flood/flood2.html).[1] Although there were many reasons for African Americans to leave the South, the flood and its consequences, especially the forced labor in the camps, were the final motivation for migrating for thousands (Barry 1997, 417).

Displacement, Resettlement, and Loss

For people affected by disasters and other environmental changes, displacement and resettlement constitute a second disaster in their lives. The impact of the initial disaster intensifies in the aftermath, both in that people experience losses and in the process of recovery. Serious disasters inflict terrible losses on people and communities, often breaking up families and uprooting communities to relocate in radically changed and/or new environments. Displacement both compounds and makes permanent many of the losses incurred in disaster. Those who can reconstruct in situ, even in much diminished circumstances, stand a better chance at recovery.

The destruction or loss through uprooting of livelihood and community require those impacted to engage in a process of reinvention. As social creatures, people's reinvention of self will be intimately linked to the reinvention of social bonds and community. This process of reinvention or recovery will have both material and social aspects. Material and social losses compound each other. Those who are uprooted, having suffered almost complete loss, like political refugees, must migrate with fewer resources with which to reconstruct their lives.

Material elements such as housing, material possessions, infrastructure, services, health care, transportation, communication, and nutrition can all be endangered, damaged, or destroyed in serious disasters and/or lost in displacement. In addition to physical damage, material losses resonate profoundly in the social world, affecting the economic, social, and cultural life of survivors. For example, material damages frequently mean the loss of livelihoods, whether through destruction or loss of worksite, tools and equipment, land or common property resources, or physical injury. Loss of livelihood and the capacity to sustain oneself endanger individual and social identity, producing a loss of status and resulting in marginalization and social disarticulation through the fragmenting of social networks such as kin, religious, and other collaborative groups and associations. The loss of a house is also the loss of a social "place." The entire community or social world is thus endangered by such individual losses. The dispersal, displacement, or death of family members fragments not only a household, but erodes the social cohesion of a community as well. Disaster-caused deaths erode networks of relationships that form the basis of personal and social identity, setting people adrift without those ties that anchor the self in the social world. Survivors of serious disasters in which there is great loss of life and prolonged devastation and displacement also suffer a loss of personal identity and partial loss of self. The loss of significant others in high-mortality disasters is also a loss of the self, since that the part of the self the other played is lost. Thus, the loss of a child means that one has lost that part of the self that was a parent. To reinforce this point, the loss of status, the social leveling, the reduction to a common level of misery, can constitute an assault on the sense of the self.

Cultural identity is at risk in uprooted communities. The loss and destruction of important cultural sites, shrines, and religious objects, and the interruption of important sacred and secular events and rituals, undermines the community's sense of itself. Disasters and displacement may endanger the identification with an environment that may once have been seen as nurturing and central to cultural identity but is feared and distrusted in the aftermath (Oliver-Smith 1992). Displacement for any group can be a crushing blow, but for indigenous peoples it can prove mortal, considering that land tenure is an essential element in the survival of indigenous societies and distinctive cultural identities.

These losses of community, family, and self compound each other to create a cumulative loss of meaning. Events and prolonged conditions of deprivation and displacement can shake the foundations of personal worldview and identity. They challenge the culturally constructed vision where the world is a place imbued with logic and life makes sense, even if it can be unfair. Major disasters rob people of the social context in which they live meaningful lives that are considered significant by others. This loss of personal relationships and the social context in which they were expressed and in which the individual was affirmed may leave people bereft of a sense of meaning and purpose in life. In this context, religion may cease to provide solace and can itself become a casualty in the aftermath of disasters.

Obviously, people forced from their known environments are separated from the material and cultural resource base upon which they have depended for life as individuals and as communities. Perhaps not so obvious is the fact that a sense of place plays an important role in individual and collective identity, in the way time and history are encoded and contextualized, and in interpersonal, community, and intercultural relations (Altman and Low 1992; Escobar 2001; Maalki 1992; Rodman 1992). A sense of place is crucial in the creation of what Giddens calls an "environment of trust" in which space, kin relations, local communities, cosmology, and tradition are linked (Giddens 1990, 102 as cited in Rodman 1992, 648).

In short, removal from a most basic physical dimension of life can mean removal from life. The disruption in individual or community identity and stability in place, resulting in resettlement in a strange landscape, can baffle and silence people like a strange language can (Basso 1988 as cited in Rodman 1992, 647). Culture loses its ontological grounding and people must struggle to construct a world that can clearly articulate continuity and identity as a community again. The human need for "environments of trust" is fundamental to the sense of order and predictability implied by culture.

Environmentally Induced Displacement and Recovery

Recovery and reconstruction for forcibly uprooted people take place in a new setting, generally far from familiar environments and people. In other words, getting to where they are going does not solve the problem. They may have stopped moving, but that is just the beginning of another process, that of resettlement. In all too many cases, resettlement, particularly when done at the community level, ends up becoming a secondary disaster. Therefore, when disasters damage or destroy communities, the uprooting of people and displacement of them far from homes and jobs makes the process of recovery doubly complex. These events/processes often set people on the road, breaking up families and communities. Uprooted people generally face the daunting task of rebuilding not only personal lives, but also communities—those relationships, networks, and structures that support people as individuals. The social destruction wrought by these phenomena takes place at both the individual level and at the community level. In some cases, survivors of these events resettle themselves individually or as families in new environments, facing the challenges of integration in new areas.

In the United States the government entrusts refugee resettlement to nongovernmental organizations (NGOs) with little attention paid to the idea of community of origin.[2] This lack of attention to community is perhaps due to Americans' ideological distance from the idea of community as something people need. Americans are portrayed as eminently mobile, able to adapt easily to new homes, new jobs, and new networks (Bellah et al. 1985). The degree to which that contention is true for Americans may be debated, but it is certainly not the case for many of the world's people. However, the discourse of displacement and resettlement in American society, that is, the choice of terminology and the scale or unit of analysis most frequently addressed, is at the level of individuals and families, whereas most large-scale displacement very frequently involves communities. This is significant because what is often lost in displacement and resettlement is the community network that enabled people to access material resources and the social and emotional support critical to survival. The community is more than the sum of the total number of individuals, and the loss of community for displaced people, particularly when the loss is the outcome of aid policies that do not take community into account, can be devastating.

To compound the issue further, when a community is resettled, it is not simply lifted up and set down whole in a new site. In most cases the community is reconfigured in specific ways. Most resettlement projects, particularly in the developing world, directly or indirectly further two additional processes: the expansion of the state and integration into regional and national market systems (Scott 1998). Neither of these processes of inclusion is particularly simple or straightforward, but in most cases they produce a restructuring of social, economic, and political relationships toward the priorities of the larger society. Resettlement itself does not always necessarily destroy "local cultures," but rather appropriates and restructures them in terms of the values and goals of distant interests, far beyond the local context. Such a process involves the reduction of local culture, society, and economy from all their varied expressions to a narrow set of institutions and activities that make them compatible with the purposes of the larger society (Garcia Canclini 1993).

The process of reconstruction, formally or informally, must address these losses, not only to reconstruct the community in a material sense, but also to support the community's efforts to rebuild itself and reconstruct the fabric of community life. For the most part, the process of reconstruction has been approached largely as a material problem. The forms of aid and assistance that are marshaled to assist displaced peoples have generally focused on issues of material need in the form of housing, nutrition, and health care. There is no denying that the often-excruciating material needs of the displaced must be addressed, but the question that is often not satisfactorily answered is *how* they should be addressed. Material aid is often donor designed and largely a transfer process. It is frequently delivered in content and form in ways that compound the social and psychological effects of destruction and displacement by undermining self-esteem, compromising community integrity and identity, and creating patterns of dependency.

In material terms, the needs of individuals, households, communities, and the extralocal systems of which they are a part are numerous, diverse, and interconnected. The procedures put in place to cope with emergency needs, however, are rarely linked to key features of community organization, although they can determine the longer-term rehabilitative system. This results in very negative impacts on the long-term viability of the community. For example, donor-driven housing forms may endanger the connection that people establish with their built environment, violating cultural norms of space and place, inhibiting both the reformation of social networks and the re-emergence of community identity (Oliver-Smith 1991). The experience of millions of displaced peoples and the failures of public housing clearly show that the built environment can seriously undermine and even prevent community from emerging, instead exaggerating the social tensions and conflict that often plague such uprooted populations. Such plans and structures are more focused on donor efficiency and cost rather than on the needs of the displaced population to reconstitute a community. In the long run the social disarticulation they foster undermines the productivity and self-sufficiency of the group.

Despite the obvious fact that employment after resettlement is essential, joblessness or lack of livelihood is one of the most common failings of resettlement projects (Cernea and McDowell 2000). From both a material and a psychological standpoint, economics drives the process of reconstruction. Employment provides needed income to replace or improve upon those personal and household needs not provided by aid, but it is also a form of action that enables people to return to being actors rather than being acted upon as disaster victims.

Disaster and displacement also causes people to lose their means of production, be it land, tools, or resource access, preventing them from engaging in normal activities. In this context it is important to establish livelihoods on the basis of traditional products, skills, and technology, allowing people to continue with known practices particularly for the initial period of adjustment. Tensions can become acute when the displaced seek to relocate in existing communities and may compete with a dense host population for scarce social and economic resources.

Key to the process of reconstruction and vulnerability reduction is the negotiation of tension due to the availability of jobs locally and in the broader and distant region. Generally, disasters markedly impact labor markets either by limiting the supply of jobs because of economic dislocation or by limiting the supply of labor because of displacement (Button and Oliver-Smith 2008). Disaster-induced displacement alters the distance and time relationships between homes and jobs. Until people resume employment, they remain dependent on external resources, and reconstruction remains incomplete.

UNDERSTANDING DISPLACEMENT AND RESETTLEMENT

If we are to understand potential mass displacements from global climate change, and respond effectively to them, we need to identify those pertinent sources of theory and information that can inform appropriate policy formation and practice. The process of resettlement in cases of involuntary uprooting has proven to be particularly challenging, and cases of successful resettlement projects are few (Scudder 2005). The vast majority of these projects, whether from disasters, development, or conflict-driven displacement, leave local people permanently displaced, disempowered, and destitute. For the vast majority of the displaced, the causes of dislocation and the uprooting process itself are nothing less than catastrophic both at the personal and the community level. These forces—natural and technological disasters, political conflict, and large scale development projects—can all be considered "totalizing phenomena" in their capacity to affect virtually every domain of human life (Oliver-Smith 2006).

The social scientific literature on displacement and resettlement is clustered around three themes: disasters, civil and military conflicts, and development projects. The scant literature from disaster-driven displacement suffers from an excessive focus at the individual and organizational levels and a relatively short temporal scale. The research from conflict-driven uprooting focuses largely on temporary camps, repatriation, and individual and family refugee resettlement to foreign countries. The literature on development-forced displacement and resettlement offers an important source of perspectives and models for furthering the understanding of global climate change–driven migration, particularly of the large-scale displacements projected for the not-too-distant future. This research is also being complemented by a growing concern regarding Internally Displaced Persons (IDP) (Deng and Cohen 1999; Koser 2007).

Although there are clear differences between disaster-forced and development-forced resettlement, particularly in terms of the "triggering" event and the question of intentionality, these differences diminish in the aftermath. Following the actual cause, there is a greater degree of commonality in the challenges people face, whether they are victims of development-forced displacement or disaster. In that sense, the victims of Hurricane Katrina or the South Asian Tsunami suffer the same persisting challenges as people displaced by development projects: homelessness, unemployment, marginalization, the loss of neighborhood and community, mental and physical health challenges, and powerlessness (Button in press; Cernea 1996b).

Regardless of the cause, displaced people, disaster victims, and refugees face a complex series of events involving dislocation; homelessness; unemployment; the dismantling of families and communities; adaptive stresses; loss of privacy; political marginalization; a decrease in mental and physical health status; and the daunting challenge of reconstituting one's livelihood, family, and community (Cernea 1990, 1997; Colson 1971; Scudder 1981, 1991). This array of challenges that displaced people face serves as a basic outline of "family resemblances" (Button in press). The "family resemblances" of disaster- and development-forced displacement and resettlement suggest that the literatures associated with these fields should enrich one another. The gap between the two fields, according to Turton (2006), is purely a social construct that inhibits a more complete understanding of displacement and resettlement processes (Button in press; Cernea and McDowell 2000).

If, in fact, the uprooted are resettled in some systematic way, the quality of the resettlement project itself may play a major role in the capacity of the community to recover from the trauma of displacement. Such projects are really about reconstructing communities after they have been materially destroyed and socially traumatized. Reconstructing and reconstituting community is an idea that needs to be approached with a certain humility and realism about the limits of current planning capacities. To date, such humility and realism have not always characterized the planners and administrators of projects dealing with uprooted peoples to any major extent. Instead, the goals of such undertakings frequently stress efficiency and cost containment over restoration of community. Such top-down initiatives have a poor record of success because of a lack of regard for local community resources. Planners often perceive the culture of uprooted people as an obstacle to success, rather than as a resource.

Communities do not construct themselves; they evolve. Even intentional communities, self-organized around a common ideology and possessing a high degree of homogeneity, are not long lasting. Reconstructing/reconstituting a community means attempting to replace, through administrative routine, an evolutionary process in which social, cultural, economic, and environmental interactions develop and enable a population to achieve a mutually sustaining social coherence and material sustenance over time. The social systems that develop are not perfect, are often far from egalitarian, and do not conform to some imagined standard of efficiency. The kind of community that sustains individual and group life is not a finely tuned mechanism or a well-balanced organism, but rather a complex, interactive, ongoing process composed of innumerable variables that are subject to the conscious and unconscious motives of its members. The idea that such a process could be the outcome of planning is ambitious, to say the least. One of the best outcomes that might be imagined for resettlement projects is to work out a system in which people can materially sustain themselves while they begin the process of social reconstruction. The least that could be hoped for might be that resettlement projects not impede the process of community reconstitution. However, if the level of impoverishment experienced by most resettled peoples is an indicator, it remains beyond the will and/or the capabilities of most contemporary policy makers and planners to provide even adequate systems of material reproduction. This fact does not bode well for the victims of potential mass displacements from processes associated with global climate change.

Although several models of the nature of voluntary planned settlement processes were developed in the late 1960s (Chambers 1969; Nelson 1973), there were few attempts at theory or model building of involuntary forms of displacement and resettlement until the late 1970s and early 1980s. There are two major theoretical approaches dealing with involuntary resettlement that policy makers and researchers have found helpful in understanding outcomes. The first was Scudder's four-stage framework that modeled first and second generation responses of resettlers (Scudder 1981; Scudder and Colson 1982). The second was Michael Cernea's Impoverishment Risks and Reconstruction model (1990). Two subsequent efforts (de Wet 2006; McDowell 2002) have also advanced our understanding of the displacement and resettlement process. Although these approaches were formulated primarily for cases of development-forced displacement and resettlement, these concepts also have application in contexts of postdisaster and postconflict resettlement (Button in press; Cernea 1996b; Scudder and Colson 1982).

First, Scudder and Colson developed an approach based on the concept of stress to describe and analyze the process of involuntary dislocation and resettlement (Scudder 1981; Scudder and Colson 1982). The Four Stage Framework, as Scudder now calls it, emphasizes how most resettlers can be expected to behave during each of the four stages necessary to the completion of a successful resettlement project (Scudder in press). Scudder and Colson posited that three forms of stress resulted from involuntary relocation and resettlement. *Physiological stress* is seen in increased morbidity and mortality rates. *Psychological stress* has four manifestations: trauma, guilt, grief, and anxiety. *Sociocultural stress*, in the form of fragmented social networks, economic deprivation, loss of power, and social dissension, is manifested as a result of the economic, political, and cultural effects of relocation. Affected people experience these three forms of stress, collectively known as *multidimensional stress*, as they pass through the displacement and resettlement process.

The process itself is represented as occurring in four stages. *Recruitment* refers to the decisions taken by authorities regarding the population to be relocated, particularly those that influence the length and severity of the stressful transition stage. The *transition* stage begins when the population to be relocated is first affected. Generally speaking, the transition stage is the longest and when affected people experience the most severe multidimensional stress. The general attitude of people during the transition stage is conservative in order to avoid the possibility of further risk and stress. The stage of *potential development* begins when people begin to abandon their conservative risk-avoidance strategies and express greater initiative and risk-taking behavior. Scudder and Colson emphasize that this stage is often never realized since many Development-Forced Displacement and Resettlement (DFDR) projects remain trapped in the transition stage by inappropriate policy and inept implementation. Equally difficult to attain is the final stage of *handing over* or *incorporation*. Achieving the incorporation stage signifies that the DFDR project has been successful. Scudder and Colson define success as achieving the local management of economic and political affairs and the phasing out of external agencies and personnel from day-to-day management of the community. The community has become able to assume its place within the larger regional context that includes host communities and other regional systems.

At roughly the same time that Scudder and Colson were developing their model, a complimentary approach emerged in political ecology, linking the ideas of *vulnerability* and *risk*. Vulnerability was initially employed in disaster research to understand how different societies experienced vastly divergent losses from similar disaster agents. An alternative perspective on human-environment relations, emphasizing the role of human interventions in generating disaster risk and impact, found that these sets of relations coalesced in the concept of vulnerability (Hewitt 1983). Vulnerability and risk, therefore, refer to the relationships between people, the environment, and the sociopolitical structures that frame the conditions in which people live. The concept of vulnerability thus integrates not only political and economic but also environmental forces in terms of both biophysical and socially constructed risk. This understanding of vulnerability enabled researchers to conceptualize how social systems generate the conditions that place different kinds of people, often differentiated along axes of class, race, ethnic, gender, or age, at different levels of risk.

As these concepts gained currency, Cernea began his writing about the risks of poverty from water projects displacements (1990), from which he developed his now well-known Impoverishment Risks and Reconstruction (IRR) approach to understanding (and mitigating) the major adverse effects of displacement. Herein he outlines eight basic risks to which people are subjected to by displacement (Cernea 1996b; Cernea and McDowell 2000). The model is based on the three basic concepts of risk, impoverishment, and reconstruction. Deriving his understanding of risk from Giddens's (1990) notion of the possibility that a certain course of action may produce negative effects, Cernea models displacement risks by deconstructing the "syncretic, multifaceted process of displacement into its identifiable principle and most widespread components": landlessness, joblessness, homelessness, marginalization, food insecurity, increased morbidity, loss of access to common property resources, and social disarticulation (2000). He further asserts that these risks will further produce serious consequences in badly or unplanned resettlement. Cernea's IRR model is designed to predict, diagnose, and resolve the problems associated with DFDR.

Two years later, Christopher McDowell combined Sustainable Livelihoods research and Cernea's IRR approach to develop a methodological framework for research on postdisaster resettlement (2002). His approach is based on the assertion in Sustainable Livelihoods research that social institutional processes are central to livelihoods in the ways they influence households' access to resources, whether natural or social. One of the principal risks in displacement and resettlement is social disarticulation, including the scattering of kinship groups and informal networks of mutual help

(Cernea and McDowell 2000). The disarticulation of spatially and culturally based patterns of self-organization, social interaction, and reciprocity constitutes a loss of essential social ties that affect access to resources and compound the loss of natural and man-made capital. Thus, in displacement and resettlement, peoples' adaptations to social disarticulation produces new dynamics that influence people's access and control over resources and often lead to further impoverishment. Therefore understanding institutional processes in resettlers' adaptive strategies is crucial for identifying the socioculturally specific nature of the risks Cernea identified as inherent in forced displacement and helps explain why displacement and resettlement so often results in greater impoverishment of affected households.

Asking why resettlement so often goes wrong, Chris de Wet has recently sought to incorporate Cernea's important insights and sees two broad approaches to responding to the question. The first approach is what he calls the "inadequate inputs" approach, which argues that resettlement projects fail because of a lack of appropriate inputs: national legal frameworks and policies, political will, funding, predisplacement research, careful implementation, and monitoring. Optimistic in tenor, the inadequate inputs approach posits that the risks and injuries of resettlement can be controlled and mitigated by appropriate policies and practices. On the other hand, De Wet finds himself moving toward what he calls the "inherent complexity" approach, arguing for "the interrelatedness of a range of factors of different orders: cultural, social, environmental, economic, institutional and political—all of which are taking place in the context of imposed space change and of local-level responses and initiatives" (de Wet 2006). These changes take place simultaneously in an interlinked and mutually influencing process of transformation and are influenced by and respond to the imposition from both external sources of power and initiatives of local actors. Therefore, the resettlement process emerges out of the complex interaction of all these factors in ways that are not predictable and that do not seem amenable to a linear-based, rational planning approach. De Wet argues that a more comprehensive and open-ended approach is necessary to understand, adapt to, and take advantage of the opportunities presented by the inherent complexity of the displacement and resettlement process. The fact that authorities are limited in the degree of control they can exercise over a project creates a space for resettlers to take greater control over the process. The challenge thus becomes the development of policy that supports a genuine participatory and open-ended approach to resettlement planning and decision-making (de Wet 2006).

Conclusion

The effects of global climate change will increase the number and scale of forced migrations in the relatively near future. For example, other cities that face threats similar to those that devastated New Orleans include Mumbai,

Caracas, Cape Town, Dar es Salaam, Manila, and Darwin. Increasing population and poverty and occupation of hazardous and vulnerable areas increases the probability of forced migrations. Socionatural disasters that trigger technological disasters make environments uninhabitable and force people to migrate. While increasing state and market integration have resulted in more resilient infrastructures in some world regions, they have also undermined traditional adaptations of rural populations to natural hazards. The specific effects of global climate change—including increased risks of flooding, storms, deforestation, desertification, soil erosion, and sea level rise—increase the probability of disasters contributing to internal and international forced migration. The catastrophic losses from Hurricane Katrina demonstrate the urgent need to develop the conceptual, strategic, and material tools to confront the increasing challenges of population displacement and resettlement from natural hazards made even more potent and complex by climate change, increasing population densities, and environmental degradation in the twenty-first century.

Over the past twenty-five years, colleagues from the field of development-induced displacement, refugee studies, and disaster research have learned that the displaced peoples we work with share many similar challenges (Cernea 1996a; Cernea and McDowell 2000; Hansen and Oliver-Smith 1982; Oliver-Smith in press; Turton 2003). Although the places and peoples are geographically and culturally distant and the sociopolitical environments and causes of dislocation often dissimilar, there remain a number of common concerns and processes. All displaced people must cope with multiple stresses and the need to adapt to a new or radically changed environment. All may experience privation, loss of homes, jobs, and the breakup of families and communities. All must mobilize social and cultural resources in their efforts to reestablish viable social groups and communities and to restore adequate levels of material life. These are important similarities that anthropologists must recognize and understand both to minimize displacement and to assist in the material reconstruction and the social reconstitution of communities. Without doubt there are important differences between disaster- and development-forced resettlement projects. However, according to Turton (2006), it is also clear that the differences between the two forms of displacement are "scientifically unsound." There are sufficient shared characteristics between the two research fields to provide a rich field for productive cross-fertilization and the development of a kind of "systematic, comparative long-term research required for improving a policy relevant theory" (Scudder in press) that is sorely needed in displacement and resettlement studies research, policy, and practice (Button in press).

As discussed earlier, although the issue of environmental refugees has generated significant debate over the last twenty years, appropriate policies pertaining to environmentally displaced peoples or other internally displaced populations have yet to attain legal status. Moreover, according to the International Federation of Red Cross and Red Crescent Societies (IFRC 2004), "there are no well-recognized and comprehensive legal instruments which identify internationally agreed rules, principles and standards for the protection and assistance of people affected by natural and technological disasters. . . . As a result, many international disaster response operations are subject to ad hoc rules and systems, which vary dramatically from country to country and impede the provision of fast and effective assistance—putting lives and dignity at risk." (IFRC 2004, 1).

The category "refugee," with all its attendant rights, still applies only to a very specifically defined group of people who, in fleeing for their lives, have crossed an international border. Although over the last decade there has been increasing concern regarding internally displaced persons and their rights, and there is increasing recognition that the causes of displacement and resettlement are far wider than wars and civil conflicts, there are still no nationally or internationally binding agreements or treaties that guarantee the rights of people who have been uprooted by other causes such as environmental disruption, disasters, or development projects. The Guiding Principles on Internal Displacement define internally displaced persons as "persons or groups of persons who have been forced or obliged to flee or leave their homes or places of habitual residence, in particular as a result of or in order to avoid the effects of armed conflict, situations of generalized violence, violations of human rights or natural or human made disasters, and who have not crossed an internally recognized state border" (http://www.UNHCHR.ch/htm/menu2/7/b/principles.htm). However, although widely recognized as an international standard, and certainly helpful in guiding NGOs and other aid organizations in assisting IDPs, the guiding principles have not been agreed upon in a binding covenant or treaty and have no legal standing.

The very real potential for global climate change to increasingly generate displacements and migrations across international borders, in combination with the dearth of appropriate policies for internally displaced persons, makes urgent the need for developing adequate legal protections and assistance programs for populations facing potential displacement by forces generated by global climate change. Current estimates for the number of environmentally displaced people around the world are highly debated for reasons discussed earlier, but the Office of the United Nations High Commissioner for Refugees estimated in 2002 that approximately 24 million people around the world had been displaced by floods, famines, and other environmental causes (UNHCR 2002, 12). These numbers could be dwarfed by the potential displacements caused by global climate change. It is both urgent and incumbent upon anthropology and the social sciences in general to contribute to the development of both theory and method that will generate legally binding policies and informed practice to address the massive displacement and resettlement that global climate change is and will increasingly cause.

Notes

1. See www.mvs.usace.army.mil/pa/esprit/2002/esp0203.pdf
2. There is, however, a debate between NGOs that favor resettling refugees with coethnics so that they can help each other with language and employment issues and those that believe in dispersing the displaced so that they will assimilate and learn English faster (Hansen 2005).

References

Altman, I. and S. Low. 1992. *Place attachment.* vol. 8: *Human Behavior and Environment: Advances in Theory and Research.* New York: Plenum.

Barry, J. M. 1997. *Rising Tide: The Great Mississippi Flood of 1927 and How it Changed America.* New York: Simon and Schuster.

Basso, K. 1988. 'Speaking with Names:' Language and Landscape Among the Western Apache. *Cultural Anthropology* 3:2: 99–130.

Bellah, R. N., R. Madsen, W. M. Sullivan, and A. Swidler. 1985. *Habits of the heart: Individualism and commitment in American life.* Berkeley: University of California Press.

Black, R. 2001. Environmental refugees: Myth or Reality? *UNHCR Working Papers* (34): 1–19.

Blaikie, P., T. Cannon, I. Davis, and B. Wisner. 1994. *At risk: Natural hazards, people's vulnerability and disasters.* London: Routledge.

Button, G. V. In press. Family resemblances between disasters and development-induced displacement: Hurricane Katrina as a comparative case study. In *Development & dispossession: The anthropology of development-induced displacement and resettlement*, ed. A. Oliver-Smith. Santa Fe, NM: School for American Research.

Button, G. V. and A. Oliver-Smith. 2008. Disaster, displacement and employment: Distortion in labor markets in post-Katrina reconstruction. In *Capitalizing on catastrophe: The globalization of disaster assistance*, eds. N. Gunewardena and M. Schuller, 123–45. Walnut Creek, CA: AltaMira.

Castles, S. 2002. Environmental change and forced migration: Making sense of the debate. *UNHCR Working Papers* (70): 1–14.

Cernea, M. 1990. Poverty risks from population displacement in water resources development, *HIID Development Discussion Paper* no. 355. Cambridge, MA: Harvard University Press.

———. 1996a. Bridging the research divide: Studying refugees and development oustees, In *In search of cool ground: War, flight and homecoming in Northeast Africa,* ed. T. Allen, 293–317. London: James Currey/Africa World Press.

———. 1996b. *Eight main risks: Impoverishment and social justice in resettlement.* Washington, DC: World Bank Environment Department.

———. 1997. The risks and reconstruction model for resettling displaced populations. *World Development* 25(10): 1569–88.

Cernea, M. and C. McDowell. 2000. *Risks and reconstruction: Experiences of resettlers and refugees.* Washington, DC: The World Bank.

Chambers, R. 1969. *Settlement schemes in tropical Africa: A study of organizations and development.* London: Routledge and Kegan Paul.

Colson, E. 1971. *The social consequences of resettlement.* Manchester: Manchester University Press.

De Wet, C. 2006. Risk, complexity and local initiative in involuntary resettlement outcomes. In *Towards improving outcomes in development-induced involuntary resettlement projects*, ed. C. de Wet, 180–202. Oxford and New York: Berghahn Books.

Deng, F. and R. Cohen. 1999. Masses in flight: The global crisis of internal displacement; and: The forsaken people: case studies of the internally displaced. *Human Rights Quarterly* 21(2): 541–44.

Dyer, C. L., D. A. Gill, and J. S. Picou. 1992. Social disruption and the Valdez oil spill: Alaskan natives in a natural resource community. *Sociology Spectrum* 12: 105–26.

El-Hinnawi, E. 1985. *Environmental refugees.* Nairobi: United Nations Environmental Programme.

Escobar, A. 2001. Culture sits in places: Reflections on globalism and subaltern strategies of localization. *Political Geography* 20: 139–74.

Garcia Canclini, N. 1993. *Transforming modernity*. Austin: University of Texas Press.

Giddens, A. 1990. *The consequences of modernity*. Cambridge: Polity Press.

Hansen, A. 2005. Black and white and the other: International immigration and change in metropolitan Atlanta. In *Beyond the gateway: Immigrants in a changing America*, eds. E. M. Gozdziak and S. F. Martin, 87–109. Lanham, MD: Lexington Books.

Hansen, A. and A. Oliver-Smith. 1982. *Involuntary migration and resettlement: The problems and responses of dislocated peoples*, Boulder, CO: Westview Press.

Hewitt, K. 1983. *Interpretations of calamity.* Winchester, MA: Allen & Unwin.

Hugo, G.1996. Environmental concerns and international migration. *International Migration Review* 30:1: 105–31.

Huq, S., S. Kovats, H. Reid and D. Satterthwaite. 2007. Reducing risks to cities from disasters and climate change. *Environment and Urbanization* 19(3): 3–15.

Intergovernmental Panel on Climate Change (IPCC). 2007a. *Climate change 2007: The physical science basis, summary for policy makers: Contribution of Working Group I to the Fourth Assessment Report of the Intergovernmental Panel on Climate Change.* Paris: IPCC.

———. 2007b. *Climate change 2007: Climate change impacts, adaptation and vulnerability, summary for policy makers: Contribution of Working Group II to the Fourth Assessment Report of the Intergovernmental Panel on Climate Change.* Brussels: IPCC.

International Federation of Red Cross and Red Crescent Societies. 2004. *World disasters report.* IFRC: Geneva.

Kenny, M. L. 2002. Drought, clientilism, fatalism and fear in Northeast Brazil. *Ethics, Place and Environment* 5(2): 123–34.

Koser, K. 2007. The global IDP situation in a changing humanitarian context. UNICEF Global Workshop on IDPs at Brookings Institute, September 4, Washington, DC.

Macias, J. M. and G. Calderon Aragon. 1994. *Desastre en Guadalajara: Notas preliminares y testimonios.* Mexico DF: CIESAS.

Maalki, L. 1992. National geographic: The rooting of peoples and the territorialization of national identity among scholars and refugees. *Cultural Anthropology* 7(1): 24–44.

McDowell, C. 2002. Involuntary resettlement, impoverishment risks, and sustainable livelihoods. *The Australasian Journal of Disaster and Trauma Studies* 2, http://www.massey.ac.nz/~trauma/issues/2002–2/mdowell.htm (accessed 4/5/08)

McGranahan, G., D. Balk, and B. Anderson. 2007. Reducing risks to cities from disasters and climate change. *Environment and Urbanization* 19(1) 17–37.

Merryman, J. 1982. Pastoral nomad settlement in response to drought: The case of the Kenya Somali. In *Involuntary migration and resettlement: The problems and responses of dislocated peoples*, eds. A. Hansen and A. Oliver-Smith, 105–20. Boulder, CO: Westview Press.

Munich Re. 1999. *Topics: Natural catastrophes. The current position.* Munich: Munich Reinsurance Company.

Myers, N. 1997. Environmental refugees. *Population and Environment* 19(2): 167–82.

Mississippi River Gallery. (n.d.). *The Flood of '27.* http://www.mvd.usace.army.mil/MRC-History-Center/gallery/flood/flood2.html (accessed 9/2/04)

Nelson, M. 1973. *Development of tropical lands: Policy issues in Latin America.* Baltimore: Johns Hopkins University Press.

Oliver-Smith, A. 1991. Success and failures in post disaster resettlement. *Disasters* 15(1): 12–24.

Oliver-Smith, A. 1992. *The martyred city: Death and rebirth in the Andes*. Prospect Heights, IL: Waveland Press.
———. 1999. What is a disaster? Anthropological perspectives on a persistent question. In *The angry earth: Disaster in anthropological perspective*, eds. A. Oliver-Smith and S. M. Hoffman, 18–34. New York: Routledge.
———. 2002. Theorizing disasters: Nature, culture, power. In *Catastrophe and culture: The anthropology of disaster*, eds. S. M. Hoffman and A. Oliver-Smith, 23–47. Santa Fe, NM: The School of American Research Press.
———. 2006. Communities after catastrophe: Reconstructing the material, reconstituting the social. In *Building communities in the 21st century*, ed. S. Hyland, 45–70. Santa Fe, NM: School of American Research Press.
———. in press. *Development and dispossession: The crisis of development forced displacement and resettlement*. Santa Fe, NM: School of Advanced Research.
Paul, B. K. 2005. Evidence against disaster caused migration: The 2004 tornado in north-central Bangladesh. *Disasters* 29(4): 370–85.
Petryna, A. 2002. *Life exposed: Biological citizens after Chernobyl*. Princeton, NJ: Princeton University Press.
Quarantelli, E. L., ed. 1998. *What is a disaster? A dozen perspectives on the question*. New York: Routledge.
Rodman, M. C. 1992. Empowering place: Multilocality and multivocality. *American Anthropologist* 94(3): 640–56.
Renaud, F., J. J. Bogardi, O. Dun, and K. Warner. 2007. Control, adapt or flee: How to face environmental migration? *InterSecTions No. 5*. United Nations University: Institute for Environment and Human Security.
Scott, J. 1998. *Seeing like a state: How certain schemes to improve the human condition have failed*. New Haven, CT: Yale University Press.
Scudder, T. 1981. What it means to be dammed: The anthropology of large-scale development projects in the tropics and subtropics. *Engineering & Science* 44(4): 9–15.
———. 2005. *The Future of large dams*. London: Earthscan.
———. In press. Resettlement theory and the Kariba case: An anthropology of resettlement. In *Development and dispossession: The anthropology of development-forced displacement and resettlement*, ed. A. Oliver-Smith. Santa Fe, NM: SAR Press.
Scudder, T., and E. Colson. 1982. From welfare to development: A conceptual framework for the analysis of dislocated people. In *Involuntary migration and resettlement*, eds. A. Hansen and A. Oliver-Smith, 267–87. Boulder, CO: Westview Press.
Shipton, P. 1990. African famines and food security: Anthropological perspectives. *Annual Review of Anthropology* 19: 353–94.
Torry, W. I. 1986. Economic development, drought and famines: Some limitations of dependency explanations. *Geojournal* 12(1): 5–18.
Turton, D. 2003. Refugees and 'other forced migrants': Towards a unitary study of forced migration. Paper presented at the Workshop on Settlement and Resettlement in Ethiopia, January 28–30. Addis Ababa.
———. 2006. Who is a forced migrant? In *Development-induced displacement: Problems, policies, and people*, ed. C. de Wet, 13–37. Oxford: Berghahn Books.
United Nations High Commission for Refugees (UNHCR). 2002. A critical time for the environment. *Refugees* 127: 2.
United Nations Millennium Project. 2005. *Environment and human well-being: A practical strategy*. London: Earthscan/James & James.
Wood, W. B. 2001. Ecomigration: Linkages between environmental change and migration. In *Global migrants, global refugees*, eds. A. R. Zolber and P. M. Benda, 42–61. New York: Berghahn.

PART 2
Anthropological Encounters

Chapter 5

Gone the Bull of Winter? Contemplating Climate Change's Cultural Implications in Northeastern Siberia, Russia[1]

Susan A. Crate

Introduction

The bull of winter is a legendary Sakha creature whose presence explains the turning from the frigid winter to the warming spring. The legend tells that the bull of winter, who keeps the cold in winter, loses his first horn at the end of January as the cold begins to let go to warmth, then his second horn melts off at the end of February and finally, by the end of March, he loses his head as spring is sure to have arrived. It seems that now with the warming, perhaps the bull of winter will no longer be. . . .

—*male Sakha elder, b. 1935*[2]

Sakha, Turkic-speaking native horse and cattle breeders of northeastern Siberia, Russia, personify winter in the form of the *Jyl Oghuha* (Bull of Winter), a white bull with blue spots, huge horns, and frosty breath. *Jyl Oghuha*'s legacy explains the extreme 100°C annual temperature range of Sakha's sub-Arctic habitat.[3] Accordingly, in early December the *Jyl Oghuha* arrives from the Arctic Ocean to hold temperatures at their coldest (–60° to –65°C; –76° to –85°F) for December and January. Although I had heard the story many times while working with Viliui Sakha[4] since 1991, in the summer of 2005 it had an unexpected ending. The realization that a cultural story, which for centuries had explained the annual temperature event of sub-Arctic winter, could perhaps become a story of how things *used to be*, alerted me to the cultural implications of global climate change. This elder's new way of recounting Sakha's age-old story of *Jyl Oghuha* was my "ethnographic moment"[5] to enter the field of climate change research. In this chapter I explore Viliui Sakha observations of global climate change, bring to light the cultural implications of global climate change, and highlight anthropology's privileged approaches to understanding different ways of knowing to move anthropologists from impartial observers into the realm of action-oriented researchers.

Sakha's Turkic ancestors migrated from Central Asia to southern Siberia around 900, and then northward to their present homeland beginning in the 1200s. They inhabit a sub-arctic region characterized by continuous permafrost with annual temperature fluctuations of 100° Celsius from –60°C (–76°F) in winter to +40°C (104°F) in summer. Viliui Sakha have adapted their southern agropastoralist subsistence to an extreme sub-Arctic environment and adapted to the throes of Russian colonization and Soviet and post-Soviet forces (Crate 2002, 2003a, 2006b). Today the majority of rural Viliui Sakha communities practice household-level food production via a system termed "cows and kin," focused on keeping cows and exchanging labor and products with kin (Crate 2003a, 2006b). They also rely heavily upon other subsistence production including gardens and greenhouses, forage (hunting, fishing, and gathering) and other domesticates including horses, pigs, and chickens. Theirs is a mixed cash economy, with most of their cash originating from state transfer payments in the form of state salaries, subsidies, and pensions.

Encountering Global Climate Change in Viliui Sakha Communities

In 2004,[6] 90 percent of the Viliui Sakha participants in my field research expressed their concerns about local climate change,[7] saying that they were seeing unprecedented change in their local areas and that they were concerned it threatened their subsistence (Crate 2006a).[8] In response to this result, in summer 2005 we worked with village youth, already engaged in our project's elder knowledge initiative (Crate 2006c), to interview thirty-three elders about their local observations of climate change. We asked a simple set of questions about what elders observed, how their lives were affected, what the causes were, and what they thought the future would bring. The elders impressed upon us that they possess ecological knowledge about how the climate was and how it has changed. In lieu of availability of comprehensive local climatic data,[9] village elders' knowledge is vital. Most elders offered testimony similar to this:

> The climate is definitely different from before. When I was little, the winters were very cold, minus 50–60 degrees. When we spit, it froze before it hit the ground and flying birds sometimes would freeze and die. The summer was a wonderful hot temperature and the hay you just cut would dry very quickly. In the last few years the climate has changed. We have rain, rain, rain all the time and winter comes late and so does spring. For people who live with a short summer when there needs to be the right weather to accomplish all for the winter and there is cool rainy times so that the hay does not dry and has to sit and sit and the quality is bad because of that. It is the right time for haying but the conditions are all wrong. (male Sakha elder, b. 1938)

What are the changes people are observing? Sakha elders reported that they cannot read the weather anymore: "From long ago we could read the weather and know what weather would come according to our "Sier-Tuom"

[Sakha sacred belief system]. But we can't do that anymore" (female Sakha elder, b. 1942). This is particularly urgent in the extreme environment of the Arctic where each day of summer is crucial to winter survival. Elders also commented that the timing of the seasons had changed, further jeopardizing winter survival. For the last decade, spring and fall have come several weeks late.

Elders also said that the climate had softened, referring again and again to *Jyl Oghuha*: "Winters have warmed and summers are not so warm. All is softer. The north is especially warming. It will be cold in winter and suddenly get warm in winter. It was never like before. Strong cold held for months. We have the legend about the bull of winter losing its horns" (male Sakha elder, b. 1925). Additionally, two qualities of the climate, both critical to survival in the north, are reported to be different: a tendency toward long periods of calms and a relative lack of humidity. The summer heat is no longer dry, but laden with humidity that stifles in high temperatures: "Before it got very hot also, like it does now, but there was air—now it gets hot and you can't breathe [humidity]." Both the lack of calms and the humidity make the Viliui Sakha's environment that much more challenging to negotiate. Although these barriers are still surmountable, elders report that family members spend more time in the seasonal tasks, most notably haying and winter activities such as hunting and wood hauling, due to the increased challenge that these climate changes pose.

Several elders commented on the loss of familiar species and the arrival of new species from the south, including a variety of insects that prey on many of the garden and forage plants that Sakha depend on. They talked about other changes in their local environment, including increased rain during the haying season, too much winter snow, increased occurrences of thunder, and a change in the quality of sunlight. Many also correlated these changes with poorer health and more diseases among their people.

We next asked elders how climate change was affecting people's daily lives. First and foremost, they talked about the effects on harvesting forage for their animals:

> It ruins the hay harvesting when it rains for two months solid. There is no winter forage for our cows and horses. Even if you plan to work every day at the hay, the weather keeps you from it. Every day it is raining. The land is going under water and the hay lands are smaller and smaller and if you keep a lot of animals, it is very hard. The hay itself has less nutrition and then when it is cut and lays and gets wet and dries many times, it also loses its nutritious quality. (male Sakha elder, b. 1932)

They talked about the negative impact on their ability to raise enough food to see them through the long winter: "So much water is bad for the garden. Potatoes rot in the ground and there are many new insects. Gardens are very late. The water and cold mean we plant potatoes a month late and some not until July" (female Sakha elder, b. 1930).

Next elders talked about how difficult it has become for their horses, which spend all winter outside and dig through the snow to find fodder. In the last decade elders have witnessed increasing amounts of snow due to warmer winter temperatures[10] and an impervious ice layer beneath the snow from a freeze/thaw that occurs commonly in the fall with warming and prevents the horses from reaching fodder. They also expressed concern about hunting, a supplemental source of food for many contemporary households, especially in the post-Soviet context: "We hunters can't hunt. I go trapping in January when the snow is thinner. But as the snow is deeper I can't go and the deep snow is bad because dogs can't run and horses can't walk. In spring and fall hunters also can't hunt because there is so much mud and boggy land" (male Sakha elder, b. 1933). Not only are hay, hunting, and foraging areas diminished due to flooding, all land areas are threatened. In one of our four research villages, there is deep concern about how water is inundating the grazing and gardening areas in the village center, another source of sustenance in these communities: "All the water ruins the usable areas near our homes—it diminishes all our land—with all the water, no one has any land anymore."

Elders also mentioned that they noticed the land was sinking in places: "The flat fields are sinking in and we want to know why—perhaps the permafrost is melting?" The most graphic of these accounts of sinking land tell of how an island near the village of Kuukei is submerging: "We have an island on the lake but now it has fallen. I have been watching for the last ten years and I see this happening" (female Sakha elder, b. 1933). However important it is to understand whether the island is in fact sinking because of melting permafrost, and whether the melting is in fact due to climate change,[11] when I heard these testimonies I was more concerned and curious about how the perception of the land actually sinking is affecting how Viliui Sakha orient themselves to their environment. Their sense of place and their understanding of "homeland" are both tied directly to an ecosystem dependent on water in its solid state. Although feeling "at home" in such icy confines is foreign to most of us, it is the familiar and the understood territory of comfort for northern inhabitants (Nuttall 1992). This was clear when we asked, "Isn't it good that it is not so cold in winter and not so hot in summer?" In response, elders unanimously argued to the opposite:[12]

> It is not bad to have warm winters, being an old person, it is great! But as Sakha people, we need strong cold here. It is how our lives are organized and how the nature works here. The big cold is good. The diseases are gone. When it is warm it snows too much and it is not warm or cold. The winter warmth affects people's blood pressure. And the heat in the summer is different, humid and very hard for people to go. It is bad for the way of life here and for survival. The nature, people, animals, and plants here are supposed to have very cold winters and very hot dry summers. That is the best for all life here. (female Sakha elder, b. 1929)

Figure 5.1: Village inhabitants take their herds down to the river to drink in the depths of the Siberian winter.

When we asked elders how they thought these changes would affect the future, all felt that conditions would progressively get worse: "As it gets warmer and warmer, the permafrost will melt and our land will be a permanent swamp and we won't be able to do anything—no pastures, no hay fields, just the high areas will remain. If it continues, then the permafrost areas will stop being frozen and it will all melt" (male Sakha elder, b. 1936). Many also made the connection between warming and its effects on health: "The worst part is that diseases will multiply in the future if it continues to get warmer and warmer. People's lives will get shorter with all the disease and no one will be able to keep animals here anymore" (female Sakha elder, b. 1944).

Some elders made a link between the local effects of global climate change and the breakdown of their contemporary social fabric: "People's attitudes will get worse and worse and things will go crazy. People's character and the way they relate has changed and I think it is because of the climate change. The way people are so violent these days I think is connected to the change in air and climate" (female Sakha elder, b. 1930). Making such connections is not unfounded. Similar cases of contextualization, the ways in which people associate changes in the natural environment with changes in their social environment, can be found in different local settings in northern Russia (Karjalainen and Habeck 2004; Simpura and Eremitcheva 1997). There are also studies in the field of biometeorology that are making such correlations in other cultural contexts.[13]

We also solicited elders' perceptions of the causes of global climate change. Many cited the presence of the reservoir of the Viliui hydroelectric station—constructed in the 1960s to supply electricity mostly for the then nascent diamond mining industry (Crate 2003b). However, studies have shown that the presence of the reservoir only results in a microclimatic change that would not include the extent of changes observed by the elders. Most elders agreed that the climate is changing due to a host of other reasons:

> They go into the cosmos too much and are mixing up the sky. When I was young they didn't go into the cosmos and we knew the weather. It rained when it was supposed to. Now it is all mixed up. Maybe from the mining activity and the electricity makers, the hydro stations, it all affects. They say the Sea [hydro station reservoir] affects us, but I don't agree. The natural climate is all mixed up. (male Sakha elder, b. 1933)

Elders commented that climate change is due to both natural and human-induced causes. When they talk about the human causes, it is important to remember that in their lifetimes they have seen the introduction and the widespread use of technology. They were born and raised on remote homesteads without electricity and now live surrounded by most varieties of technology. It is an easy step to relate the changes in their physical environment with the entry and advancement of this technology. Explaining the changes as "caused by nature" also makes sense given that they live in a highly variable climate to begin with and also know there have been climatic changes in the past.

Natural causes elders talked about included changes from nature itself, the changing direction of the Earth and all planets, each with a magnetic pull that is affecting us, changing sky and clouds, and the melting of the ice on the Arctic Ocean, bringing lots of clouds and rain. Human-induced causes included the "breaking" of the atmosphere by rockets and bombs that go up into the sky, and by humans going into the cosmos too much and mixing up the sky; the changing of the atmosphere by something in the atmosphere that makes it all very warm, and by all the "technika" people are using that fouls the air; the holes in the ozone and the other wreckage done with all our technology; and too many atom bombs. Although at first consideration some of the contributing factors these elders mention seem irrelevant to Western scientific thought on the subject, their ideas are both relevant and culturally provocative: the former because many of their ideas are related to the anthropogenic drivers known to be partially causing global climate change, the latter because so much of their attention is focused on activities in the sky and outer space that have to do with Soviet technologies introduced in their lifetimes.

Some elders provided explanations that related to phenomena other than global climate change. One commented, "The elders said it was like this last century also and they say that every century the same conditions come around—one hundred years ago also the land was under water." Sakha also have a cultural understanding of there being dry and wet years:

> They said that we would be having dry years now, but it is the opposite. Very wet years have come, lots of rain. Not in the spring when we need it, but in the summer when it gets in the way. There are many times as much water as there should be in the wet years, and if it continues like this, we will all go under water. We had the wet years and so it should be dry by now. (male Sakha elder, b. 1932)

These are important historical events that need further investigation in order to tease out just how Sakha's ancestors adapted to and survived these cyclical changes prior to the Soviet period. Additionally, several elders explained that the waterlogged fields had more to do with Sakha's negligence to work the land as they did in centuries past: "Before—in the Soviet time and before that—since our ancestors first came to these parts, we would make the fields so they were free of water, but not now." However, understanding the inundation of fields by water in the context of other observed changes attributed to climate change refutes these explanations.

Many of the elders' testimonies reveal that they seek to understand local climate change not only based on their observations, but also by integrating knowledge from other sources. One source was the ancient Sakha proverb "Tiiiekhtere ool uieghe, khachchagha Buus baiaghal irieghe," meaning, "They will survive until the day when the Arctic Ocean melts." Several elders recollected this proverb when they heard of the 2005 summer catastrophic flooding that occurred of the Yana River in the north of the Sakha Republic. Three villages were so heavily flooded that the residents had to permanently relocate. Reporting of this incident substantiated it not as an isolated phenomenon but directly related to the "fact" that the Arctic Ocean is no longer freezing up completely in the winters, resulting in increased water regimes for the entire republic.

In the summer of 2005, I identified only two media sources addressing global climate change that reached the villages. One was the British Broadcasting Company (BBC)'s airing of *The Day After Tomorrow*, the 2004 action/adventure, science fiction/fantasy thriller, on midday local television several times that summer. It is likely that many of the elders' comments about the global implications of local climate change were based on images and sound bites from this film.

The second media source was an article in the republic-wide *Komsomolskaia Pravda v Yakutii* by a Dr. Trofim Maksimov, a biologist and climate scientist in the capital city, Yakutsk (Ivanova 2005).[14] His extensive research in the Sakha Republic shows that average temperatures have risen by 2–3.5°C in the last one hundred years and that average winter temperatures for the same time period are 10°C warmer. This correlates directly with the elders' observations. His findings also document the movement of floral species

Figure 5.2: Cow-keeping households need to harvest two tons of hay per cow (approximately the size of the pictured stack, once it is finished) to fodder them through the nine-month winter.

northward and more temperate species coming into the republic from the south. Again, elders have made similar observations. Despite Maksimov's outspokenness and comprehensive information, only a handful of Viliui inhabitants subscribe to this newspaper or have received either his message or other outside information about the extent and causes of global climate change.

The elders' testimonies reveal no debate about *whether* climate change is occurring. Like most indigenous cultures practicing subsistence, they are, by default, ethnoclimatologists. With a continuous stream of experiential data, they know things are changing. Working with these communities to facilitate adaptive responses to these physical changes is critical. Anthropologists have a unique role, as interpreters of culture, to understand and act in response to global climate change's cultural effects.

The Cultural Implications of Global Climate Change and Indigenous Peoples

Transformations of both symbolic cultures[15] and subsistence cultures, such as the changes described here, reframe the implications of unprecedented global climate change. Global climate change, in causes, effects, and amelioration, is intimately and ultimately about culture: Global climate change is caused by the multiple drivers of our global consumer culture, transforms

Figure 5.3: Hay fields like this one are increasingly inundated with water due to the local effects of climate change.

symbolic and subsistence cultures, and will only be forestalled via a cultural transformation from degenerative to regenerative consumer behavior. Accordingly, anthropologists are strategically well suited to interpret, facilitate, translate, communicate, advocate, and act both in the field and at home, taking action and responding to the causes of change and communities facing and adapting to change.[16] As the cases presented in this volume show, climate change is forcing not just community adaptation and resilience, but also relocation of human, animal, and plant populations. Lost with those relocations are the intimate human-environment relationships that not only ground and substantiate indigenous worldviews but also work to maintain and steward local land-scapes. In some cases, moves also result in the loss of mythological symbols, meteorological orientation, and even the very totem and mainstay plants and animals that ground a culture.

We need not be over confident in our research partners' capacity to adapt. Although it seems completely plausible that highly adaptive cultures will find ways to feed themselves even if their main animals and plants cannot survive the projected climactic shifts, as anthropologists we need to grapple with the cultural implications of the loss of animals and plants that are central to daily subsistence practices, cycles of annual events, and sacred cosmologies (Crate 2008). The cultural implications could be analogous to the disorientation, alienation, and loss of meaning in life that happens when any people are removed from their environment of origin, like Native Americans moved onto reservations (Castile and Bee 1992; Prucha 1985; White 1983). The only difference is that while in some cases communities themselves will move, in other cases it is the environment that is moving.[17] As the earth literally changes beneath their feet, it is vital to understand the cognitive reverberations within and cultural implications for a people's sense of homeland and place.

If we agree, as Keith Basso convincingly argues, that human existence is irrevocably situated in time and space, that social life is everywhere accomplished through an exchange of symbolic forms, and that wisdom "sits in places" (1996, 53), then we need to grapple with the extent to which global climate change is and will increasingly transform these spaces, symbolic forms, and places. It follows that the result will be great loss of wisdom, of the physical make-ups of cosmologies and worldviews, and of the very human-environment interactions that are a culture's core (Netting 1968, 1993; Steward 1955).

Exploring Anthropological Research Approaches to Address Global Climate Change

Anthropologists can be most effective by using the tools of applied, advocacy-oriented, and public anthropology (Borofsky 2006; Chambers 1985; Gould and Kolb 1964; Kirsch 2002; Nagengast and Vélez-Ibáñez 2004; Rylko-Bauer et al. 2006). Advocacy is key not only in our collaborative relationship with communities but also in representing their best interests in policy and other advocacy contexts. In many parts of the world, indigenous peoples are actively advocating for themselves. However, there are places, such as northern Russia, where civil society and self-advocacy do not have the legacy that exists in Canada, Greenland, and Alaska, where indigenous groups are proactive on issues such as global climate change.[18] In such places anthropologists can work as communicators both to our indigenous research partners (what information they need about global climate change and in what proper form[s]) and by seeking out the local, regional, and national channels through which local voices can affect policy. Similarly, we can link our research partners with other communities who have gone through similar experiences (Cutter and Emrich 2006; Hoffman and Oliver-Smith 2002; Oliver-Smith 1996, 2005; Thomalla et al. 2006).

Research on climate change, the bulk of which to date is in the Arctic, does not address global climate change's cultural implications. Observations and perceptions of local effects of climate change, such as those of the Viliui Sakha presented here, reveal a need to develop research projects focusing on the cognitive/perceptual orientations of communities. Our research agendas must first investigate how our research partners perceive change (Crate 2008), and then use their understandings to encourage positive change. Now that many of our research partners are actively listening to their elders, the time is ripe for those elders' messages to inform the world and for anthropologists to take to heart and fully fathom the cultural implications and our innate responsibilities to act on all of our behalf. In the end we discover that each culture has its own *Jyl Oghuha* that is not only central to how that culture orients their daily/seasonal activities, worldview, and cultural identity, but is also part of the amalgamation of ethnodiversity that, like biodiversity, is intrinsic to the robust health and continued human, plant, and animal habitation of the planet.

Notes

1. I would like to acknowledge the people of the Viliui regions of western Sakha, Russia, with whom my ongoing research is possible. For constructive comments I thank Mark Nuttall. The research documented in this article was supported by the National Science Foundation under Grant No. 0532993. Any opinions, findings, conclusions, and recommendations expressed in this material are those of the author and do not necessarily reflect the views of the National Science Foundation. This chapter is largely based on parts of an article for *Current Anthropology*. The material was originally published in *Current Anthropology*, vol. 49, issue 5, page #. To these ends, I also gratefully acknowledge Wenner-Gren.
2. All quotes are anonymous, except for birth year and gender.
3. There are several portrayals of the Bull of Winter in classic Sakha ethnographic texts (Ergis 1974: 123–24; Kulakovskii: 1979: 45–46; Seroshevski 1993:26; Sivtsev 1996: 131).
4. Refers to Sakha inhabiting the Viliui River watershed of western Sakha Republic.
5. A term I use based on Stuart Kirsch's use (Kirsch 2002:175) and for which he gives credit to Marilyn Strathern (1999).
6. In context of my 2003–2006 NSF project entitled "Investigating the Economic and Environmental Resilience of Viliui Sakha Villages: Building Capacity, Assessing Sustainability, Gaining Knowledge," engaging local Viliui Sakha communities in defining sustainability and identifying barriers preventing them from realizing those definitions.
7. We administered surveys to a stratified sample of 30% (Elgeeii: n=63, Kutana: n=24) of all households surveyed by Crate in 1999–2000 (Elgeeii: n=210, Kutana: n=79). The survey instrument was developed based upon both the communities' definitions of sustainability generated during the first field season of the project and standardized questions used in the Survey of Living Conditions in the Arctic project (http://www.arcticlivingconditions.org/).
8. This was a collaborative project involving myself, one research assistant from the US, a research assistant in each of the four villages, and the direct involvement of the communities themselves. Hence, my use of the pronoun "we."
9. There are regional stations that provide data on a republic-wide level. However, these data are not translated into public information specific to the villages where these elders live.
10. Typically it snows in these areas from mid-September to mid-November and then again from mid-February to mid-March. In the deep winter it is too cold to snow. In the last decade or so, as winter temperatures are milder, it tends to snow for longer periods in both the fall and spring and the cold period of no snow is increasingly briefer.
11. Many of the pastures of the Viliui Sakha communities are located in thermokarst depressions known under the local name *alaas* (Crate 2006b: 9–11). *Alaas* are characterized by very specific processes of freezing and thawing, permafrost degradation

but also permafrost build-up. See Washburn (1979: 274) for an illustration of *alaas*' development cycle.

12. Granted, shorter winters may actually be beneficial for cattle and horse breeding. Horses and cattle will spend less time in the stables and barns (and more time on the pastures) if the annual average temperature increases. However, more precipitation (snow) and a higher frequency of freezing/thawing events will have an adverse effect.
13. For a broader context for the influences of climate on psychological factors, see *International Journal of Biometeorology* and http://biometeorology.org/.
14. The *Suntaar Sonunnaar*, the regional paper that most inhabitants subscribe to if they get any paper, was lacking in information on climate change from 2003–2005.
15. In this article I use the term *culture* to refer to both the series of prescribed human activities and the prescribed symbols that give those activities significance; both the specific way a given people classify, codify, and communicate experience symbolically and the way that people live in accordance to beliefs, language, and history. Culture includes technology, art, science, and moral and ethical systems. All humans possess culture and the world is made up of a diversity of cultures. Accordingly, I use the term in both its singular and plural forms.
16. Although it is beyond the principle focus of this article to discuss the multiplicity of causes for and effects of the transformation of culture resulting from unprecedented global climate change, I do want to mention these larger implications as I see them. The causes and effects of global climate change are about people and power, ethics and morals, environmental costs and justice, and cultural and spiritual survival. Scholars are beginning to address the equity and justice implications of climate change. See, for example Thomas and Twyman (2005). On a temporal scale, the effects of global climate change are the indirect costs of imperialism and colonization—the "nonpoint" fall-out for peoples who have been largely ignored. These are the same peoples whose territories that have long been dumping grounds for uranium, industrial societies' trash heaps, and transboundary pollutants. This is environmental colonialism at its fullest development—its ultimate scale—with far-reaching social and cultural implications. Global climate change is the result of global processes that were neither caused by, nor can they be mitigated by, the majority of climate-sensitive world regions now experiencing the most unprecedented change Thus indigenous peoples find themselves at the mercy of and adapting to changes far beyond their control.
17. I take poetic license here by saying that "the environment moves." It works well within the analogy. I fully acknowledge that the environment cannot move but that it changes.
18. I am not implying that it is necessary to install "civil society" in Viliui Sakha communities from scratch. I am emphasizing here that Inuit and other northern communities are far more successful when it comes to expressing their concerns and interests in the wider (global) public. Since the fall of the Soviet Union there has been a gradual increase in existing political institutions, NGOs, and researchers-cum-advocates in the Russian North. For example, in the case of the Eveny, local elites can articulate their concerns—at least to some extent—via RAIPON (the Russian Association of Indigenous Peoples of the North). Vasilii Robbek and his team of researchers in Yakutsk have been trying to address several politically relevant issues in their research, and at least with some success (see http://www.sitc.ru/ync/narod1.htm). Places and spaces for self-determination in Sakha and the Russian North in general are very different from those in Alaskan or Canadian Northern communities. Local educational institutions, such as schools, libraries, houses of culture, etc., do play a significant role in ecological/ environmental education and campaigning, and these institutions should be considered and included in the process of local "capacity building."

References

Basso, K. 1996. Wisdom sits in places: Notes on a western Apache landscape. In *Senses of place*, eds. K. Basso and S. Feld. Santa Fe, NM: School of American Research Press.

Borofsky, R. 2006. Conceptualizing public anthropology. http://www.publicanthropology.org/Defining/definingpa.htm (accessed May 30, 2007).

Castile, G. P and R. L. Bee, eds. 1992. *State and reservation: New perspectives on federal Indian policy*. Tucson: University of Arizona Press.

Chambers, E. 1985. *Applied anthropology: A practical guide*. Englewood Cliffs, NJ: Prentice Hall.

Crate, S. 2002. Viliui Sakha oral history: The key to contemporary household survival. *Arctic Anthropology* 39(1): 134–54.

———. 2003a. Viliui Sakha post-Soviet adaptation: A sub-Arctic test of Netting's smallholder theory. *Human Ecology* 31(4): 499–528.

———. 2003b. Co-option in Siberia: The case of diamonds and the Vilyuy Sakha. *Polar Geography* 26(4) (2002): 289–307.

———. 2006a. Investigating local definitions of sustainability in the Arctic: Insights from post-Soviet Sakha villages. *Arctic* 59(3): 115–31.

———. 2006b. *Cows, kin and globalization: An ethnography of sustainability.* Walnut Creek, CA: AltaMira Press.

———. 2006c Elder knowledge and sustainable livelihoods in post-Soviet Russia: Finding dialogue across the generations. *Arctic Anthropology* 43(1): 40–51.

———. 2008. Gone the bull of winter? Grappling with the cultural implications of and anthropology's role(s) in global climate change. *Current Anthropology* 49(4): 569–95.

Cutter, S. and C. Emrich. 2006. Moral hazard, social catastrophe: The changing face of vulnerability along the hurricane coasts. *The ANNALS of the American Academy of Political and Social Science* 604(1): 102–12. http://ann.sagepub.com/cgi/content/abstract/604/1/102 (accessed May 30, 2007).

Ergis, G. U. 1974. *Ocherki pa Yakutskomy folklory* [Essays on Yakut folklore]. Moscow: Science Publishers.

Gould, J. and W. L. Kolb, eds. 1964. A dictionary of the social sciences. New York: The Free Press of Glencoe.

Hoffman, S. M. and A. Oliver-Smith, eds. 2002. *Catastrophe and culture: The anthropology of disaster*. Santa Fe, NM: School of American Research.

IPCC. 2007. *Climate change 2007: Impacts, adaptation and vulnerability. Working Group II summary for policymakers*. Geneva: IPCC Secretariat. http://www.ipcc.ch/SPM13apr07.pdf (accessed May 30, 2007).

Ivanova, E. 2005. *Khvatit gasovat'* [Enough gassing]. *Komsolovkaia Pravda v Yakutii.* 28 July–4 August: 1.

Karjaleinen, T. P. and J. O. Habeck. 2004. "When 'the environment' comes to visit: local environmental knowledge in the far north of Russia. *Environmental Values* 13(2): 167–86.

Kirsch, S. 2002. Anthropology and advocacy: A case study of the campaign against the Ok Tedi mine. *Critical Anthropology* 22: 175–200.

Kulakovskii, A. E. 1979. *Nauchni trude* [Scientific works]. Yakutsk: Yakutskoe Knizhnoe Izdatel'stvo.

Nanengast, C. and C. G. Velez-Ibanez. 2004. *Human rights: The scholar as activist.* Oklahoma City, OK: Society for Applied Anthropology.

Netting, R. M. 1968. *Hill farmers of Nigeria: Cultural ecology of the Kofyar of the Jos Plateau*. Seattle: University of Washington Press.

———. 1993. *Smallholders, householders: Farm families and the ecology of intensive, sustainable agriculture*. Stanford, CA: Stanford University Press.

Nuttall, M. 1992. *Arctic homeland: Kinship, community and development in Northwest Greenland*. Toronto: University of Toronto Press.

Oliver-Smith, A. 1996. Anthropological research on hazards and disasters *Annual Review of Anthropology* 25: 303–28.
———. 2005. Communities after catastrophe: Reconstructing the material, reconstituting the social. in *Community building in the 21st century*, ed. S. Hyland, 45–70. Santa Fe, NM: School of American Research Press.
Prucha, F. P. 1985. *The Indians in American society: From the Revolutionary War to the present.* Berkeley: University of California Press.
Rylko-Bauer, B., M. Singer, and J. van Willigen. 2006. Reclaiming applied anthropology: Its past, present, and future. *American Anthropologist* 108(1): 178–90.
Seroshevskii, V. L. 1993 [1896]. *Yakuti* [The Yakut]. Moscow: ROSSPEN.
Siegel, J. 2006. Arctic National Wildlife Refuge divides Americans. *Anthropology News* 47(3): 26–27.
Simpura, J. and G. Eremitcheva. 1997. Dirt: Symbolic and practical dimensions of social problems in St. Petersburg. *International Journal of Urban and Regional Research* 21 (3): 476–80.
Sivtsev, D. K. 1996 [1947]. *Sakha fol'klora: Khomyyrynn'yk* [Sakha folklore Collection] Novosibirsk: Nauka.
Steward, J. H. 1955. *Theory of culture change.* Urbana: University of Illinois Press.
Strathern, M. 1999. The ethnographic effect II. In *Property, substance, and effect: Anthropological essays on persons and things*, 229–61. London: Athlone.
Thomalla, F., T. Downing, E. Spanger-Siegfried, G. Hand, and J. Rockstrom. 2006. Reducing hazard vulnerability: Towards a common approach between disaster risk reduction and climate adaptation. *Disasters* 30(1): 39–48. http://www.blackwell-synergy.com/doi/abs/10.1111/j.1467–9523.2006.00305.
Thomas, D. and C. Twyman. 2005. Equity and justice in climate change adaptation amongst natural-resource-dependent societies. *Global Environmental Change* 15: 115–24.
Washburn, A. L. 1979. Geocryology: A survey of periglacial processes and environments. London: Edward Arnold.
Watt-Cloutier, S. 2004. Climate change and human rights. In *Human Rights Dialogue*, special issue on "Environmental rights," Carnegie Institute. http://www.cceia.org/viewMedia.php/prmTemplateID/8/prmID/4445.
White, R. 1983. *The Roots of dependency: Subsistence, environment, and social change among the Choctaws, Pawnees, and Navajos.* Lincoln: University of Nebraska Press.

Chapter 6

Sea Ice: The Sociocultural Dimensions of a Melting Environment in the Arctic

Anne Henshaw

Introduction

In the mainstream media, images of polar bears perched on pans of melting ice have become emblematic of climate change and its immediate threats. While visually effective, do polar bears serve as an appropriate symbol or proxy for highlighting human vulnerabilities and adaptive capacity to climate change, especially for the 400,000 indigenous people who call the Arctic their home? In part, the answer to this question depends on who you ask and the scale of your analysis.

For Inuit of the eastern Canadian Arctic, sea ice, in the context of warming, becomes a window into the social, economic, and political forces that define how climate change is experienced, documented, perceived, and portrayed on multiple spatial scales. At the local level, sea ice greatly influences travel and mobility. Its concentration on a seasonal basis has direct links with the distribution of marine mammals and sea birds on which Inuit and many Arctic groups depend for their cultural livelihoods and to meet their socioeconomic needs. The impacts of shorter sea ice seasons and unpredictable conditions vary and individuals and communities are coping with these changes using a variety of strategies. At a regional and global level, a melting environment underscores the political dimensions that climate change holds for Inuit in an age of empowerment. On the one hand, Inuit are working at the national and international level to highlight the threats that climate change holds for them. On the other, Inuit are positioning themselves to take advantage of the economic development opportunities that a reduced sea ice environment will offer.

In this chapter, I explore the human dimensions of climate change relative to sea ice, a key indicator of warming that represents critical habitat not just for polar bears but for people. Using the results from the *Arctic Climate Impact Assessment* (ACIA 2005) and related studies in concert with my own climate change research among Inuit living on southwest Baffin Island, Nunavut, I explore how these issues manifest themselves in the context of social science research. With this background I discuss the important role that anthropology should continue to play in building community-based research models that foster collaboration between scientists and indigenous people, groups both concerned with the growing effects of climate change in the Arctic.

Background to the Study Area

Over the last ten years I have collaborated with the community of Kinngait (Cape Dorset) on a long-term environmental knowledge project. Historically, Inuit from Kinngait[1] inhabited a land use area encompassing the southwest corner of Baffin Island, including expansive coastline and islands that border the north side of Hudson Strait and eastern Hudson Bay. Today Kinngait is one of twenty-six communities that make up Nunavut, Canada's newest territory. The community is a medium-sized hamlet with approximately 1,200 residents who spend considerable time traveling along the well-established routes and areas from Foxe Peninsula and extending as far east as Kimmirut (Lake Harbour). The region has a rich archaeological record dating back at least four thousand years with Pre Dorset, Dorset, and Thule sites scattered across the landscape. The Hudson's Bay Company opened a post in the present-day community in 1913, followed by a Roman Catholic mission that operated from 1938–1960, and a Royal Canadian Mounted Police (RCMP) attachment that opened in 1965. Many Inuit were forced to permanently settle in the community in the 1960s with the advent of government housing, health, and educational programs and the West Baffin Cooperative, the main distributor of soapstone carvings, prints, drawings, and engravings (Kemp 1976, 125). Many of the eldest members of the com-munity interviewed as part of my research grew up in outpost camps located along the coastline[2] where they moved regularly across vast areas of the land and sea to hunt and trap. This chapter will draw on their experiences as well as the experiences of Inuit living in other similar communities in the eastern Arctic.

Understanding Human-Sea Ice Dynamics at the Local Level

Impacts

Across the Arctic, anthropologists and other social scientists have addressed the relationships between indigenous groups and sea ice in a number of studies (Aporta 2002; Boas 1964; Kassam 2000; Krupnik 1993; Nelson 1969; Okakok 1981). These relationships have been reconsidered and reinvestigated in recent years in the context of rapidly changing physical, social, and political environments (Ford et al. 2006a, 2006b; Gearheard et al. 2006; Henshaw 2003a; Krupnik and Jolly 2002; Laidler and Elee 2006; Tremblay et al. 2006). These studies highlight three overarching themes related to impacts, coping strategies, and vulnerabilities, which illustrate what Inuit currently experience in relation to shorter sea ice seasons.

Sea ice dominates the Inuit seascape for most of the calendar year and is highly sensitive to weather conditions as well as longer-term climate change. Landfast ice (i.e., ice that remains attached to land) forms first along the interior bays and inlets in the early winter (November/December) and usually reaches a maximum thickness of two meters (Jacobs, Barry, and Weaver 1975, 523). However, significant amounts of interannual and regional variability exist in the timing of freeze-up and ablation of landfast ice (Jacobs and Newell 1979; Jacobs et al. 1975). For Inuit, late-forming sea ice and early spring break-ups have far-reaching impacts as these phenomena not only affect the availability and distribution of marine mammals but also determine how and when the Inuit can travel to hunt and fish.

For Inuit, mobility is more than a transitional activity getting from one place to another: it forms a fundamental part of who they are and how they experience and learn about their environment (Aporta 2004; Henshaw 2006a). As the sea ice environment changes, the risks associated with travel increase as conditions become less reliable (Henshaw 2006b). As Qimeata Nungusuituq, an experienced hunter from Kinngait, recounts, changing wind and snow patterns have created rough conditions that make it more difficult to travel across the ice:

> Since I was a child, [ice] would freeze up in the fall when it was supposed to but nowadays ice conditions are not the same as [when I was growing up]. . . . [Normally] it would have been nice and flat [smooth] for a dog team or a snow mobile to be traveling but now it is all rough ice going out. . . . So much south wind coming in the last few years, instead of coming from inland, I think that is why the ice keeps breaking up because the wind keeps coming from the south. Before, the snow wouldn't fall when the ice was forming, but nowadays it snows and deteriorates freezing of the ice and makes it rough. (Nungusuituq 2000)

Additionally, as Kinngait elder Kananginak Putuguk[3] explains, the sea ice is not freezing as solid, making travel especially dangerous

> In springtime the ice is going out a lot sooner than it used to, but in fall time the last couple of years I've noticed even though it is very cold outside the ice is not forming as it should, even though it is not very windy and even though it is still very cold, everything else is freezing, the ice around here is not freezing like it should and these *paunai*, they are forming in more areas I think than they used to, there used to be not as many, today there is more. (Putuguk 2000)

The regular routes used by Inuit to traverse the ice are also sensitive to changes in the location of the floe edge and early spring break-ups. The floe edge is the place where the fast ice ends and the open water or moving ice begins. The floe edge is important because it is biologically rich area and serves as a hunting platform for Inuit to harvest a variety of marine mammals and birds. Although traveling along the floe edge is inherently dangerous, risk associated with such travel increases as thinning and unstable ice conditions, especially in spring, become more common (Ford et al. 2006a, 2006b; Laidler and Elee 2006; Tremblay et al. 2006).

While most of the recent research in the Canadian Arctic focuses on the negative impacts associated with climate change, there are some communities in other circumpolar areas where the impacts from a later freeze-up have both pluses and minuses. For example, in Kotzebue, Alaska, positive impacts include better whitefish, clamming, arctic fox, and spotted seal harvests, and better access to caribou and driftwood, while negative impacts involve poor access to Kotzebue for people living outside the community, rough ice conditions, more danger from thin ice, erosion, and flooding (Huntington and Fox 2005).

It is also important to note that sea ice is one of many changes in weather and climate noted by indigenous peoples in the Arctic. In interviews in 2000, experienced Inuit hunters explained to me, based on their environmental observations and related indicators, that the weather seemed to be "changing faster" and in unpredictable and sudden ways. As Kananginak Putuguk and Alika Parr describes:

> When I was growing up watching . . . they could tell what the weather would be like then. [Today] I can't really tell what it is going to be [like], what the clouds are telling me because the wind changes just like that. Clouds forming nowadays just come from nowhere and you don't know if it is going to be windy or not, or if it is going to be snowing. You can't tell what it is going to be like nowadays. (Putuguk 2000)

> Now, today, we get . . . blizzards all of the sudden . . . when it seems like it'll be a nice day. But we get a blizzard all of the sudden. . . . It wasn't like that before. (Parr 2000)

This increasing unpredictability has many implications, foremost being that the forecasting techniques Inuit have come to depend on are no longer reliable, adding more risk to land, sea, and ice travel. As Oshuituq Ipeelie describes, "Nowadays in the last two to three years, I try to forecast [using] whatever sign [clouds, sun, wind, moon] I used when I was growing up. I try to use [the] same things I used before today, but if I tried to use that, I am never really right anymore, I'm not really accurate anymore" (Ipeelie 2000).

Loss of their ability to predict using local indicators puts Inuit in a vulnerable position, but one not without options and ways to cope. For example, most Inuit living in this region today rely on radio weather broadcast from Environment Canada in combination with a specialized weather vocabulary and knowledge of local indicators passed down through generations of Inuit who learned how to travel safely on the land, sea, and ice. In this sense, Inuit draw from as many sources of knowledge as possible to make decisions that can often mean the difference between life and death. It is not unusual for Inuit to observe local indicators, listen to the weather report, and use the single sideband radio to talk to other families out traveling on the land before they decide where and when to travel.

Ways of Coping

Recent studies show that Inuit are responding to increased risk and unpredictability of changing sea ice conditions in a number of other ways as well. Responses vary by community and are dictated by particular histories, perceptions of change, and the viability of options available to groups (Ford and Smit 2004). In Sachs Harbor, Canada, responses include individual adjustments to the timing, location, and methods of harvesting animals and adjusting the overall mix of animals harvested to minimize risk (Berkes and Jolly 2001). In Kinngait, Laidler and Elee documented a number of ways Inuit use to cope with the unstable ice conditions, including the use of boats to travel over open water areas, avoiding travel near the floe edge where currents and winds are strong (especially after recent snowfalls), traveling closer to town, and using dog teams to detect (un)safe ice conditions (2006, 166). In the community of Arctic Bay, Ford and others (2006a, 142) report that, in addition to the use of boats and the kind of avoidance behavior noted by Laidler and Elee, Inuit (who can afford it[4]) are taking along new safety equipment like immersion suits, personal locator beacons, and satellite phones to minimize their exposure to risk while hunting at the floe edge during spring. I think this changing behavior, reflects the fact that Inuit are thinking about risk in a different way, not only because conditions are more unpredictable, but also because many individuals lack extensive experience of traveling on the sea ice.

Vulnerabilities

Despite these coping strategies, Inuit remain vulnerable to climate change on a number of fronts. Foremost, Inuit vulnerability is in large part due to the fact they live in sedentary communities with infrastructures that rely on fossil fuels for heat and transportation. Prior to the 1960s government-mandated settlement, Inuit and their ancestors were able to cope with a range of sea ice extremes by relying on opportunistic and mobile lifeways involving resource sharing, flexible group size and settlement patterns, and diversified subsistence strategies that took advantage of seasonally available resources (Henshaw 2003b). Although most Inuit communities still engage a flexible use of resources, a detailed understanding of the environment, and an active intercommunity sharing network, sedentism, a reduced spatial access to resources, and increased exposures to contaminants make Inuit especially vulnerable to rapid climate change (Ford et al. 2006a; Nuttall et al. 2005).

Additionally, most Arctic communities represent a mix of market and subsistence-based economies affected by changes in a globalized marketplace

(Wenzel 2000). In Arctic Bay, for example, Ford and community residents (Ford et al. 2006a, 141) report that floe edge hunting in spring is a relatively recent phenomenon resulting from government-mandated quotas and the commercial exploitation of narwhal for its ivory tusk: "Traditionally, hunters avoided this time, waiting for the narwhal to migrate closer to the community where they could be hunted more safely close to the shore, often taken by kayak on open water" (Wilkinson 1955, as quoted in Ford et al. 2006a, 143). The use of new technologies (GPS units, immersion suits, satellite phones, beacons) are helping hunters who can afford them to manage the dangers of sea ice travel while leaving others to assume more risk without adequate experiential knowledge (Aporta and Higgs 2005; George et al. 2004). Many elders voice their concern that Inuit youth today, educated more in village schools and less on the land and sea, are not gaining the experience they need to understand the inherent dangers of arctic travel. Programs like the Canadian Junior Rangers help in this regard by taking high school students out on excursions on the land, sea, and ice to learn basic Inuit survival skills. However, many youth who do not participate in such programs are unprepared for arctic travel and often have a false sense of security with the technology they bring along (Henshaw 2003b).

Climate Change at the Regional and Global Scale: Threat or Opportunity?

At the same time as communities and researchers are documenting the climate change impacts Inuit experience, the vulnerabilities they are exposed to, and the responses they make to such change, the overarching theme of melting sea ice has served as a catalyst for Inuit to work at the national and international level to bring attention to their plight and to highlight climate change as a threat to their cultural survival. At the national level, Nunavut[5] was the first jurisdiction to sign a climate change Memorandum of Understanding (MOU) with the Canadian federal government agreeing to "explore cooperation in areas such as climate change and education, energy efficiency and the development of renewable and alternative energy programs. In addition, the governments will explore cooperation to advance knowledge of climate change impacts and adaptation strategies in Nunavut through *Inuit Qaujimajatuqangit* (traditional knowledge) and research; and support pan Northern and multilateral initiatives related to emission reduction and adaptation."[6] This commitment has fostered an increase in climate change studies, including the ones discussed earlier. Exploration into alternative energy programs in Nunavut is still in its early stages, although increasing costs of diesel fuel will certainly provide incentive for more investment in this area.

At the same time, Inuit in Nunavut are actively working with national and international nongovernmental organizations (NGOs) such as the Inuit Circumpolar Council (ICC) and the national Inuit organization, Inuit Tapiriit Kanatami (ITK), to profile the tangible ways Inuit across the Arctic are being impacted by climate change. Sheila Watt-Cloutier, 2002–2006 chair of the ICC and native of Kuujuaq, Nunavik, was nominated for the 2007 Nobel Peace Prize for her work in bring climate change and global contaminants to the attention of the UN on behalf of Inuit people. The eventual award of the Nobel Peace Prize did recognize work done on global climate change, and was shared by former US vice-president Al Gore and the Intergovernmental Panel on Climate Change (IPCC). Even though Watt-Cloutier ended up not being awarded the prize, her work continues to be heralded by Inuit leaders and others. Acknowledging the importance of Watt-Cloutier's nomination, Mary Simon, president of the ITK reports:

> The Arctic is the barometer of global environmental health, and this nomination for Sheila Watt-Cloutier acknowledges that fact. Sheila Watt-Cloutier has worked tirelessly as an environmental activist on the international stage to bring the issue of persistent organic pollutants and global climate change to the attention of leaders and legislators at the United Nations and throughout the globe." (ICC press release)[7]

Currently, the Inter-American Commission on Human Rights will be holding a hearing to determine whether climate change is a human rights violation based on a petition filed by Sheila Watt-Cloutier and sixty-five other Inuit working in conjunction with the Centre for International Environmental Law and Earth Justice. The petition to the UN stated the US is in violation of Inuit human rights by refusing to sign any international treaties to cut its greenhouse gas emissions (CBC February 6, 2007)[8]. In addition, Watt-Cloutier is working through the United Nations Environmental Program's (UNEP) polar center (GRID-Arendal) in Norway called "Many Strong Voices." The purpose of the program is to foster connections between "vulnerable arctic communities and those of small island developing states whose future is in danger due to rising sea levels."[9] According to a recent article in Nuanvut's weekly newspaper, *Nunatsiaq News*, about the program, Watt-Cloutier states:

> Climate change in the Arctic is a human issue, a family issue, a community issue, and an issue of cultural survival. The joining of circumpolar peoples with Pacific Island and Caribbean States is surely part of the answer in addressing these issues. Many small voices can make a loud noise. As we melt, the small developing island states sink. (George 2005)

Clearly, Inuit, like Watt-Cloutier and her successor at the ICC Patricia Cochran, are advocating for their communities using international bodies in a strategic way to highlight the negative impacts climate change poses for Inuit and other indigenous people in the north.

At the same time that Inuit are voicing their concerns over the threats of climate change, they are actively pursuing economic development opportunities that can result from a warmer, more ice-free environment. As perennial summer sea ice concentration continues to shrink in the Arctic (ACIA 2005), circumpolar nations are actively asserting their territorial rights in the north to ensure their access to new shipping lanes and increased marine transport. Growing access to the Arctic Ocean will increase conflicts among competing users of arctic waterways (ACIA 2005). Governments, including Canada, are asserting their sovereignty in Arctic waters by increasing patrol vessels and reserve troops that will include a military training center to be located somewhere in Nunavut. As part of this increased presence, the federal government was working toward building a deep water joint military-civilian port, something the Nunavut Territorial government and local communities were actively encouraging and supporting. Such a port would not only offer a more cost-effective means to transport cargo to the Arctic but would also boost tourism from cruise ships, and enable fishing vessels to unload their catch at a port in Nunavut, instead of Newfoundland as they do now. For communities in Nunavut, the benefit from such a port would offer new investments for the future. In February 2007, the Canadian government backed off their promise of the seaport, to the dismay of Nunavut governmental officials (Thompson 2007). However, in October 2007, Canadian Prime Minister Stephen Harper asserted his commitment to the Arctic vis-à-vis the "Speech from the Throne," so the idea will likely gain momentum in the coming years. As he stated: "New opportunities are emerging across the Arctic, and new challenges from other shores. Our government will bring forward an integrated northern strategy focused on strengthening Canada's sovereignty, protecting our environmental heritage, promoting economic and social development, and improving and devolving governance, so that northerners have greater control over their destinies."[10] From the regional vantage point, a melting environment amplifies the multiple voices of climate change, reflecting both the threats and promises it holds for arctic societies trying to cope with an uncertain future.

Role of Anthropology

Anthropology in Canada has a varied track record in its relationship with indigenous peoples (Dyck and Waldram 1993). Of late it is increasingly characterized by research that fosters collaboration between scientists and communities. Currently, anthropologists in collaboration with other social scientists are calling for more research that engages indigenous partners and recognizes them through formal academic channels, through both granting agencies like the National Science Foundation and coauthorship of publications (Brook, M'Lot, and McLachlan 2006; Ford et al. 2006a, 2006b; Gearheard et al. 2006; Huntington and Fox 2005; Krupnik et al. 2005; Laidler and Elee 2006; MacDonald, Arragutainaq, and Novalinga 1997; Ooseva et al. 2004; Tremblay et al. 2006). Many of these projects, some of which focus on sea ice, emphasize the importance of training and collaboration. For example, in Nunavik, Tremblay et al. (2006) train Inuit to monitor ice conditions and conduct their own interviews with experienced Inuit hunters on how routes are affected by changing ice conditions, in an attempt to link traditional and scientific knowledge. Although collaborative-based research seeking to link different knowledge structures is not without significant challenges (Brook et al. 2006; Gilchrist, Mallory, and Merkel 2005; Nadasdy 1999), its most productive applications come when there are overlapping interests and time-tested trust among participants (Huntington 1998, 2000; Huntington et al. 2002).

In my own research, such overlapping interests revealed themselves in the context of an environmental knowledge project focused on the gathering of place names. For Kinngaitmiut, place names are part of a rich cultural heritage that defines their history and knowledge of the land, sea, and ice. Until recently Inuit place names did not appear on any published maps, yet they are an integral part of how Inuit navigate, monitor, and share stories about the world they inhabit. Today, these maps not only recognize claims Inuit have to their homeland but also serve an immediate need relative to climate change, as they are used by search and rescue teams who increasingly are called on to locate stranded travelers.

From my perspective as an environmental anthropologist, toponyms represent a form of indigenous knowledge that can be mapped in cartographic space and serve as a basis of comparison to other environmental phenomena sensitive to climate change, including changing sea ice. As Hunn (1996, 19) has pointed out, toponyms indicate *where things happen*, and by definition those happenings occur not once but over the course of seasons, years, or even generations. Even place names that may refer to an event take on new meanings as successive generations of Inuit traverse the land. When the nature of activities or meanings associated with certain named places change, toponyms provide an important benchmark for understanding and documenting changing people and landscapes over time. For example, the several Kinngaitmiut place names that referred to locations where Inuit gathered in winter or once built snow house villages become a proxy for the location of the floe edge during historic times. In this example, these places serve as historical markers of past activities (building snow houses) and environmental features (location of the floe edge) that are sensitive to changing sea ice conditions (Henshaw 2006a). Because the floe edge is particularly dynamic, its location is an important indicator of instability in the marine environment, especially during early spring break-ups.

Conclusion

In a time of political empowerment, Inuit themselves, not anthropologists, are their own best advocates when it comes to climate change. Whether they are highlighting the threats of climate change to their cultural livelihood or seeking economic development opportunities as waterways open up, Inuit are exercising their rights to self-determination. Inuit speak with not one voice, but many.

This is not to say anthropology and other social sciences do not have a role to play. The collaborative, community-based work that defines much of the research in the Arctic can serve as an important model for other regions where indigenous rights are not recognized to the same extent as they are in a liberal democracy like Canada. However, this kind of research is not without dispute. Here I return to the image of the polar bear perched on a pan of melting ice. As environmental nongovernmental organizations petition the US Fish and Wildlife Service to make the polar bear an endangered species under the US Endangered Species Act, Nunavut Environmental Minister Patterk Netser recently traveled to Washington, DC to challenge the petition, which he claims "shows a disturbing lack of respect for indigenous knowledge" (George 2007). In this context, the polar bear quickly becomes "an inconvenient truth" as it is in embroiled in the different perceptions of climate change and its immediate threats. While the causes of climate change are now well documented, the complex and rapidly changing social and political environments that define the issue are continuing to unfold. As anthropologists, we thrive in such complexity and are in a unique position to use our anthropological lens to carry out the kinds of in-depth and compassionate investigations an issue of this magnitude deserves.

NOTES

1. Historically, Inuit from Kinngait are known as the Sikusiilarmiut or "people of the place where there is open water in winter" (Boas 1964).
2. Despite government attempts to have Inuit settle permanently in the 1960s and 1970s, many continued to live in outpost camps during these decades in order to hunt, trap, and fish through the year.
3. Kanaginak Putuguk's surname is also spelled Pootoogook.
4. Many Inuit can not afford such safety devices due to the high cost of living in the North and the limited employment opportunities.
5. Nunavut Territory was created in 1999. It was created as part of the Nunavut Land Claims Agreement Act and also represents a form of public governance within Canada.
6. See the ITK website http://www.itk.ca/environment/climate-change-index.php and Government of Canada website http://www.climatechange.gc.ca/default.asp.
7. See ICC website for press release: http://inuitcircumpolar.com/index.php?auto_slide=&ID=397&Lang=En&Parent_ID=¤t_slide_num=.
8. See CBC website for full article: http://www.cbc.ca/canada/north/story/2007/02/06/climate-hearing.html.
9. See UNEP website for details on the program see URL: http://polar.grida.no/news.cfm?pressReleaseltemID=1002.
10. For the full speech given by Prime Minister Harper see URL: http://www.sft-ddt.gc.ca/eng/index.asp.

REFERENCES

ACIA. 2005. *Arctic Climate Impact Assessment: Scientific report*. Cambridge, UK: Cambridge University Press.

Aporta, C. 2002. Life on the ice: Understanding the codes of a changing environment. *Polar Record* 38(207): 341–54.

Aporta, C. 2003. New ways of mapping: Using GPS mapping software to plot place names and trails in Igloolik, Nunavut. *Arctic* 56(4): 321–27.

———. 2004. Routes, trails, and tracks: Trail breaking among Inuit of Igloolik. *Etudes/Inuit/Studies* 28(2): 9–38.

Aporta, C. and E. Higgs. 2005. Satellite culture: Global positioning systems, Inuit wayfinding and the need for a new account for technology. *Current Anthropology* 46(5): 729–53.

Berkes, F. and D. Jolly. 2001. Adapting to climate change: Social-ecological resilience in a Canadian Western Arctic community. *Conservation Ecology* 5(2): 18. http://www.consecol.org/vol15/iss2/art18.

Boas, F. 1964. *The Central Eskimo*. Lincoln: University of Nebraska Press.

Brook, R., M. M'Lot, and S. McLachlan, S. 2006. Pitfalls to avoid when linking traditional and scientific knowledge. In *Climate change: Linking traditional and scientific knowledge*, eds. R. Riewe and J. Oakes, 139–46. Winnipeg: University of Manitoba, Aboriginal Issues Press.

Dyck, N. and J. B. Waldram, eds. 1993. *Anthropology, public policy, and native peoples in Canada*. Montreal: McGill-Queen's University Press.

Ford, J. and the community of Arctic Bay. 2006a. Hunting on thin ice: Risks associated with the Arctic Bay narwhal hunt. In *Climate change: Linking traditional and scientific knowledge*, eds. R. Riewe and J. Oakes, 139–46. Winnipeg: University of Manitoba, Aboriginal Issues Press.

Ford, J. and the community of Igloolik. 2006b. Sensitivity of Iglulingmiut hunters to hazards associated with climate change. In *Climate change: Linking traditional and scientific knowledge*, eds. R. Riewe and J. Oakes, 139–46. Winnipeg: University of Manitoba, Aboriginal Issues Press.

Ford, J. D., and B. Smit. 2004. A framework for assessing the vulnerability of communities in the Canadian Arctic to risks associated with climate change. *Arctic* 57(4): 389–400.

Gearheard, S., W. Matumeak, I. Angutikjuaq, J. Maslanik, H. P. Huntington, J. Leavitt, D. Matumeak Kagak, G. Tigullaraq, and R. G. Barry. 2006. "It's not that simple": A comparison of sea ice environments, uses of sea ice, and vulnerability to change in Barrow, Alaska, USA, and Clyde River, Nunavut, Canada. *Ambio* 35(4): 203–11.

George, J. C., H. Huntington, K. Brewster, H. Eicken, D. Norton, and R. Glenn. 2004. Observations on shorefast ice dynamics in Arctic Alaska and the responses of the Inupiat hunting community. *Arctic* 57(4): 363–74.

George, J. 2005. Many small voices make loud noise in Montreal. *Nunatsiaq News*, December 16, 2005.

———. 2007. Polar bear threat "hysterical" Nester tells U.S. last week. *Nunatsiaq News*, March 16, 2005.

Gilchrist, G., M. Mallory and F. Merkel. 2005. Can local ecological knowledge contribute to wildlife management? Case studies of migratory birds. *Ecology and Society* 10(1): 20. http://www.ecologyandsociety.org/vol10/iss1/art20/(accessed April 4, 2008).

Henshaw, A. 2003a. Climate and culture in the north: The interface of archaeology, paleoenvironmental science, and oral history. In *Weather, climate and culture*, eds. S. Strauss and B. Orlove, 217–32. New York: Berg Publications.

———. 2003b. Polynyas and ice edge habitats in cultural context: Archaeological perspectives from southeast Baffin Island. *Arctic* 56(1): 1–13.

———. 2006a. Learning landscapes: Pausing along the journey: Learning landscapes, environmental change and place names amongst the Sikusilarmiut. 2006. *Arctic Anthropology* 43(1): 52–66.

———. 2006b. Winds of change: Weather knowledge amongst the Sikusilarmiut. In *Climate change: Linking traditional and scientific knowledge*, eds. R. Riewe and J. Oakes, 177–88. Winnipeg: University of Manitoba, Aboriginal Issues Press.

Hunn, E. 1996. Columbia Plateau Indian place names: What can they teach us? *Journal of Linguistic Anthropology* 6(1): 3–26.
Huntington, H. P. 1998. Observations on the utility of the semi-directive interview for documenting traditional ecological knowledge. *Arctic* 51(3): 237–42.
———. 2000. Using traditional ecological knowledge in science: Methods and applications. *Ecological Applications* 10: 1270–74.
Huntington, H. and S. Fox (Lead Authors). 2005. The changing arctic: Indigenous perspectives. In *Arctic climate impact assessment*, 62–98. Cambridge: Cambridge University Press.
Huntington, H. P., P. K. Brown-Schwalenberg, M. E. Fernandez-Gimenez, K. J. Frost, D. W. Norton, and D. H. Rosenberg. 2002. Observations on the workshop as a means of improving communication between holders of traditional and scientific knowledge. *Environmental Management* 30(6): 778–92.
Ipeelie, O. 2000. Interview transcript, July 2000. Interpreter Pootoogook Elee. Transcript on file with the author.
Jacobs, J. and J. P. Newell. 1979. Recent-year-to-year variations in seasonal temperatures and sea ice conditions in the eastern Canadian Arctic. *Arctic* 32(4): 345–54.
Jacobs, J., R. Barry, and R. Weaver. 1975. Fast ice characteristics, with special reference to the eastern Canadian Arctic. *Polar Record* 17(110): 521–36.
Kassam, K. A. 2000. Subsistence harvesting, sea ice, and climate change: A case study of Wainwright, Alaska. NSF project report, The Arctic Institute of North America and the University of Calgary (draft), 22 pp.
Kemp, W. 1976. Inuit land use in south and east Baffin Island. In *Inuit land use and occupancy project*, ed. M. Freeman. Ottawa: Department of Indian and Northern Affairs.
Krupnik, I. 1993. *Arctic adaptations: Native whalers and reindeer herders of northern Eurasia*. Hanover, NH: University Press of New England.
Krupnik, I. and D. Jolly, eds. 2002. *The earth is faster now: Indigenous observations of Arctic environmental change*. Fairbanks: Arctic Research Consortium of the United States.
Krupnik, I., M. Bravo, Y. Csonka, G. Hovelsrud-Broda, L. Müller-Wille, B. Poppel, P. Schweitzer, and S. Sörlin. 2005. Social sciences and humanities in the International Polar Year 2007–2008: An integrating mission. *Arctic* 58(1): 91–101.
Laidler, G. J. and P. Elee. 2006. Sea ice processes and change: Exposure and risk in Cape Dorset. In *Climate change: Linking traditional and scientific knowledge*, eds. R. Riewe and J. Oakes, 55–176. Winnipeg: University of Manitoba, Aboriginal Issues Press.
MacDonald, M., L. Arragutainaq, and Z. Novalinga. 1997. *Voices from the bay: Traditional ecological knowledge of Inuit and Cree*. Canadian Arctic Resource Committee and Environmental Committee of Municipality of Sanikiliuaq, Ottawa, Ontario and Sanikiluaq, N.W.T.
Nadasdy, P. 1999. The politics of TEK: Power and the "integration" of knowledge. *Arctic Anthropology* 36: 118.
Nelson, R. K. 1969. *Hunters of the northern ice*. Chicago: University of Chicago.
Nungusuituq, Q. 2000. Interview transcript, June, 2000. Interpreter Pootoogook Elee. Transcript on file with the author.
Nuttall, M., F. Berkes, B. Forbes, G. Kofinas, T. Vlassova and G. Wenzel 2005. Hunting, herding, fishing, and gathering: Indigenous peoples and renewable resource use in the Arctic. In *Arctic climate impact assessment*, 650–90. Cambridge: Cambridge University Press.
Okakok, L., ed. 1981. *Puiguitkaat, the 1978 elder's conference at Barrow, Alaska*. Barrow, AK: North Slope Borough, Inupiat History, Language and Culture Commission.
Ooseva, C. C. N. G. N., C. Alowa, and I. Krupnik, eds. 2004. *Watching ice and weather our way*. Washington, DC and Savooga, AK: Arctic Studies Center, Smithsonian Institution and Savoonga Whaling Captain Association, Washington.

Parr, A. 2000. Interview transcript, July 2000. Interpreter Pootoogook Elee. Transcript on file with the author.

Pitseolak, P. and D. Harley Eber. 1993. *People from our side: A life story with photographs and oral biography*. Montreal and Kingston: McGill-Queen's University Press.

Putuguk, K. 2000. Interview transcript, July 2000. Interpreter Pootoogook Elee. Transcript on file with the author.

Thompson, J. 2004. Mayor fears deepwater port flip flop. *Nunatsiaq News*, February 23, 2007.

Tremblay, M., C. Furgal, V. LaFortune, C. Larrivée, J.-P. Savard, M. Barrett, T. Annanack, N. Enish, P. Tookalook, and B. Etidloie. 2006. Communities and ice: Linking traditional and scientific knowledge. In *Climate change: Linking traditional and scientific knowledge*, eds. R. Riewe and J. Oakes, 123–38. Winnipeg: University of Manitoba, Aboriginal Issues Press.

Wilkinson, D. 1955. *Land of the long day*. Toronto: Clarke, Irwin & Company Limited.

Wenzel, G. W. 2000. Sharing, money, and modern Inuit subsistence: Obligation and reciprocity at Clyde River, Nunavut. In *The social economy of sharing: Resource allocation and modern hunter-gatherers*, eds. G. W. Wenzel, G. Hovelsrud-Broda, and N. Kishigami. *Senri Ethnological Studies* 53: 61–87.

Chapter 7

Global Models, Local Risks: Responding to Climate Change in the Swiss Alps

Sarah Strauss

In 1989, Renato Rosaldo wrote of "Grief and the Headhunter's Rage," describing the moment at which his own personal experiences of death, specifically, the loss of his wife Michelle, finally allowed him to understand what his Ilongot friends had been telling him for years regarding the intimate connection between headhunting, rage, and grief. The key to this understanding, argued Rosaldo, was the recognition that "the ethnographer, as a positioned subject, grasps certain human phenomena better than others. He or she occupies a position or structural location and observes with a particular angle of vision" (1989, 19). When a researcher has not had the kinds of life experiences that inform the explanations and behaviors of his or her research subjects, it is difficult to grasp the significance of what they are saying. Headhunting is arguably quite a specialized cultural practice, and one with which few anthropologists would have intimate knowledge. The experience of death, however, is universal, and it was this experience that allowed Rosaldo insight into this aspect of the Ilongot world.

In this volume, authors have been asked to describe their encounters with the local effects of climate change for the communities with whom they have been conducting research. While perhaps not invoking the emotional force that discussions of headhunting can incur, climate change is a politically charged issue that generates both moral outrage and cries for action. In particular, many self-designated environmentalists (including Nobel laureate Al Gore) have directed their energies toward morally charged mandates for action—Use Less Carbon!—which have sometimes resulted in an equal and opposite reaction by other communities or nations. In its varied and extensive manifestations, climate change is certainly a global phenomenon, but one that it is locally experienced by specific cultures and within circumscribed environments. The ability of people to adapt to the changes wrought by a variable climate is often constrained by external factors that complicate seemingly simple solutions.

As anthropologists, we seek to understand and translate, helping make the experiences of one place/time/people intelligible to those who inhabit different lifeworlds. In this chapter, I present the case of Leukerbad in

Switzerland, where I have been working to understand the "social lives" (cf. Appadurai 1986) of water and weather over the past decade. While my research was initially focused only on Leukerbad's water in its myriad forms, the impacts of climate change on those water resources soon forced me to broaden the scope of the project, examining climate variability and modeling in the wider context of the Swiss Alps.

Leukerbad is situated in the Dalatal, a side valley off of the Rhone River valley in the Swiss canton of Wallis (also known as "Valais" in French; it is a bilingual canton). Leukerbad has been known for its therapeutic thermal baths since well before the Romans settled in this area. For three hundred years after the founding of the Catholic parish in 1501, until avalanche control became more effective in the nineteenth century, avalanches and fires continually dismantled the ever-growing infrastructure of the village. For most of the eighteenth and nineteenth centuries, the population of Leukerbad was quite stable at about five hundred people. Since the late 1950s, there has been rapid growth in both health- and tourism-related industries, from medical clinics to ski areas, accompanied by a great deal of real estate development to accommodate the changing economic opportunities. The population in 2001 was around 1700, though it has since declined by about 200; these numbers reflect a distribution of about equal numbers of natives, "Uesserschwiiz" or other Swiss, and foreigners, mostly Portuguese service workers or other European health workers.

As with many other locations in the Alps, the retreat of glaciers following the 1850 highstand that marked the end of the Little Ice Age has been a dominant feature of the Leukerbad landscape. Like other Alpine communities, Leukerbad has few natural resources besides water, and its inhabitants made do with minimal-wage labor and small subsistence farms until the 1960s. While all of the villages in the region had historical books of local transactions, most have been destroyed by fire or avalanche; Leukerbad was very fortunate in that regard, and has an excellent historical record. Despite the shift from a subsistence-oriented economy, weather and climate have continued to be important concerns for the economic health of the valley. The short to midrange consequences of altered precipitation patterns for the ski area, as well as the long-term impact of global environmental change on the glaciers that permit a stable flow of water through the Dalatal, both depend on reliable precipitation.

Glacier Stories

Glaciers have historically been important to Alpine culture for many reasons: as reservoirs for drinking water and power generation, as raw materials for commerce (in the time before refrigeration), as tourist destinations, as visible markers of environmental change, and as repositories for lost souls. In many villages throughout Wallis/Valais, tales about glaciers abound. In Ausserberg, a community on the Sudhang (literally meaning "south-facing wall of the Valley") known for its dry climate, they tell this story about "The Artificial Glacier": The people of Ausserberg were unhappy because, unlike neighboring villages, they didn't have a glacier of their own. So they went up into the mountains, cut blocks of ice from a nearby glacier, and brought them back to Ausserberg. Once they had stacked up enough chunks of ice, the climate in their community changed and the grapes froze on the vines. The dismayed villagers hurriedly disassembled their newly minted glacier, saying, "Better wine than water!" (Jegerlehner 1989 [1906], my translation/paraphrase).

In the Loetschental, one valley to the east of Leukerbad, siren tales of entrapped souls who call out to the living are told, cautioning people to stay close to home. The dangerous nature of glaciers is clear in these tales, which warn simultaneously about the terrors of nature and the importance of following religious traditions to the fullest in order to avoid the glacial purgatory. From Loetschental, too, comes the story of the creation of the Langgletscher—somewhat similar to the Ausserberg story, but with a rather different ending: Seven hundred years ago, they say, the glaciers in the Loetschental were very far up on the walls of the valley and higher, by Concordiaplatz. The lower valley was filled with flowering meadows and grassy slopes, but received very little rain, and so the villagers often had to travel quite far to get water for themselves and their animals. One day, a stranger came through, and told them that if a pure young virgin girl went up to the high mountains surrounding the valley at dawn, pulled seven shards off of seven different glaciers, and returned to the valley to stack them up in the desired location, that a new glacier would grow. This task was accomplished, and from that day, the new glacier began to grow, eventually covering over the hut where the young girl had lived, and surging across the entire valley. The river that came down from the new glacier provided all of the water that the villages could use (Jegerlehner 1989 [1906], my translation/paraphrase).

In some versions of this story, the name of the stranger who gave the advice was Lonza, and the new river was named after him. The regional hydroelectric power company that serves both the Leukerbad and Loetschental valleys today is called the Lonza. In the century after this tale was recorded, the surging of the Langgletscher—300 meters in the period 1918–1924, 160 meters in 1980–1989 (Kanton Wallis 2000)—has shown that the glacier surge depicted in the folktale, though seemingly exaggerated, fits with local experience (cf. Cruikshank 2005).

Glaciers have been an integral feature of life in the upper Rhone region, yet their rapid demise in both length and overall mass suggests the need to consider an ice-free future. Glaciers in the Alps have lost at least 50 percent of their total mass since 1850, and are projected to lose between 50 to 90 percent of what remains, certainly by the end of the twenty-first century (Haeberli and Beniston 1998), and perhaps even by 2050, depending on the size of the glacier (Alcarno et al. 2007). The range is extreme, depending on the initial size, location, and orientation of the glacier in question. What will happen if these glaciers disappear? In Leukerbad, local people have varying opinions, from "Nothing at all," to "We will have to leave the valley where our families settled over five hundred years ago." Within a couple of generations (by 2050), this community will have to make difficult decisions about water resource distribution and energy supplies that may have implications extending well beyond the reaches of their narrow valley. Projected regional temperature increases averaging 4°C combined with significant shifts in timing, amount, and intensity of precipitation (Beniston 2006) will affect ski area operation, hydroelectric power generation, avalanche and flooding hazards, and glacier retreat itself.

Since 1998, I have spent one year of time conducting field research on water, weather, and climate in Leukerbad. Methods included household surveys, individual interviews, archival research, and participant-observation. In a 2001 survey of 95 residents (representing approximately 75 out of 400 households in the village), 93 percent asserted that the glaciers above Leukerbad were in retreat, and nearly 100 percent felt that global climate change was taking place.

Because Leukerbad has long been known as the home of many valuable varieties of water: thermal springs; freshwater springs; rivers, lakes, and glaciers; and more recently, a ski area, it is an excellent laboratory for examining the impact of glacial retreat. Leukerbad, and the other four communities that comprise the Dalatal, receive about 75 percent of their total electrical energy from a small hydroelectric plant powered by the river Dala, which itself is fed by the three small glaciers described previously. Since 1958, when the first large rehabilitation clinic was built, Leukerbad residents have increasingly depended on health and leisure tourism to support their local economy. Their previous self-sufficiency has been all but lost. Before 1960, most families had retained a subsistence garden and a few livestock to help them survive the year; now, fewer than ten families have any livestock at all, and most of the large gardens are gone. Until recently this community has had easy access to clean, fresh water. The impact of global environmental change and specifically of glacier retreat on this community will most likely affect that access.

But what does it mean to inhabitants of Leukerbad that their glaciers are disappearing? Discussions about perceived impacts of glacial retreat on the local community show that many people were well aware of the significance of the retreat. The comments of two men, one of whom grew up in the valley and another who moved there for professional reasons in his thirties, illustrate this well:

> OK, fine, I only know what others have said [about climate change and glacier retreat]. There will be more rock avalanches—the glaciers keep the region and

> the mountain together. The resources are already low, the water resources. If the glaciers aren't there we might not have any more water in July—it would certainly have a big influence on our life. (Int. 47)

> Yes, so what will happen then? That is in any case not rosy, if we know that life requires three elements: light, air, and water. Now they take one away. The hope is that the time might come again, one hopes that perhaps the time will come again that there is a cold period, and enormous snowfall. That's wishful thinking, but that won't come, it is always continuing to get warmer. (Int. 31)

Many individuals, particularly older residents with extensive family connections to the valley, were more phlegmatic about the effects of climate change, and less willing to be pessimistic about the future, as a prominent ninety-year-old man indicated:

> Yes, I would say so, people talk a lot about climate. Isn't it true that . . . the temperature of the earth will get two degrees warmer, I believe. Or it has already shifted two degrees since the turn of the century. . . . So there were more glaciers one thousand years ago or so, right? If it keeps going that way, yes, well, thank God we humans don't know everything. It could always change somewhat in a way that one just can't predict. (Int. 37)

Another well-known citizen in his eighties commented that "the scientists speak about 1.5 degrees or something, that is not noticeable. You can't say you noticed that, you never notice that" (Int. 46).

When I spoke with people, a more frequent response was to talk about the nature of glaciers as storage banks or reservoirs for water that is needed throughout the year. As the folktales I relate above make quite clear, in the past, villagers clearly recognized both the dangers as well as the advantages of having glaciers nearby:

> I would say so, right, that if the glaciers are natural reservoirs, then we would have to build more of these reservoirs and then we can again keep the water if it is coming down in the form of rain [instead of snow]. The reservoirs [dammed lakes that have already built] would no longer be for electricity, then, but for drinking water. . . . (Int. 31)

> Yes, it's already happening, and it makes one a bit afraid. Speaking of drinking water, we indeed obtain it more or less directly from the glaciers around here, and they are becoming ever smaller. If the warming that we are afraid of happens, I don't know . . . it can change again, it has always happened that way. First ice ages, and then earlier it was tropical here. . . . The wheel will always turn again, though perhaps we won't be around to experience it. . . . The glacier is a water bank, or reservoir, and if it goes away, that will certainly create a lack; we will have go to groundwater [for drinking water], or I don't know. (Int. 49)

About 75 percent of the survey respondents said that they felt that the local climate has become not only warmer, but also wetter than in earlier decades, and this perception is borne out by recent IPCC projections (Alcarno et al. 2007):

> Probably, if the glaciers disappear, I can imagine that a really big [climate] change must have taken place. . . . Precipitation has stayed pretty stable from previous times, but the more the glacier melts, the more precipitation [rain showers] there will have been. Probably then the seas will rise, how that will affect things, what kind of effects there will be, I just can't say, it's too hard to say. But the precipitation, yes, it will rain more and it will be more difficult for people to hold back the rainwater. The glaciers are a natural bank, and that will go away, if all that I have mentioned happens. (Int. 45)

On one hand, Leukerbadners live with the visible risks and benefits of glaciers on a daily basis. From the middle of the village on a good day, the evidence of the rapid retreat of their local glaciers can be observed and monitored. And, like many Swiss, they have a great concern for safeguarding the environment. Glaciers continue to fascinate the urban (now eco-) tourist and activist, as they have since the heyday of the nineteenth-century "grand tour." During the mid-nineteenth century, the peak period of glacial advance invited the construction of numerous hotels and chapels adjacent to the tongues of Alpine glaciers (e.g., Gletsch by the Rhonegletscher and Fluehkapelle above Leukerbad). But locals do not tend to see glaciers as quite so significant a marker. The same glacier may have both surged and retreated several times within a single human lifespan, and these changes, while often correlated with climatic processes, may also be linked to other geophysical forces. The attitudes of mountain people in Leukerbad, accustomed as they have been to scraping out a living from a difficult landscape, reflect a far greater concern with sudden and proximal threats like avalanches and floods than with the relatively gradual and somewhat remote risk of glacial retreat. But the reality is that all of these events may be causally related, as with the severe weather event in Leukerbad last summer that caused extensive flooding, and ultimately a water shortage. The preferred response is to shore up the defenses, and to work within the context of the challenges that appear, but not to attempt to predict, or avert, what the long-range future might bring.

Climate models and other kinds of expert knowledge are well publicized in Switzerland, and people are aware of the national and international scientific discourse on these environmental issues. Even so, there is still the sense that, like the souls trapped in the purgatory of the glacier, the water bank that comprises the glaciers is stored at the mercy of God, and released as is His will. Most of the Leukerbadners I have spoken with over the years—whether overtly religious or not—feel that they can deal with whatever situation comes to pass. They have had a hard life before, and may well again, and if they have to leave their homes, moving to the urban or lower areas to find work—well, none of these things are really in their control, so there is no point to worrying about them. No matter if they believe that climate change is happening, that it is anthropogenic in origin, or that it will drastically affect their own and their children's lifestyle—there is still the feeling that nothing preventative can be done. With such a fatalistic attitude, how can such a community be moved to action on behalf of reducing the impacts of climate change in their backyard? And, more importantly, are all communities in the Valais region of Switzerland similar to Leukerbad in their attitudes about climate change?

Lessons from Public Health: Models, Moral Certitude, and "Compliance"

Climate models do not have very high resolution; even the most detailed regional models only manage about a thirty-square-kilometer grid. The Dalatal, with Leukerbad as its main population center, and the Loetschental, with its string of small villages, are adjacent valleys; they are a scant ten kilometers apart as the crow flies. They share many geographical and geological features, and have similar weather and climate. In terms of the ability of regional climate models to forecast impacts, they are virtually identical. They are also similar in terms of sociocultural research questions: the main language is German, tourism is important to both, and subsistence farming has all but disappeared.

But the local histories of these valleys are quite divergent, in part because of the influence of significant natural features: Leukerbad is home to one of the largest thermal spring systems in Europe, which has influenced both settlement patterns and land use. It has also attracted the wealthiest and most notable citizens from several continents (including Mark Twain and Johann Goethe) to "take the waters"; accessibility to the Dalatal since the Neolithic was enhanced by the fact that the Gemmipass (connecting Leukerbad to the Berner Oberland region) was not fully glaciated in the last glacial period, so it was more easily traversed than most other north/south routes across the Alps (and offered the chance for a hot bath, surely news that travels fast in any age). In contrast, the Loetschental was, until the completion of the railway tunnel in 1913, quite isolated due to the narrow, closed off topography of the lower valley. This valley is embraced by one of the largest glacier systems in the Alps, in contrast to Leukerbad's minimal glaciation. Kippel, a main village of the Loetschental, has been the focus of extended anthropological study (Friedl 1974), as has Toerbel, a previously inaccessible village high on the slopes above the Rhone to the south (Netting 1981). From these brief details, it is clear that the histories of these adjacent valleys have produced very different communities, and although they share many physical and cultural features, there is every reason to expect the people of these valleys to react quite differently to natural hazards and risks in their local environment.

At this point, it may be useful to invoke Ulrich Beck's distinction between *risks* and *threats*: in earlier industrial and preindustrial societies, real environmental *threats* (e.g., specific extreme weather events) existed. But climate change is not a threat, it is a socially constructed *risk*, because it is "1) not limitable, either socially or temporally; 2) not accountable according to the prevailing rules of causality, guilt, and liability; and 3) neither compensable nor insurable" (Beck 1995, 2). To be concerned about climate change in a more than "tsk-tsk" sort of way is to be possessed of a particular kind of consciousness that depends on an assumption of its inherent risks. To do that, as Beck points out, we must buy into the narratives of technical expertise that currently describe this situation; we must be fully "scientized." "Perceiving an ecological crisis is only possible if, for instance, chemical formulas become 'forms of experience' in the Kantian sense, historical a prioris, which therefore determine everyday perception" (Beck 1995, 124). Taking a stance of cultural relativism, as is the habit of anthropologists, requires us to assume first that each culture develops its own logic, and follows a reasonable course of action in relation to that peculiar logic, which may or may not bear any resemblance to the expected rationalism of the Euro-American West.

If it is the case that two adjacent regions have developed sufficiently different local cultures, as can be argued for Leukerbad and the Loetschental, we might expect that their perceptions of risks related to climate change, and their subsequent responses, might also be quite different. While climate models do not distinguish difference at that local a level, and would offer the same degree of risk for each community—and thus might invoke a similar response from the policymaking community as to the most effective steps suggested for mitigating that risk—an anthropological analysis requires us to examine the evidence with a different eye, and can help us see why such broad-scale efforts may not succeed in "retraining" local communities to respond in a way that the scientific community finds appropriate.

Here, we might take a lesson from another major scientific discourse with the lay community—that of public health. Since the rise of international and community-based public health programs in the 1950s, anthropologists have often been called upon to solve problems relating to these public health efforts. The major lesson that has been learned from such encounters is that *knowledge does not equal behavior change*; people everywhere have contingencies in their lives that cannot be addressed by public health messages that use words of one syllable. Mark Nichter, a medical anthropologist who has worked extensively on community health issues across Asia, wrote an entire article entitled "Drink Boiled Water" (Nichter 1985) that is devoted to all of the reasons why people in a Sri Lankan village understood the message, yet did not boil their water; they did not "comply" with the simple directive. Environmental and climate scientists need to take note of this lesson, and not expect that rationales deriving from a scientific understanding of a problem and its most efficient solutions will be sufficient to change the behavior of people whose lives and histories are otherwise constrained. In our zeal to convince governments and lay publics alike of the rightness of our message, we must not fall into the trap of blaming the victims, assuming that noncompliance is a sign of either stupidity or malevolence. It may rather be simply an indication that life is complicated, for everyone. The problems faced by local communities in adapting to global environmental changes are not simple, and their solutions are almost guaranteed to be both diverse and complex. Anthropologists, living in our own communities and often having the privilege of traveling to our research locations, are in a unique position to understand Renato Rosaldo's insight regarding grief, rage, and headhunting as we can exchange translations of our own localized experiences of climate change with those of our consultants in the field: we are all feeling the effects, both long and short term, of a changing climate, but the solutions that will be applicable to this global problem cannot be cast from a single mold.

REFERENCES

Alcarno, J., et al. 2007. Europe. In *Climate change 2007: Impacts, adaptations, and vulnerability.* Contribution of WG II to the Fourth Assessment Report of the IPCC, eds. M. L. Parry, et al. Cambridge: Cambridge University Press, 541–80.

Appadurai, A. 1986. *The social life of things.* Cambridge: Cambridge University Press.

Beck, U. 1995. *Ecological enlightenment.* Trans. Mark Ritter. Atlantic Highlands, NJ: Humanities Press International.

Beniston, M. 2006. Mountain weather and climate: A general overview and a focus on climate change in the Alps. *Hydrobiologia* 562: 3–16.

Cruikshank, J. 2005. *Do glaciers listen?* Vancouver: University of British Columbia Press.

Friedl, J. 1974. *Kippel: A changing village in the Alps.* New York: Holt, Rinehart and Winston Publishers.

Haeberli, W. and M. Beniston. 1998. Climate change and its impacts on glaciers and permafrost in the Alps. *Ambio* 27(4): 258–65.

Jegerlehner, J. 1989 [1906]. *Walliser Sagen.* Zurich: Edition Olms AG.

Kanton Wallis. 2000. Brochure. Kandidat UNESCO Weltnaturerbe: Jungfrau/Aletsch/Bietschhorn region.

Netting, R. M. 1981. *Balancing on an Alp: Ecological change and continuity in a Swiss mountain community.* Cambridge: Cambridge University Press.

Nichter, M. 1985. Drink boiled water. *Social Science & Medicine* 21(6): 667–69.

Rosaldo, R. 1989. *Culture and truth.* Boston: Beacon Press.

Chapter 8

Storm Warnings: The Role of Anthropology in Adapting to Sea-Level Rise in Southwestern Bangladesh

Timothy Finan

Nature of the Problem

In all corners but the most intractable, the debate over climate change is over. The Fourth IPCC Assessment has established a consensus position for the science community complete with carefully and transparently calculated levels of confidence (IPCC 2007). Anthropogenic activity has led to an unprecedented warming of the planet, and the consequent impacts are projected to be disastrous for large segments of the world's population.[1] Moreover, the world's poorest are the most exposed and vulnerable, and they are poised to experience the direct and pernicious impacts of global climate change.

The public discourse of climate change has begun to permeate most levels of human activity and interest, and it is now part and parcel of the public domain. As often happens when ideas emanate from laboratories or oceanic buoys or academic walls, climate change has experienced its own transformation, passing through stages of rejection, skepticism, and contestation before emerging as an accepted reality and awakening a global awareness. This awareness has generated two general foci of attention: one on the human causes of global warming (our collective *carbon footprint*, to use a current trope) and the other on how best to navigate along the twisting trail of impacts.

This chapter focuses on the latter and asks what the role of anthropology is and can be in understanding the impacts of climate change. Natural scientists have adequate tools at their disposal to describe the anticipated physical impacts of warming with some precision, and have done so in the series of IPCC reports and elsewhere. These reports, however, do not address the impacts on *human* systems. They are restricted to so many hectares of agricultural land lost, temperature increases and desiccation in this or that region, disappearing water supplies, vegetation change, decreases in biodiversity, loss of fish stocks, and so on. This is climate change without a human face.

Anthropology, however, provides the theoretical concepts and the methodologies that can shift the focus to the dynamic interface of natural and human systems under change.

As climate anthropology creates space in the academic and practitioner landscape, it has engendered a set of theoretical approaches based on a core set of interrelated concepts—some drawn from a hallowed anthropological tradition, others adopted from the biological sciences. One such foundational cornerstone of this theory is *adaptation*, described as "the decision-making process and the set of actions undertaken to maintain the capacity to deal with future change or perturbations to a social-ecological system without undergoing significant changes in function, structural identity, or feedbacks of that system while maintaining the option to develop" (Nelson, Adger, and Brown 2007, 113). Adaptation has long been used in anthropology to describe successful or "functional" interactions of human cultures in localized environments as part of a long-term evolutionary process (e.g., Cohen 1974, 46). More current definitions, as exemplified above, focus on the response side to outside perturbations and tend to have shorter-term timeframes.

The concepts of *vulnerability* and *resilience* explain variations in the success of the adaptation process. As Füssel (2007) points out, vulnerability has meant different things to many thinkers—and has undergone a steady process of refinement in each, more complex application. Today there is consensus in the climate research community that vulnerability is primarily a social characteristic rather than a physical one (Eakin and Luers 2006), defined as "the present and future capability of a social group to withstand a socio-physical stress." Vulnerability is not about the outcomes of past responses but the potential of future ones, determined by a range of current factors such as asset entitlement and resources, environmental and institutional contexts, and the severity and duration of exposure to the stress or shock.

Resilience, borrowed from ecology, is considered the mirror image of vulnerability since it explains the capacity of a household or community to "absorb" stress without a change in structure and function. Adger (2000) distinguishes between ecological and social resilience and also highlights their synergy by suggesting that a resilient ecosystem may reinforce the resilience of the social system (and vice versa). In effect, the resilience concept becomes more attractive to climate change anthropologists due to its implicit recognition of human agency.

As I shall seek to exemplify below, anthropology provides an appropriate lens to assess the nature of adaptation, vulnerability, and resilience. Oliver-Smith (in Hoffman and Oliver-Smith 2002, 40) employed the term *human-environment mutuality* when writing on disasters to describe the interaction of a dynamic social system with a dynamic natural system. From an anthropological perspective, adaptation, vulnerability, and resilience are social phenomena that reflect an accumulation of localized decisions.

Human beings in institutional and environmental contexts make decisions regarding the mobilization and allocation of resources aimed at achieving welfare goals of one sort or another. These decisions are sometimes constrained by such factors as powerlessness, inequity, and suppression (similar to what Blainkie et al. call "root causes"), but they also demonstrate an ingenious ability to cope and survive. As a fundamental human process, then, adaptation—seen as the reduction of vulnerability and the enhancement of resilience—is a function not only of natural system adjustment but also of power, culture, race, class, gender, ethnicity, and the other building blocks of anthropological theory.

The current livelihoods framework so pervasive in development practice is an organizational device that formally incorporates natural system change (abrupt or cumulative) into a dynamic human system defined by its multiple asset packages (human, social, political, economic, and physical capitals), the sets of decisions that mobilize and allocate these resources, and the outcomes of these decisions (Scoones 1998). Germane to my argument here, the livelihoods approach is widely considered the accepted empirical methodology for assessing vulnerability and resilience in communities (e.g., Brocklesby and Hobley 2003; Little et al. 2001), and it is quintessentially holistic in facilitating the integration of culture, power, gender, class, etc. into an analysis of adaptation to climate change.

The anthropology mission I propose here is to establish the conceptual credentials and assemble the appropriate toolkit needed to understand how changes in the natural system will revise current terms of engagement at the level of communities and households. In some cases, global climate change will generate abrupt events, such as floods and storms; in other cases, the change will be experienced as trend lines, only suspected at first but then perceptible from afar. In either case, human populations in time and space will sense the pressure and will begin to respond either proactively or reactively. To understand this dynamic is the gauntlet at the feet of anthropology. Let me illustrate with an example.

Shrimp Aquaculture Livelihoods and Sea-level Rise in Coastal Bangladesh

The projected rise in sea level, particularly in "contained" water bodies such as the Bay of Bengal, is one aspect of global change that now appears inevitable. The most conservative projection reported in the IPCC Fourth Assessment (2007) is estimated at 40 cm by the end of the twenty-first century. More significantly, however, the rise in sea level in the northern Bay of Bengal combined with warmer sea surface temperatures will likely result in more severe storm and cyclone activity with accompanying high water extremes, surges, and salt intrusion. In Bangladesh, the magnitude of these impacts, especially for coastal populations, is beyond comprehension. Nearly 100 million people could be affected; 10 percent of the fertile agricultural land could be destroyed; farming and fishing livelihoods could be completely compromised; and the fragile Sundarbans mangrove forests, with their unique biodiversity, are in danger of disappearing altogether (Ahmed, Alam, and Rehman 1999; Ali 1999). On few stages will the drama of adaptation to climate change be played out with such high stakes as in this highly vulnerable region.

Bangladesh is one of the world's most densely populated and impoverished countries. More than half its estimated population of 149 million live on less than $1 a day, and its total annual per capita income (GDP/per capita) is less than $400. One of the world's mightiest riverine systems, the Ganges-Brahmaputra-Meghna (GBM), carves its way through Bangladesh, forming a large delta that covers around one-third of the country's area and is home to 70 percent of the population. Annually, strong southwestern monsoons move across the Bay of Bengal dumping an average of two meters of rainfall on most of the country. The GBM—relatively low in the dry winter months—fills with water accumulated in the vast Himalayan watershed and floods the deltaic lowlands as it moves to the sea. From a livelihood perspective, moderate annual flooding restores the fabled fertility of the country's paddy land and distributes and replenishes the freshwater fish stocks so critical to the food security of the population, especially the poor. However, excessive flooding leads to catastrophic levels of damage, including high levels of mortality and livelihood disruption. Thus, the natural and human systems have negotiated an uneasy balance, where livelihoods are dependent upon the annual renewal of the resource base, yet anomalies and extremes can have devastating consequences. The livelihood stress that is introduced to this deltaic system by global warming is manifest, at least initially, in more severe cyclonic events, storm surges, excessive flooding, and backwater effects (saltwater intrusion).

The inhabited coast of Bangladesh is 36,000 km^2 in area and densely populated due to its fertile agriculture and fishing. The southern coastal region, in its entirety from Khulna to Chittagong, is a highly complex and dynamic hydrological system driven by a delicate interplay of natural and human influences. Such factors as seasonal flooding, water salinity, and tidal movements affect the diversity and distribution of valuable species and the quality and quantity of agricultural land, which in turn determine the organization and outcomes of local livelihood systems. Over several decades, the national government invested in large dikes and polders intended to "control" the force of nature and to reclaim land for agricultural production. The primary staple crop, paddy rice, is grown in freshwater and the earthen dams were to keep out the saline water from the bay and rivers near the mouth of the delta. In the 1980s, however, the patterns of land use along the coast were altered. Bangladesh discovered the tremendous economic potential of the shrimp export markets to Europe, Japan, and the US, and well-heeled Bangladeshi capitalists (with political backing) acquired many of the coastal areas behind the polders for shrimp aquaculture, forcing out the farmers and replacing the paddy fields with the saline water enclosures.

As will be emphasized below, the vulnerability of coastal livelihoods is determined not only by the impacts of climate change, but also by the characteristics of the social, political, and economic systems distributed across the vulnerable landscape. The region is densely populated, a fact that by itself increases exposure to extreme climate events. The vulnerability of coastal populations increases due to the high level of resource concentration, pervasive social inequity, lack of political voice and representation, embedded corruption, high rates of illiteracy, alarmingly low levels of child nutrition, widespread exposure to arsenic toxicity in the drinking water, and the social exclusion of women from public life.[2] Half of all agricultural workers are landless, crushing debt is common, employment is seasonal and low paid, and the very poor are forced to reside in inadequate shelters located in areas prone to flooding and storms. Thus, the process of adaptation to climate change cannot be viewed as a mechanical adjustment to natural perturbation, but as a more profound sociocultural confrontation with root causes of vulnerability.

The Freshwater Beels and Shrimp Aquaculture

To illustrate the above points, I now turn to the local. Beels are small depressions located at various distances from a coastline which are connected to the complex hydrology of a region through canals and minor rivers. Throughout southwest Bangladesh there are hundreds of such beels, and they vary in surface area from several to several hundred acres and in quality from brackish and saline to freshwater. I focus here on freshwater beels. During the dry winter months (November to February) beel water levels recede into small isolated pools and drier areas are cultivated. During the monsoon season (roughly May to September), the beels expand into small lakes. Although beel land is privately owned and managed—mostly by the villagers who reside on the edges—the water in the beel is considered *khas* land, or public lands and open to all (Barkat, Zaman, and Raithan 2001).[3]

Historically the beels are used for *boro* rice cultivation and pasture for livestock during the dry months. During the monsoon months, when the beels fill up with rain and flood waters from nearby rivers and canals of the delta, they provide a public access fishing source for the surrounding communities, particularly the landless laborers. Thus, two distinct groups—with different livelihood characteristics and vulnerabilities—have coexisted and managed the beel resources. However, the emergence of the prawn export industry in the last two decades has transformed this resource management system.[4] The value of prawn exports in Bangladesh is now estimated at $350 million, behind only garments and remittances as the major sources of foreign exchange. In the case of the beels, the introduction of the freshwater *macrobrachium* species (*golda*) prawn has the local economy.

There are now over 100,000 golda producers, and many more households benefit from provisioning inputs to this system, particularly from the supply of post-larvae and labor.

Golda aquaculture was introduced by an enterprising individual who wondered if shrimp could be raised in freshwater as it was in the saline coastal ponds. The golda prawn spend their post-larvae stage in brackish water, then migrate up rivers to grow to adulthood in freshwater before returning to the sea. This innovative farmer obtained a sample of post-larvae and grew them out in a small pond near his home. Based on his success, he introduced a golda production technology in the beel areas. Farmers with land rights to the dry beel areas built rectangular enclosures with areas ranging from a fraction to several acres. They did so by raising earthen walls called *ghers,* trapezoidal in shape and about one meter high and a half meter across the top. These enclosures keep rainfall in and flood waters out. During the dry season, boro paddy land is cultivated as was always done; after the monsoon season starts, farmers put post-larvae in a small flooded trench along the inside of the gher. As the rains fill up the trench, the post-larvae spill out into the open field where they grow to adulthood. A monsoon rice crop (*aman*) is transplanted inside the gher, with most farmers also integrating fish aquaculture. Additionally, vegetables and other crops for sale and home consumption are grown along the dike walls.

From a livelihood perspective, the gher technology has diversified income, intensified the productivity of scarce resources (land), and tapped greater quantities of abundant labor resources. Unlike *bagda* shrimp production (in saline beels),[5] gher production is the economic mainstay of mainly small-scale farmers. The intensive management requirement of gher production precludes the construction of large ghers, precluding large capital investment. The "vertical" intensification and diversification of production and the integration of resource-scarce farmers into an international market have greatly increased income-earning opportunities for extremely poor farm households. This transformation has particularly benefited women, who now grow vegetables on dike walls and participate in the ancillary income-generating activities that provide the post-larvae fry and feed to the system (Finan and Biswas 2005). The introduction and widespread adoption of gher technology has reduced the vulnerability of many families in the coastal beel villages by increasing household income and assets, and it has also had a major regional impact. Ancillary support activities generate seasonal employment for an estimated 400,000 fry collectors and for a large number of feed suppliers and marketing agents.

Despite this seeming success story of sustainable and appropriate development, this livelihood system is extremely vulnerable to the impacts of global climate change and associated sea-level rise. The nature of this vulnerability lies in the physical configuration of the gher structures, the biology of prawn production, and in the vicissitudes of international markets.

For example, gher technology dramatically alters the landscape of coastal freshwater beels. Farmers living on the edge of beels have moved away from (single crop) rice and pasture production (and fishing) by creating irregular checkerboard patterns of ghers toward the center of the beel. In some cases, the embankments and enclosures actually cross natural (and public) water canals that flush the hydrologic system. The disruption of the natural hydrological cycle (resulting in waterlogging) and the interference with the movement of fish species have already been identified as both social and environmental problems.[6] The intensification in the use of beel systems has favored those with access to land (either through ownership or rental) and has hurt local fishing livelihoods. Whether the decline in fishing is compensated by increases in gher labor demand has not yet been determined.

Sea-level rise will likely result in a much larger volume of saline water moving into the canals that feed the beel hydrology, contaminating water resources and eroding gher embankments. Case in point: an extreme flood in 1998 destroyed thousands of ghers, resulting in an emergency response from the national government and international aid agencies. Another likely result of sea-level rise is saltwater intrusion through groundwater flows. Such consequences bring us to question what form successful adaptation can take. In lieu of abandoning their livelihood altogether, the challenge is to determine the ways in which the beel communities can enhance their resilience and adjust to these pressures. In what ways can anthropology meet this challenge?

The Role of Anthropology

In a given beel, there are multiple communities in the perimeter, carrying out a diversity of livelihood activities. The large landless population fishes and provides labor to the local agricultural economy; others offer services including rickshaw pulling, sewing, petty commerce, and the like. There is great variation in the beels in terms of land size and land tenure relations, but most gher farmers are land limited, capital scarce, and vulnerable to market forces. Considering the impacts of global climate change and associated sea-level rise, the options for adaptation are stark. Saltwater intrusion and the backwater effect may completely compromise the gher system by disrupting the delicate ecology and hydrology of the beels. Severe flooding due to increased storm and cyclone activity could destroy the physical infrastructure. Both could mean the abandonment of the golda livelihood. On the other hand, households and communities could potentially devise the necessary institutional arrangements and technology to lessen the impacts of these natural systems stressors. I examine here how anthropology and its unique toolkit can promote this second alternative of mitigation and adaptation.

A previous study of adaptation to climate extremes in southern Arizona (Finan et al. 2002), suggests that the long-term process of adjustment requires two major inputs—technology and social reorganization. In the middle of the nineteenth century, Mormon settlers used flood irrigation, requiring a community effort to build diversion dams across small rivers to channel water to the fields. As water became increasingly scarcer in the river system, a deep-well technology was introduced to lift water from the riverbed alluvium. The introduction of this technology, however, required the reorganization of rules and regulations regarding the distribution and use of alluvial water in order to assure that all farmers would be served.

I argue that the dynamic relationship between technology adoption and institutional reorganization similarly applies to the beel dwellers of coastal Bangladesh. An anthropological contribution to climate change in this region has three dimensions. The first is a *distributive* dimension. Anthropology is well equipped to describe the distribution of resource access and resource management strategies among the communities surrounding the beels. Livelihood analysis (e.g., Bebbington 1999; Scoones 1998) provides a framework for the organization of essential household and community data categories such as the environmental stressors; household assets (human capital, physical capital, social capital, economic capital, etc.); strategies of resource mobilization and allocation; and outcomes in terms of income, food security, health, education, and the like. A livelihood assessment—using households as analytical units—specifies the patterns of resource access and use across a social landscape made up of farmers, laborers, fisher people, and shopkeepers, each analyzed as a prototypical livelihood system. The output of the livelihood assessment then is a profile of the distribution of vulnerabilities across livelihood systems and, from a community perspective, how these livelihood systems are articulated, interdependent, and synergistic. In effect, the livelihood approach defines the "response space" (Thomas et al. 2005) of the beel communities—that is, their own capacity to adjust to changes in the natural system.

The second dimension is *institutional*. In anthropology, institutional analysis ties the local household with broader forces and influences such as markets, political systems, and government agencies. In part, the institutional analysis specifies power relationships, demonstrating forms of dependency, local and global market distortions, and the impacts of outside interventions both private and public. This dimension also focuses on the set of resources and interventions that are beyond the capacity of the community itself to mobilize, including infrastructure works, availability of and access to new technologies, sources of information, and networks. In the case of Bangladesh coastal regions, the institutional analysis identifies the government agencies responsible for water management (e.g., the Bangladesh Water Development Board—BWDB); local, national, and international NGOs with programs in the region; local universities and research centers; the national disaster response and management system; and other institutions that link coastal communities to broader networks. Whereas the distributive dimension enhances an understanding of the local capacity, the institutional dimension provides access to appropriate information, technology, and alternative models of organization.

The third dimension is the *empowering* of local management. Anthropologists have contributed significantly to models of community-driven development and common-property management strategies. This dimension has immediate relevance to the process of adaptation to global climate change in the beel regions, since it is increasingly clear that beel communities will need to reorganize local institutions (or devise new ones) to deal with increased environmental uncertainty. Currently beels are managed privately, and the ghers are constructed without any broader constraints to their location other than a recognized land right. The gher in effect reduces the amount of common property that is available to the community during the monsoon system when the beels become open access water bodies. Furthermore, beels are dynamic hydrological systems influenced by the larger riverine delta dynamics. Thus, decisions made by one gher owner might adversely impact the entire beel system and the other gher farmers. Beel dwellers and environmentalists already recognize and are concerned about waterlogging (due to blocking of canals) and its impact on health. To reduce the vulnerability associated with sea-level rise and to enhance community resilience, organizational adjustment will be necessary. The direction of reorganization will likely be toward a form of common property. This does not mean that the ghers themselves would become units of collective production, but that the beel, as a hydrological system, would be managed under collective agreement, similar to the management of lobster stocks by fishermen in the northeastern US. Such an institutional reorganization effectively places beel management under a framework of participatory community decision-making. This institutional reorganization better positions the beel communities to seek the technological solutions that will be required to mitigate the impacts of saltwater intrusion and sea surge.

Conclusion

The projections of sea-level rise impacts in the Bay of Bengal for the millions that inhabit the coastal areas of Bangladesh are dire. The government is contemplating large-scale infrastructural interventions to counteract a rising sea level. Such grand earthworks will not completely mitigate the litany of threats associated with global climate change. Community and locally specific adaptations are also necessary. An anthropological approach can localize the adaptation process and seek to understand how local communities can adjust to and reorganize for increased environmental uncertainty. Anthropology can provide the human face to climate change—to translate it into real-time households and communities managing their own resource systems. An anthropological toolkit is appropriate to this problem-solving challenge. In the case of the beel systems of coastal Bangladesh, anthropology can assess the distribution of local resource access, use and livelihood profiles, identify institutional relationships that link communities to broader systems (including access to new technology and knowledge), and facilitate the process of local resource management and problem solving. To achieve such urgent goals, it is critical that anthropology "gear up" for the imminent challenge that climate change will create as its impacts are manifest among local populations. It is well positioned and well equipped to do so.

NOTES

1. The ten recorded hottest years have occurred since the beginning of 1990s (Adger 1999; Adger et al. 2003). Recent ice core analysis shows that the earth has never undergone any comparable episodes of "natural" warming over the last 650,000 years.
2. In traditional rural society in Bangladesh, Muslim women are not permitted to participate in public affairs without the permission or presence of the husband or male relative. Even going to the market is frowned upon in conservative parts of the region.
3. The larger saline beels are more permanent and state-owned, but leased for use by private individuals.
4. There is a major livelihood difference between the capitalist production of "shrimp" in the large saline ponds and the production of freshwater "prawns" in the beels. This distinction is not, however, the focus of the present chapter.
5. *Bagda* refers to the peneid varieties of shrimp grown in large saline beels.
6. Other problems involve the by-catch loss in postlarvae collection and the diminishing numbers of the apple snail, whose flesh has been a major source of feed for the prawn (see Finan and Biswas 2005).

REFERENCES

Adger, W. N. 1999. Social vulnerability to climate change and extremes in coastal Vietnam. *World Development* 27: 249–69.

———. 2000. Social and ecological resilience: Are they related? *Progress in Human Geography* 24(3): 347–64.

Adger, W. N., S. Huq, K. Brown, D. Conway, and M. Hulme. 2003. Adaptation to climate change in the developing world. *Progress in Development Studies* 3(3): 179–95.

Ahmed, A. U., M. Alam, and A. A. Rahman. 1999. Adaptation to climate change in Bangladesh: Future outlook. In *Vulnerability and adaptation to climate change for Bangladesh*, eds. S. Huq, M. Asaduzzaman, Z. Karim, and F. Mahtab, 125–43. The Dorcrecht, Netherlands: J. Kluwer Academic Publishers.

Ali, A.1999. Climate change impacts and adaptation assessment in Bangladesh. *Climate Research* 12: 109–16.

Barkat, A., S. Zaman, and S. Raihan. 2001. The political economy of Khas land in Bangladesh, Dhaka: Association for land reform and development.

Bebbington, A. 1999. Capitals and capabilities: A framework for analyzing peasant viability, rural livelihoods and poverty. *World Development* 27(12): 2021–44.

Blaikie P., T. Cannon, I. Davis, and B. Wisner. 1994. At risk: Natural hazards, people's vulnerability and disaster. New York: Routledge Press.

Brocklesby, M. A. and M. Hobley. 2003. The practice of design: Developing the chars livelihood programme in Bangladesh. *Journal of International Development* 15: 893–909.

Cohen, Y. A. 1974. Culture as adaptation. In *Man in adaptation: The cultural present*, ed. Y. A. Cohen, 45–70. Chicago: Aldine Press.

Eakin, H. and A. L. Luers. 2006. Assessing the vulnerability of social-environmental systems *Annual Review of Environment and Resources* 31: 365–94.

Finan, Timothy J., C. T. West, T. Mcguire, and D. Austin. 2002. Processes of adaptation to climate variability: A case study from the US Southwest. *Climate Research* 21: 299–310.

Finan, T. J. and P. Biswas. 2005. Challenges to adaptive management: The GOLDA project in southwest Bangladesh. Report presented to the SANREM CRSP, University of Georgia, Athens.

Füssel, H.-M. 2007. Vulnerability: A generally applicable conceptual framework for climate change research. *Global Environmental Change* 17: 155–67.

Intergovernmental Panel on Climate Change. 2007. Assessment of adaptation practices, options, constraints and capacity. *In* Contribution of Working Group II to the Fourth Assessment Report of the Intergovernmental Panel on Climate Change, eds. M. L. Parry, O. F. Canziani, J. P. Palutikof, P. J. van der Linden, and C. E. Hanson, 976. Cambridge: Cambridge University Press.

Little, P. D., K. Smith, B. A. Cellarius, D. L. Coppock, and C. B. Barrett. 2001. Avoiding disaster: Diversification and risk management among East African Herders. *Development and Change* 32: 401–33.

Nelson, D. R., W. N. Adger, and K. Brown. 2007. Adaptation to environmental change: Contributions of a resilience framework. *Annual Review of Environment and Resources* 32(11): 10.

O'Brien, K., R. Leichenko, U. Kelkar, H. Venema, G. Aandahl, H.r Tompkins, A. Javed, S.Bhadwal, S. Barg, L Nygaard, and J. West. 2004. Mapping vulnerability to multiple stressors: Climate change and globalization in India, *Global Environmental Change* 14(4): 303–13.

Oliver-Smith, A. 2002. Theorizing disasters: Nature, power, and culture. In *Catastrophe and culture: The Anthropology of disaster*, eds. S. Hoffman and A. Oliver-Smith, 23–48. Santa Fe, NM: School of American Research Press.

Scoones, I. 1998. Sustainable rural livelihoods: A framework for analysis. Working Paper 72, Institute for Development Studies, Brighton, UK.

Thomas, D., H. Osbahr, C. Twyman, N. Adger, and B. Hewitson. 2005. Adaptive: Adaptations to climate change amongst natural resource-dependant societies in the developing world: Across the Southern African climate gradient. Technical Report 35. Tyndall Centre for Climate Change Research, East Anglia University, Norwich, UK.

Chapter 9

Salmon Nation: Climate Change and Tribal Sovereignty

Benedict J. Colombi

Introduction

Nez Perce interactions in the Snake and Columbia River drainages are a matrix of water, salmon, labor, ceremony, and place. Thus, to know this cultural history is to understand a particular river system and the people who have changed it. In this chapter I examine how humans have shaped the natural history of this region and how, in turn, nature shapes human relationships. Rivers in this context are "organic machines," an energy system that is at once an abstract idea and a concrete form, like migrating salmon, the flow of the river, and large dams (White 1995).

I emphasize Nez Perce culture, salmon, and nation building because each are situated in and interact according to their various biological, social, and political contexts. Nez Perce have fished for salmon for at least five thousand years, and salmon deliver power and energy that has improved the everyday lives and well-being of this indigenous people. Nation building, on the other hand, is largely a sociopolitical construct that reacts to and takes advantage of an encroaching modern world-system, and, moreover, is built on the doctrines of reserved rights, self-determination, and sovereignty. I consider how climate change challenges Nez Perce cultural and ecological sustainability, and explore ongoing tribal efforts to build an autonomous salmon nation in the twenty-first century.

In studying climate change I underscore how my encounters are situated within shifting notions of place and in understanding the impact that this has on Nez Perce culture and environment. In other words, global climate change and its regional impacts have the overwhelming potential to change Nez Perce homelands and sense of place, as well as transform contemporary political, legal, and economic contexts. A changing river, changing salmon returns, and how people respond to them, all contribute to changing ideas and understandings of place.

Salmon are the crucial link to Nez Perce cultural and ecological sustainability. For Nez Perce, salmon bring the ocean's energy inland to plants, animals, and people. Without migrating salmon, Nez Perce say, the river would die. More precisely, energy from salmon is pumped upstream from marine environments in the form of carbon, nitrogen, phosphorus, and other nutrients. Salmon, most importantly, are a keystone species and more than 140 Pacific Northwest species of animals and plants, including species of bears, trees, birds, and insects, rely on salmon as a major source of survival and energy (Cederholm et al. 2001).

Recently questions have emerged about how global climate change will trigger new ecosystem regimes and impact cultural and subsistence rights, framing global climate change to be a threat not only to the biodiversity of salmon, but also to the maintenance and reproduction of indigenous rights. Locally climate change is prompting the adoption of different adaptive strategies with a change in place according to the areas of rivers and land Nez Perce once used. Intimate human-environment relations, which are primary to Nez Perce cosmology and cultural diversity, lose place in the Northwest landscape. This dilemma elicits a discussion of how Nez Perce place, environment, and salmon nation building are negotiated, especially as climate change proceeds and sustainability is challenged and transformed.

Nez Perce Place and Environment

The Nez Perce (*niimiipuu*)[1] story is about water, land, salmon, game, and roots. These are the ideological and material foundations from which Nez Perce built their indigenous culture (Marshall 2006). Without these ideological and material foundations, Nez Perce creation is nonexistent. Thus, Nez Perce respond to these traditional (*walíim*) forms of natural resources with prayer and contemporary reverence. The Creator (*haniyaw'áat*), for instance, fashioned both the world and humanity, and the foundations for life express a particular history in place and environment. Nez Perce interactions with the Columbia Basin are a matrix of labor, ceremony, and place, told in terms of salmon (*léwliks*) and water (*kúus*).

Nez Perce stories and the history of kin relations and community are also tied to salmon and water, with individual and collective identities vested in symbolic and material sources of salmon, places to fish for salmon, and water in the Columbia Basin. Nez Perce develop relations and identity in regard to family, band, tribe, and their relation to land, water, and salmon (Marshall 2006). Social cohesion and basic values are therefore enhanced and governed by these aforementioned relations.

Fish and water are widely used in important Nez Perce daily life and ceremony and are necessary for the fulfillment of individual and community life. For Nez Perce these events include: births; funerals; testimonial "giveaways" for the first anniversary marking an individual's death; weddings; name-giving ceremonies; "first salmon," "first kill," and "first roots" ceremonies marking adulthood; and pow-wows and other celebrations, including dinners conducted to share and give thanks for the joy of life (Marshall 2006). The dinners, which are both ritual feasts and nonritual meals, ideally include items

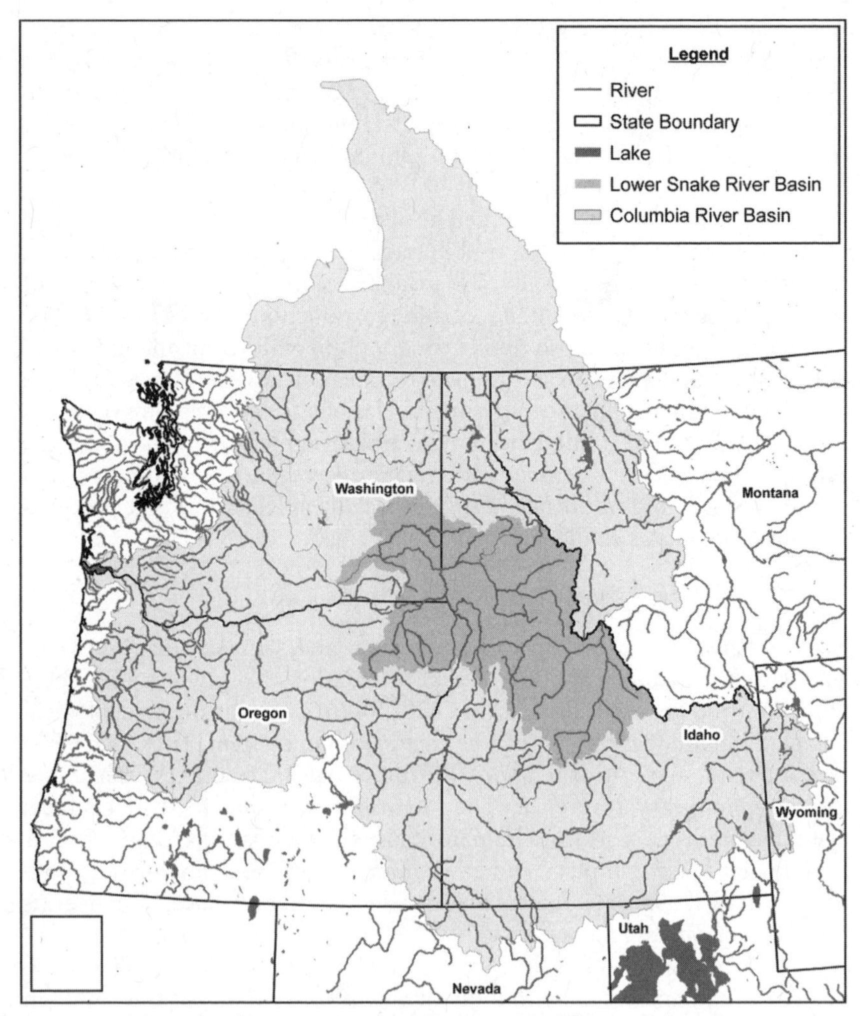

Figure 9.1: The Columbia River Basin. © The Nez Perce Tribe, Land Services Program, Jeff Cronce, 2008.

unavailable for purchase in supermarkets, including water (*kúus*); chinook salmon (*nac'óox*); meat—elk, deer, moose, and bison (*núukt*); roots (*qáaws*); and huckleberries (*cemíitx*). The capture of all these traditional foods is thought of as a gift (*pínitiní*) by the Creator, because these living beings gave up their lives so that Nez Perce can continue to prosper.

For fish, Nez Perce prefer salmon—in first place chinook, then sockeye, and lastly silver. Nez Perce also use other fish species, including eel (*hésu'*), sturgeon (*qîilex*), and steelhead trout (*héyey*), all anadromous fish, and other native fish including cutthroat (*waw'álam*) and bull (*ís'lam*) trout, northern pikeminnow, sucker, and chiselmouth. Nonnative fish species, such as carp, walleye, and bass are rarely if ever used, regarded as either culturally insignificant or unimportant. Store-bought fish is unacceptable. Chinook salmon from hatcheries is acceptable but not preferred. Fish other than chinook salmon is generally disliked (Marshall 2006).

Water (*kúus*), just like fish, has an ideological and material importance to Nez Perce cosmology and everyday survival. From an ideological perspective, water is home to powerful spirits and, materially, water is used for medicine and healing purposes. According to Nez Perce cosmology, eddies and confluences of free-flowing rivers and waterfalls are thought of as the homes of spirits. Similar to how Nez Perce regard fish, not all water sources are considered the same or equal in both importance and preference. Springs possess the purest, strongest, and most spiritually powerful water and is poured on hot rocks in the ritual sweathouse. Cold flowing water from high mountain streams is less preferred than spring water, but is considered "better" than water that runs at lower elevations, with less velocity, and at higher temperatures.

Water (*kúus*) and salmon (*léwliks*) are essential to everything that is Nez Perce and are found in streams and rivers of great cultural importance. Basic values and beliefs in water and salmon are evident as moral instruction in Nez Perce traditional stories, such as "Coyote Breaks the Fish Dam at Celilo," "The Maiden and the Salmon," "How Salmon Got Over the Falls," and "Coyote and Salmon" (Phinney 1934; Spinden 1908; Walker 1998). These stories illuminate the creation of the world and the beings that inhabit it, and include places in the Columbia and Snake River system, from Celilo Falls on the mid-Columbia to the tributaries of the Snake River, containing the Palouse, Tucannon, Clearwater, Grande Ronde, Salmon, Weiser, and Payette rivers in the Snake River Basin. Except for above the lower falls on the Palouse River, all of these rivers and streams supported annual returns of salmon, and all of the subbasins, including the Palouse River, flourished with abundant springs, cold running water, waterfalls, and deep holes and eddies.

Salmon Biodiversity and Climate Change

The scientific consensus is that global climate change is transforming the world. In February 2007, the Intergovernmental Panel on Climate Change (IPCC) reported that the earth is warming and that humans are largely responsible for the change. What are the global implications of a changing climate? It is argued that drought will intensify and create various arid regions, savanna grasslands will replace tropical forests, glaciers will diminish, rising sea levels will exacerbate coastal flooding, ocean coral will markedly decline, winter seasons will be less severe, and the Arctic permafrost will thaw (Kerr 2007). Overall, no world region is immune to the effects of global climate change. What are the projected implications of climate change in the Columbia Basin?

The greatest threat from climate change to Nez Perce includes a reduction in the biodiversity of salmon and water-related resources (Battin et al. 2007). Both salmon and water define the region with its spectacular forests, abrupt topographical and microclimate changes, and prolific salmon runs and freshwater resources. Continuing stress, however, from nearly two centuries of unprecedented commercial growth and the projected implications of global climate change is transforming this environment and the cultures dependant upon it by reducing numbers of available and harvestable salmon and spatially shifting annual snow and rainfall patterns.

At the beginning of the twenty-first century there were over 10 million inhabitants in the Pacific Northwest, three orders of magnitude greater than the indigenous population in 1750 (Bodley 2006). By the late nineteenth century the immigrant population exceeded the Native population, and by the early twentieth century the total settler population was more than one million people. Before World War II economic expansion and population growth fueled the construction of large dams, and the Bonneville and Grand Coulee dams were completed on the Columbia River in 1938 and 1942, respectively. After World War II per-capita energy consumption increased dramatically and between 1985 and 2003 the economy nearly doubled. A century of commercial growth has degraded ecosystems, diminished the opportunities for many ranching, farming, and forest-based communities, and partially destroyed the great Columbia River salmon fishery (Augerot 2005). Massive change of historic ecosystems has removed 80 to 90 percent of the old-growth coniferous forests in the Cascades. Timber cutting, grazing, and fire suppression have made the remaining forests prone to disease and fire. Ninety percent of the sagebrush steppe in Idaho and 99 percent of the Palouse Prairie steppe, a unique ecosystem in north-central Idaho and southeastern Washington, has been removed, mostly for urban and agricultural development. Dramatic human environmental change to native habitat has led to the extinction of fourteen bird and mammal species from Washington and Oregon (Iten et al. 2001), and the state of Oregon lists forty-two additional mammals and birds and as "species of concern." Researchers predict that in coming decades the social and environmental sustainability of the entire Pacific Northwest will be further challenged by the combined effects of global climate change, water shortages, and more severe stream flow fluctuations (Parson et al. 2001), and by increases in the price of fossil fuels as global peak oil production is exceeded (Deffeyes 2003).

For Nez Perce, salmon ecosystems define cosmology, labor, energy, and economy. Salmon link biodiversity and productivity because salmon are "transport vector[s]" of key processes in the movement of "materials, and energy and nutrients between marine, aquatic, and terrestrial ecosystems" (Cederholm et al. 2001, 652). Spawning salmon in the Columbia River drainage historically transported over 100 million kilograms of energy and materials from marine to terrestrial ecosystems (10 to 16 million fish) annually, making it the world's richest inland fishery (Lichatowich 1999). Before the arrival of Europeans these rich ecosystems supported some 700,000 indigenous people diversified into forty-seven cultural subareas and representing eleven language families (Kroeber 1939). Allowing for the impact of

European disease, the precontact population may have been twice this size (Boyd 1985). Indigenous peoples in the Columbia River drainage consumed nearly 36 billion kilocalories in returning salmon (Bodley 2006). The seven species of Pacific Northwest salmon (*Oncorhynchus* sp.) have presumably existed in their present form for six million years, and they have been a food source for indigenous peoples for millennia. Now in the Columbia Basin, salmon runs measure roughly 20 percent of their historic levels of 10 to 16 million before 1805, with as few as 200,000 fish returning annually (Augerot 2005). The current decline in Columbia Basin salmon can be attributed to the impacts by hydroelectric dams, irrigation projects, and overall habitat loss (Blumm 2002; Lichatowich 1999).

It is projected that global climate change will spatially redistribute stream flow scenarios and reduce the amount annual freshwater cycling (Parson et al. 2001). Sharp variations in water are affected by the reduction in annual snow pack (Mote et al. 2003). Pacific Northwest average annual air temperatures warmed by between 0.7 and 0.9 degrees Celsius in the twentieth century, and climate models suggest that additional increases from 1.5 and 3.2 degrees Celsius will occur by the mid-twenty-first century (Battin et al. 2007). These higher air temperatures could harm salmon during spawning, incubation, and rearing stages of their life. Warmer temperatures create earlier snowmelt and less moisture falling as snow. Increases in rain versus snow will lead to increased winter peak flows that scour stream and riverbeds and obliterate salmon eggs. A reduction in snow pack results in diminished flows in summer and fall, decreasing the availability of suitable spawning habitat and expediting increases in water temperatures.

Little is known about the ability of salmon to adjust to global climate change. The negative effects of climate change, however, are projected to be most pronounced in the higher and more pristine tributaries of major river systems (Battin et al. 2007). In the Nez Perce homeland, rivers and streams are markedly cooler, higher, and offer suitable habitat for spawning salmon than warmer, lower elevation streams located further downstream in the lower Columbia River basin. Perturbations in climate and increases in average temperatures will likely challenge the remaining salmon stocks in the next fifty to one hundred years. Recent research in the Pacific Northwest on chinook salmon populations suggests declines between 20 and 40 percent by 2050 (Battin et al. 2007). Indigenous peoples depend on chinook salmon for their large size and high fat content. A significant decline in chinook salmon therefore threatens indigenous sociocultural and ecological sustainability.[2]

In the Pacific Northwest the sustainability advantages of smaller-scale Native nations challenges the negative effects of modern commercial growth and the projected impacts and loss from climate change. One of the building blocks of modern-day tribal sovereignty is for Native nations to aim to increase their capacities for autonomous self-governance and sustainable community development in Native nation building (Cornell and Kalt 1998). One goal could therefore be to build a modern-day salmon nation and thereby transcend the current state of the environment with proactive policies to alter internal and external situations of colonization.

Salmon Nation Building and Climate Change

In building a salmon nation, Nez Perce and other indigenous nations actively seek a more sustainable future. Native nations aim to develop alternative systems to external policies of assimilation and pro-growth economic development. Salmon nation building includes developing indigenous strategies to fashion current and informed policy making, and to restore and implement those decisions in Native homelands (Native Nations Institute 2007). The goal in salmon nation building is to develop comprehensive and collective strategies in rebuilding societies and ecosystems that are sustainable.

Salmon nation building addresses creative cultural adaptation to the current salmon crisis and may also have the capacity to countervail the negative effects on salmon-related resources from global climate change. Drawing on the doctrine of reserved rights (Wilkins and Lomawaima 2001) and the Treaty of 1855, the Nez Perce Tribe generates policy and revenue in natural resources in fisheries, wildlife programs, and land management. In this regard, the Nez Perce effectively build a strong sovereign nation and shape a vigorous future economy that fits their own circumstances and culture.

The primary goal for Nez Perce tribal natural resource programs is to recover and restore all species and populations of anadromous salmon and resident fish in the Nez Perce traditional homeland. In collaboration with the Columbia River Inter-Tribal Fish Commission (CRITFC), the Nez Perce Fisheries program provides scientific, technical, and policy inputs to protect reserved rights in salmon and water in the Columbia River drainage at all "usual and accustomed places." The Nez Perce Tribe operates seven fish hatcheries on and off the reservation, and monitors the harvest of 50 percent of the available adult salmon migrating in the Columbia River drainage. The Nez Perce Tribe and other Native nations provide recommendations for the protection and restoration of critical habitat for salmon populations listed under the federal Endangered Species Act. Consultation between Native nations and federal agencies results in the issuance of biological opinions for the survival and recovery of listed salmon species from a federal action that is a direct or indirect alteration of critical habitat. Native nations, as a result, generally oppose large dams and other water development projects that negatively impact surviving salmon and water quality.

Water is a central concern in salmon nation building. In 2005 the Nez Perce Tribe formed an agreement between non-Native water users, the Idaho State Senate, and the US Congress in the Snake River Basin Adjudication (SRBA)—a water rights case introduced in 1986 to settle more than 150,000 outstanding claims to water in the Snake River drainage. The Nez Perce Tribe, in return, drew from their cultural connections to salmon and water, and formed an agreement in which the Bureau of Reclamation may lease up to 427,000 acre-feet of water from the state to increase flow augmentation in the Snake River drainage and help endangered salmon. Additional water will flow down the Snake River, aid salmon migration, and improve Nez Perce fish and habitat projects. Nez Perce thus use reserved rights, self-determination, and autonomous self-governance to build a salmon nation and reduce internal and external conflicts from over 150 years of pro-growth, commercial development.

Ultimately, then, the greatest potential conflict to Nez Perce culture stems from the projected impacts of climate variation and its associated consequences to salmon and water. In the Pacific Northwest, indigenous nations react to climate change with novel and innovative policies (Hanna 2007). Adjudicating water rights for salmon is a powerful tool in an environment of increasing demands and declining supplies. The federal Endangered Species Act is a valuable legal strategy for Native nations aiming to protect salmon populations from extinction, and additional legal structures, such as contract law, may provide another means by which Native nations might attempt to secure in-stream flows to protect migrating salmon. Furthermore, in the protection of salmon, indigenous policies aim to designate off-reservation landholdings as public lands in national parks and monuments, and in wild and scenic rivers.

The formation of intergovernmental and intertribal cooperation has resulted in the Columbia River Inter-Tribal Fish Commission and other collaborations with various federal agencies such as the National Oceanographic and Atmospheric Administration (NOAA) and the National Fish and Wildlife Service (NFWS). In salmon restoration, these partnerships are effective in comanaging hatchery programs and in developing long-range management strategies. Native nations develop and implement strong policies on the future of dams and related irrigation projects. Native nations, for instance, might seek to enforce dam operators to release more water when needed to improve fish passage, and, when necessary, litigate for the decommissioning of dams as a measure of last resort.

Nez Perce nation building includes on- and off-reservation carbon sequestration. The Nez Perce Tribe has committed to twenty-nine forest-restoration projects and about five thousand acres to carbon sequestration, and plantings of Douglas fir and ponderosa pine saplings are projected to absorb a year's worth of carbon dioxide from nearly 500,000 cars, trucks, and SUVs (Zaffos 2006). The Nez Perce Tribe aims to have corporations offset their greenhouse gas emissions by paying to keep trees growing and for forests to remain intact. Few American companies are presently mandated to curb greenhouse emissions with carbon sequestrations, but tribal efforts are models that demonstrate for others that the real value in forests is to keep them alive and in place. Twenty-first-century Nez Perce tribal policy is ultimately about maintaining healthy ecosystems and building a salmon nation in the Columbia River drainage.

Conclusion

This essay juxtaposes the gravity of global climate change with a sense of place for Nez Perce and other indigenous peoples inhabiting the Columbia River drainage for the last 10,000 years. The interdependencies of salmon, water, and the physical world illustrates the geography of the human experience, and how humans shape the natural history of an area and how nature shapes human relationships. The rivers of the Columbia Basin, in this context, are an "organic machine . . . an energy system which, although modified by human interventions, maintains its natural, its "unmade" qualities" (White 1995, ix).

For the past 150 years, the environmental effects of industrial production have contributed to the longer drought cycles of climate change, pollution, and radioactivity, and to diminishing salmon runs in the Columbia River Basin. I suppose the impacts from human-induced climate change will cause some to ask, "Does the country of the past transform and supplant the country of the present?" Or to ask, "What happened here? Who was involved? What was it like? And, why should it matter?" (Basso 1996, 5). For the Nez Perce, a strong sense of place provides the answers to all these questions and more.

Notions of place for the Nez Perce describe the importance of the earth, water, and salmon. How the Nez Perce collectively build a salmon nation suggests that "place-making involves multiple acts of remembering and imagining which inform each other in complex ways" (Basso 1996, 6). Nez Perce interactions with the environment in the present context are challenged by the shifting notions of place and in the transformation of contemporary political, legal, and economic domains. This essay reveals the extent to which the cultural and ecological sustainability of the Columbia River Basin and tribal efforts to build an autonomous salmon nation are both exacerbated and mitigated by the increasing regional effects and impacts of twenty-first-century global climate change.

Measurable declines in salmon and water are immediate dangers to Nez Perce rights and their continuing way of life. We can only wonder how climate change will trigger the materialization of new ecosystem regimes and impact cultural and subsistence rights in the places the Nez Perce once used. Effective adaptation to such changes will require long-term action by a greater community of citizens, including extraordinary alliances between Native nations and non-Native peoples to work together to strengthen and improve a common place and a common watershed. Necessary actions include placing limits on commercial growth, developing renewable sources of energy, and maintaining lower levels of consumption. Realizing social and environmental justice in response to the local effects of global climate change will require effective management by highly autonomous local communities, such as Native nations, and in fully representative regional democracies and effective global institutions.

Notes

1. All Nez Perce language (*nimiipuutimt*) words are in parentheses and italicized. The spellings are derived from Haruo Aoki's (1994) *Nez Perce Dictionary*.
2. I use the term *sustainability* to refer to the ability of the members of any human society to acquire the energy and materials needed for successful cross-generational maintenance and reproduction of individual households, society, and culture. "Sustainability" has different definitions for different people. One general definition is "the global problem of how to meet human needs in a world of declining material resources, persistent poverty, conflict, and resource degradation" (Bodley 2006).

References

Aoki, H. 1994. *Nez Perce dictionary*. Berkeley: University of California Press.

Augerot, X. 2005. *Atlas of Pacific salmon: The first map-based status assessment of salmon in the north pacific*. Berkeley: University of California Press.

Basso, K. H. 1996. *Wisdom sits in places: Landscape and language among the western Apache*. Albuquerque: University of New Mexico Press.

Battin, J., M. W. Wiley, M. H. Ruckelshaus, R. N. Palmer, E. Korb, K. K. Bartz, and H. Imaki. 2007. Projected impacts of climate change on salmon habitat restoration. *Proceedings of the National Academy of Sciences of the United States of America* 104(16): 6720–25.

Blumm, M. 2002. *Sacrificing the salmon: A legal and policy history of the decline of Columbia Basin salmon*. Den Bosch, Netherlands: BookWorld Publications.

Bodley, J. H. 2006. Scale, power, and sustainability in the Pacific Northwest. Paper presented at the Annual Meeting, Society for Applied Anthropology. Vancouver, British Columbia, March 27.

Boyd, R. T. 1985. The introduction of infectious diseases among the Indians of the Pacific Northwest, 1774–1874. PhD dissertation, University of Washington, Seattle.

Cederholm, C. J., D. H. Johnson, R. E. Bilby, L. G. Dominguez, A. M. Garrett, W. H. Graeber, E. L. Greda, M. D. Kunze, B. G. Marcot, J. F. Palmisano, R. W. Plotnikoff, W. G. Pearcy, C. A. Simenstad, and P. C. Trotter. 2001. Pacific salmon and wildlife-ecological contexts, relationships, and implications for management. In *Wildlife-habitat relationships in Oregon and Washington*, eds. D. H. Johnson and T. A. O'Neil, 628–84. Corvallis: Oregon State University.

Cornell, S. and J. P. Kalt. 1998. Sovereignty and nation-building: The development challenge in Indian country today. *American Indian Culture and Research Journal* 22(3):187–214.

Deffeyes, K. S. 2003. *Hubbert's peak: The impending world oil shortage*. Princeton, NJ: Princeton University Press.

Hanna, J. M. 2007. Native communities and climate change: Protecting tribal resources as part of national climate policy. Report published by the Natural Resources Law Center. Colorado Law School, University of Colorado, Boulder.

Iten, C., T. A. O'Neill, K. A. Bettinger, and D. H. Johnson. 2001. Extirpated species of Oregon and Washington. In *Wildlife-habitat relationships in Oregon and Washington*, eds. D. H. Johnson and T. A. O'Neill, 452–73. Corvallis: Oregon State University.

Kerr, R. A. 2007. Global warming is changing the world. *Science* 316(5822): 188–90.

Kroeber, A. L. 1939. *Cultural and natural areas of native North America*. Berkeley: University of California Press.

Lichatowich, J. 1999. *Salmon without rivers: A history of the Pacific salmon crises*. Washington, DC: Island Press.

Marshall, A. G. 2006. Fish, water, and Nez Perce life. *Idaho Law Review* 42(3): 763–93.

Mote, P. W., E. A. Parson, A. F. Hamlet, W. S. Keeton, D. Lettenmaier, N. Mantua, E. L. Miles, D. W. Peterson, D. L. Peterson, R. Slaughter, and A. K. Snover. 2003. Preparing for climatic change: The water, salmon, and forests of the Pacific Northwest. *Climatic Change* 61: 45–88.

Native Nations Institute. 2007. What is native nation building? Electronic Document, http://nni.arizona.edu/whoweare/whatis.php, accessed October 17.

Parson, E. A., P. W. Mote, A. Hamlet, N. Mantua, A. Snover, W. Keeton, E. Miles, D. Canning, and K. Gray Ideker. 2001. Potential consequences of climate variability and change for the Pacific Northwest. In *Climate change impacts on the United States: The potential consequences of climate variability and change*, ed. National Assessment Synthesis Team, 247–80. Washington, DC: US Global Change Research Program.

Phinney, A. 1934. *Nez Perce texts*. New York: Columbia University Press.

Spinden, H. J. 1908. *The Nez Perce Indians*. Memoirs of the American Anthropological Association vol. 2, part 3. Lancaster, PA: The New Era Printing Company.

Walker, D. E. Jr. 1998. *Nez Perce coyote tales: The myth cycle*. In collaboration with Daniel N. Mathews, illustrations by Marc Seahmer. Norman: University of Oklahoma Press.

White, R. 1995. *The organic machine*. New York: Hill and Wang.

Wilkins, D. E. and K. Tsianina Lomawaima. 2001. *Uneven ground: American Indian sovereignty and federal law*. Norman: University of Oklahoma Press.

Zaffos, J. 2006. Tribes look to cash in with 'tree-market" environmentalism. *High Country News* 13(5): 5.

Chapter 10

Global Averages, Local Extremes: The Subtleties and Complexities of Climate Change in Papua New Guinea

Jerry Jacka

Introduction

2007 marked a watershed moment in the reception of human-induced climate change around the planet. With the release in February of that year of the Fourth Assessment Report (AR4) by the United Nations Intergovernmental Panel on Climate Change (IPCC) Working Group I, the physical science evidence that humans are influencing climate change was revised to a 90 percent probability, up from a 66 percent probability just six years prior (IPCC 2007, 8). A few months later, in April 2007, the IPCC Working Group II released another report discussing impacts, adaptation, and vulnerability to climate change. In May 2007, Working Group III released their report on the potentials for mitigating climate change. Work on these reports culminated with the IPCC sharing the 2007 Nobel Peace Prize with Al Gore in October for all of their work on highlighting climate change. While these reports and accolades represent a sea change in popular perceptions of human-induced climate change, they also have a glaring deficiency: indigenous peoples are either scarcely mentioned or completely absent from any discussions regarding their perspectives and knowledge of climate change (see Salick and Byg 2007). As anthropologists, it is imperative that we serve as witnesses to global climate change for the often-neglected and remote communities where we work in order to communicate their stories of adaptation and vulnerability to these social and environmental changes.

Indigenous peoples are, of course, not powerless or voiceless and one may wonder what purpose such a volume as this one serves. I believe, however, that the ethnographic documentation of indigenous peoples' responses to climate change can serve as a powerful methodology for action in influencing policies and practices related to global climate change. Indigenous people have always responded to environmental variations (droughts, famines, etc.) via a number of resilient mechanisms, such as traditional environmental knowledge of wild fallback foods, diversity of domesticated crops, long-distance exchange networks, and temporary migrations. Moreover, in the absence of long-term (or any) climatological records, oral traditions and local knowledge of changing climatic conditions can present a more complete picture of climate change in areas without these records. One such area, and the focus of this chapter, is the Porgera Valley, home to several thousand Ipili and Enga speakers in the highlands of Papua New Guinea (PNG).

In this chapter, I document climatic changes that Porgerans describe as having occurred over the past seventy years and show that Porgerans interpret climate change along several different dimensions—moral, agricultural, environmental, and cosmological. This indicates that mitigating climate change vulnerabilities in some communities will require a number of cultural and environmental realms that Northern governments and policy makers have not considered. I also explore in more detail what exactly Porgerans mean when they talk about climate. As Sarah Strauss and Ben Orlove (2003) indicate in their introduction to the edited volume *Weather, Climate, Culture*, it is crucial to distinguish between weather, season, and climate, or in other words between short-term, medium-term, and long-term climatological forces. While climate, a long-term trend, has been receiving the most attention in the press, extreme weather and seasonal events are apt to cause much greater human suffering in the next decade. Pielke and Bravo de Guenni (2004) argue that the dominant model in climate science, like that used by the IPCC, is overly concerned with modeling future scenarios of climate change without modeling the more immediate human vulnerabilities to climate change (see also Feddema et al. 2005; Pielke 2005). Thus there is a danger in focusing on the average at the expense of the extremes, particularly for people whose livelihoods are dependent upon subsistence agricultural practices.

In this context, I also explore medium-term trends, in particular impacts coming from seasonal weather fluctuations related to the El Niño Southern Oscillation, and the implications of more frequent El Niños for Porgeran people. The IPCC's Third Assessment Report noted that El Niño events "became more frequent, persistent, and intense during the last 20 to 30 years compared to the previous 100 years" (IPCC 2001, 6). In the PNG highlands, El Niños are frequently associated with severe and prolonged droughts. The AR4 supports this by claiming that "more intense and longer droughts have been observed over wider areas since the 1970s, particularly in the tropics and subtropics" (IPCC 2007, 6). Droughts in the highlands are also associated with frosts (above 1,500 m) that destroy the fragile vines of the sweet potato, the staple of highlands populations (Brown and Powell 1974; Sillitoe 1996; Waddell 1975; Wohlt 1982). The 1997 El Niño, for instance, affected crop production throughout the country's population of 5 million people, resulting in over a half a million people receiving direct food aid from government and NGO aid groups (Ellis 2003).

Understanding the dynamics of climate change in PNG has a far greater urgency than just the impacts on the populations living in and around the

Porgera Valley. With the largest extant block of tropical rain forest in the Western Pacific and Southeast Asia (Primack and Corlett 2005, 8), the forests of the island of New Guinea deserve special attention for their role in providing carbon sinks for the increase of anthropogenic carbon dioxide in the atmosphere (Achard et al. 2002; Detwiler and Hall 1988)—one of the key greenhouse gases linked to global climate change (IPCC 2007). An increase in El Niños and droughts will undoubtedly result in more massive fires like the ones that raged through the tropical rain forests of Indonesia in 1997, 1998, and 2005. These changes in forest cover perpetuate an unfortunate feedback cycle, such that the loss of forest cover contributes to warming that contributes to more droughts and fires leading to more loss of forest cover (Feddema et al. 2005).

Changing Ecologies, Changing Landscapes

When I arrived in Porgera (see Figure 1) in late 1998 to begin fourteen months of research, the 1997 El Niño had been over for a year. The impacts and memories of the event were still prevalent, though. People talked of how all the local springs dried up, and how the nearest water to be found was over three kilometers away. The smoke from the forest fires in Indonesia had caused the sun and moon to turn red, foreshadowing apocalyptic events that the recently Christianized Porgerans had heard of from the Book of Revelation. A solar eclipse and a strong earthquake that also affected the region at this time upset fundamental cosmological principles that people linked to larger moral and social problems in the valley. Large numbers of migrants, who had come from higher-altitude areas, were still living in the Porgera Valley, waiting for their gardens to begin producing again before leaving. Many of the migrants reported leaving behind the sick and elderly who could not travel to Porgera, many of whom were now dead (see Lemonnier 2001, 166).

Elevations in the Porgera Valley range from 900 to over 3,700 meters above sea level. In the region where I worked (at about 1,800 meters), frosts had not impacted the gardens, but drought had dried up all of the sweet potato mounds, creating severe famine conditions. Above 2,200 meters, frosts and droughts had combined to completely destroy higher-altitude gardens. The migrants were all from the high-altitude grasslands to the south of the Porgera Valley, where the lowest elevations are around 2,300 meters. The severity of the famine was mitigated by the presence of a multinational mining venture in the valley, Porgera Joint Venture (PJV), operator of the Porgera Gold Mine, which annually has produced about 900,000 ounces of gold since opening in 1990. In concert with the PNG government and AusAID (see Bourke et al. 2001), bags of rice were given out throughout the valley while people's gardens withered in the sun and frost.

I eventually came to understand that the events of the 1997 El Niño were merely an exceptional climatic anomaly in an environment that Porgerans

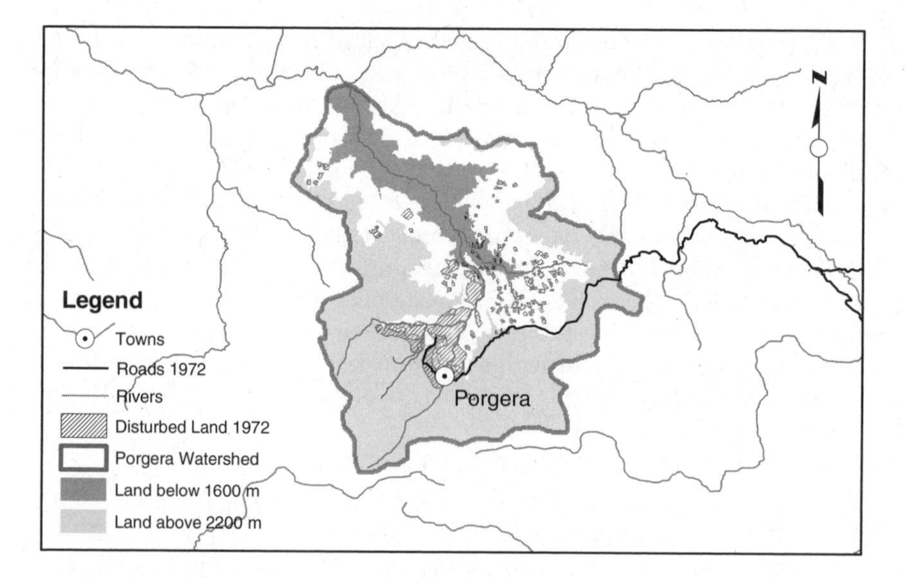

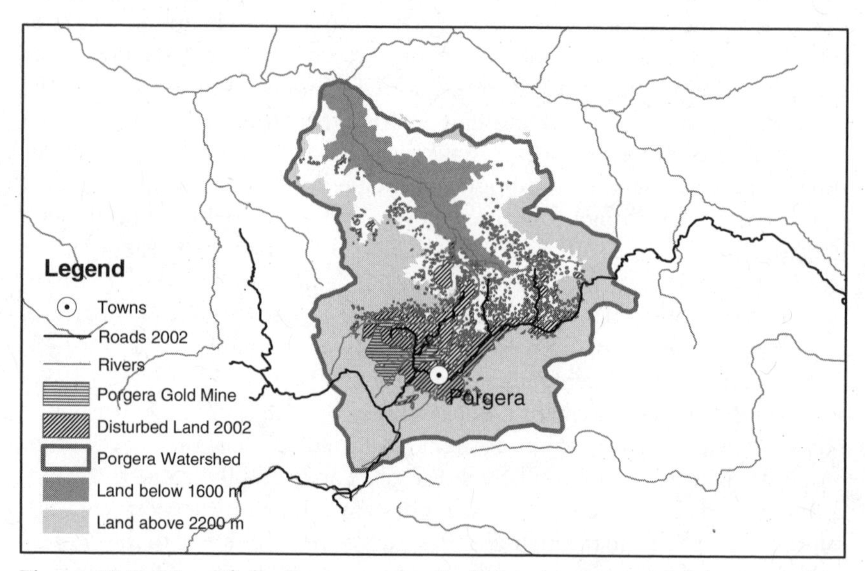

Figure 10.1: Map of the Porgera watershed showing area of land impacted by gardening and forest clearance in 1972 and 2002.

felt had been under change for some time. The most discussed transformation was the increase in temperature that everyone felt had been underway for the last few decades (cf. Salick and Byg 2007, 13). Further evidence for this was seen in the movement of species from one ecological zone to another. In an indigenous ecological perspective, Porgerans characterize their landscape in three ecological zones, *wapi*, which stretches from the lowest reaches of the valley up to around 1,600 to 1,700 meters; *andakama*, starting at the upper end of the wapi and reaching to about 2,100 to 2,200 meters; and *aiyandaka*, the highest altitude ecological zone. A number of features characterize these zones. Wapi is hot, humid, has a lot of insects, and is marked by species such as fruit pandanus (*Pandanus conoideus*) and wild bananas (*Musa* sp.). Andakama literally means "the place of houses and clearings" and is where people traditionally lived and gardened. The aiyandaka is a cold, rainy zone with waterlogged soils and marked by a particular variety of wild nut pandanus (*Pandanus brosimos*), *wapena* (see Figure 2). People also mentioned that several species of birds that only used to be found in the wapi could now be seen flying around in the andakama. Domesticated banana species that used to only grow in the lower section of the andakama could now be planted and would grow well in the aiyandaka. In essence, there was the perception that species had shifted upwards in response to the general warming trend.

Inhabitants also reported that foreign species invasions accompanied the warming temperatures. A grassy weed with detachable seeds that gets stuck in people's clothing grows rampant in fallowing sweet potato gardens. Locally called "poor man's grass," elders report seeing this for the first time when they were in their teens. A beetle that attacks and kills casuarina (*Casuarina oligodon*) trees, the most important tree species for domestic use as firewood, was reported to only have arrived in the valley in the last decade. Large stands of dying casuarina were pointed out to me around the valley by many people.

Figure 10.2: Harvesting pandanus nuts in the aiyandaka.

Extreme rainfall events are another area of concern locally. On average, Porgera shows little seasonality. Rainfall records from PJV (1974 to 2004) show a low average of about 250 millimeters of rain per month from June to September, increasing to a high of about 350 millimeters per month during the rest of the year—a pattern that one group of climatologists has termed "wet and very wet" (McAlpine et al. 1983, 61). Recent El Niños in 1982 and 1997 have fit the pattern of drought and frost described above. The 1987, 1991, and 1993 El Niños, however, while also logging some of the lowest rainfall amounts on record for Porgera, have paradoxically also resulted in some of the highest monthly rainfall totals. This could be accounted for by a pattern that some highlands researchers have noted of extremely heavy rainfalls preceding major El Niño events (Bourke 1989; Brookfield and Allen 1989).

These climatic extremes of drought/flood cycles that will be experienced globally due to climate change, not the averages that climate change models focus on, will have the greatest impacts on Porgeran livelihoods. Due to the lack of seasonality in this part of the PNG highlands, there are no set and regular planting times but rather a constant rotation. Women generally plant the same number of sweet potato mounds every week that they harvest. Sweet potatoes cannot be stored, so there must always be a harvestable source of tubers. Extreme rainfall oversaturates the ground, preventing tuber development, while drought shrivels the vines of the plant before tubers can grow.

As previously stated, Porgerans largely interpret the changes they are experiencing as moral and cosmological imbalances in their world. In the sections that follow, I examine some of these changes and trace the complex ways in which climatic changes are being experienced in the affected communities.

Development, Migration, and Land Cover Change

With the 1987 development of the Porgera Gold Mine, the four-wheel-drive road into the valley was upgraded to allow average cars access into the area. Porgeran landowners associated with the mine regularly receive significant cash payments for compensation, royalties, and wages (see Filer 1999). As a consequence, other Papua New Guineans have flooded into the valley to partake in as much of the development action as they can. From a population of around 8,000 in the late 1980s, mining officials estimate there are more than 35,000 people in the valley today (Ila Temu, personal communication, 9 December 2006). Newcomers and Porgerans alike have settled into camps along the road that services the mine. Because all of the immigrants are dependent upon subsistence gardening for their livelihoods, severe deforestation is occurring in the higher-altitude rain forest where the potential for crop losses are substantially increased due to frost (see Figure 3).

Figure 10.3: Migration caused deforestation along the road through the high altitude rain forest.

These settlements are unique in that they are located at a much higher altitude than Porgerans previously lived and gardened. During the colonial era (pre-1975), the four-wheel-drive road was constructed at an altitude of about 2,200 to 2,400 meters above sea level at the base of a high limestone massif. Prior to mining development, Porgerans lived and gardened primarily between 1,650 and 2,150 meters in the andakama. There are three factors that inhibited them from moving into their present locations along the road at the base of the mountains. The first of these is altitude: above 2,100 meters frosts are an ever-present danger. The second is that due to local weather effects with the mountains along the southern edge of the Porgera valley, there is much greater rainfall at the base of the mountains. People living along the road, today, complain about how much harder it is to grow crops due to the soil being constantly waterlogged. Third, the upper rain forest was the location where a key ritual was performed that was critical for regulating the amount of rain that fell. I discuss the implications of the demise of this ritual in the next section.

When the four-wheel-drive road into Porgera was being built during 1970–71, elder men reported to me the great fear they had of cutting trees and digging gravel for the road. Konapi Malo claimed, "We never came up to the upper rain forest without a good reason. We weren't supposed to cut down trees or disturb anything, or else the spirits would punish us. When the *kiap* (colonial officer) made us work on the road many of us ran away and hid. But then the kiap came with his policemen and shackled us all together and made us work on the road in shackles."

Satellite data track the extent to which people have moved up into the rainforest since the building of the first road in the early 1970s, the upgrading of it in the late 1980s, and within the last couple of years. In 1972, of all the disturbed lands in the Porgera watershed, only 17 percent of these lands were above 2,200 meters. This number increased to 41 percent by the end of 2002. In terms of area, there were 585 hectares of disturbed lands above 2,200 meters in 1972, which increased to 2,812 hectares by December 2002 (see Figure 1). All of this land has been converted from tropical broadleaf forest with an unbroken canopy to gardens, cleared spaces, and scraggly, second-growth forests. Earlier research on land-cover change and climate indicated that tropical deforestation would result in a warming of 1° to 2°C in deforested areas (DeFries et al. 2002). The IPCC climate change models, though, largely ignore the impact of land-cover change on regional temperatures, precipitation, and vegetation (Pielke 2005). However, when more recent research (Feddema et al. 2005) included tropical deforestation and its effect on regional climates into the IPCC climate change models, they found that deforestation would actually increase the temperature more than 2°C in tropical regions, there would be greater amounts of precipitation, but there would also be longer dry periods. Since over 40 percent of all agricultural activities now occur in the potential frost zone in Porgera, there is only greater likelihood of significant crop losses and more widespread famines. Research has also shown that trees planted in gardens and along their borders helped mitigate the damage of high altitude frosts in a 1972 El Niño (Brown and Powell 1974, 6), yet given the relationships seen with migration into the valley and deforestation, this remains an unlikely proposition for people to use to decrease their vulnerability to future frosts.

THE COSMOLOGICAL IMPLICATIONS OF CLIMATE CHANGE

On an almost daily basis in Porgera people would say to me that "the ground is ending" (*yu koyo peya*). I would ask them to explain what they meant. They would point around and remark that all one had to do was see the signs of the impending end. Women would comment that the ground no longer had grease (*yu ipane*), crops wouldn't grow as well because there was now too much rain. There were also all sorts of crops that now grew at different altitudes than in the times of their parents and grandparents.

One elder man even told me that the sun and moon no longer rose and set behind the proper mountains as the entire world had shifted and was coming apart. He was alluding to the fact that Porgeran cosmology envisions the world as a series of plates. The places where the plates come together are powerful sites where ritual activities were performed in the recent past to placate the spirits that sat at these junctures and held the world together. In Porgera, the most important ritual site was located in the high-altitude rain forest, the aiyandaka, and was destroyed during the building of the road. A guardian spirit named Lemeane was said to live at the ritual site and hold the world together to keep it from falling apart. Lemeane was also responsible for regulating the amount of rainfall and water in the Porgeran cosmos (Biersack 1998; Gibbs 1975). Rituals performed to Lemeane involved killing pigs and "feeding" the grease of pigs to him so that he would stay asleep and not upset the water balance of the world. The grease of the pigs was also believed to flow outward through roots embedded in the landscape, thereby renewing the fertility of the earth so that people, pigs, and crops would flourish. Lenape Pasia described what happened to the ritual site when the road was built:

> We were the caretakers of Auwalo Anda [the site]. We performed the rituals to control the rain and renew the earth. When the road was being built, the white men made us come to Auwalo Anda and cut down the big trees guarding the site so we could use the lumber for the road. We were terrified because this place was sacred. During the cutting of the trees, one of the trees fell wrong and killed a man. We knew this was Lemeane waking up and getting angry. Later, one of the men who cut the first tree had all of his children born with six fingers. We now wonder if we shouldn't have come up to the high rain forest and destroyed the Auwalo Anda site.

In 1972, within months of the road being finished (and in the midst of a large drought caused by the 1972 El Niño), a Catholic priest came to the site and sprinkled holy water on the stones that represented Lemeane's bones and where the pig grease was smeared. One of the men present said that everyone was "shrieking and running around in fear" that Lemeane was going to crack the ground apart and destroy the world. When some time went by and the world did not end, people started to convert to Christianity, according to this man.

While the rituals associated with Lemeane and Auwalo Anda are no longer performed, the cosmological concerns that the rituals assuaged have not gone away. Talking of the climatic anomalies that came with the 1997 El Niño many people stated that during that time they feared that Lemeane was angry and that the world was finally going to end. One man, now a Christian pastor, said that some people talked of renewing the rituals to Lemeane. "But how can we do that now," he wondered aloud to me, "we have turned our backs on the past and are now Christians." Consequently, many Porgerans understand the environmental changes as somehow linked to the impending return of Jesus. One woman said that the warming experienced in Porgera was "the fire that Jesus was bringing from heaven to destroy the world" (see Ballard 2000; Biersack 2005; Jacka 2005).

Associated with the demise of a ritual oriented toward mitigating the climatic uncertainties of rainfall, migration into the high altitude rain forest (aiyandaka) upset human and spirit relationships that were founded on ideas of reciprocity with spirits of the aiyandaka. Spirits called *tawe wanda* (sky women) were the "caretakers" of the upper rain forest. While generally benevolent toward humans, sky women could also harm individuals who wantonly destroyed resources in this ecological zone. Occasionally men would chop down wapena pandanus trees, said to be the "sweet potato" of the sky women, to get at the nut pods. Unable to climb the trees due to the mossy trunk, men would have to mourn these chopped down trees with the same emotion that would be exhibited to a deceased kinsperson. A failure to "repay" the gift of the pandanus nuts by showing contrition would anger sky women who had the ability (like Lemeane) to bring drenching rains that would cut short trips to the upper rain forest for the purposes of hunting and gathering. In some cases, the clouds and rain could confuse a person so greatly that they would fall to their deaths from cliff tops. With increased settlement into the aiyandaka by Porgerans and outsiders, the attitudes and practices of reciprocity toward sky women has largely ceased, although many people openly wondered to me if the heavy rainfalls along the highway were not somehow related to this fact.

Conclusion

The connections drawn in this chapter—climate change, deforestation, development, disrupted relations between humans and spirits, the end of the world—are not easily fit into a simplified scheme of cause and effect that appeals to Western rationality. Unlike the obvious implications of indigenous Arctic peoples trying to cope with a rapidly changing environment due to warming (ACIA 2004, and see also the essays in this volume by Broadbent and Lantto, Crate, Henshaw, Nuttall, Marino and Schweitzer, and Stuckenberger), Porgeran experiences of climate change reverberate through their society in complex and subtle ways. Climate change did not force Porgerans into living and gardening at higher altitudes, but the implications of a changing climate with more frequent and intense El Niños, and the extreme variabilities that these cause in weather in Porgera—droughts, deluges, frosts—will severely impact people's livelihoods in these higher altitudes. Likewise, climate change did not cause the increasing deforestation in Porgera, but the deforestation of the tropics has profound implications as one of the drivers of future climate change, which, as a result of increased dry periods, can lead to large forest fires.

A critical factor to consider is how Porgerans understand climate change, not due to their own impacts on the environment, or the global increase of greenhouse gases, but rather as a societal breakdown between themselves and the rituals oriented toward powerful spirits that control the cosmos. The task for anthropologists in writing about climate change will be to trace some of these less-obvious dynamics of climate change, such as are occurring in Porgera, in order to ensure that peoples on the margins of climate change research have a voice in this global phenomenon.

Focusing on the local and regional realities of climate change, and not globally modeled averages, is critical; as one climatologist notes, "People and ecosystems experience the effects of environmental change regionally, and not as global averaged values" (Pielke 2005, 1626; see also Pilkey and Pilkey-Jarvis 2007). Mitigating people's vulnerabilities to climate change will have to take the dynamics of culture and regional political-economic factors like development into consideration. Vulnerabilities will come not just from observable phenomena like rising seas, desertification, and melting permafrost. They will also be rooted in a people's understanding of the ontology and cosmology of indigenously perceived climate processes and how transformations in these ontological and cosmological notions threaten their cultural order.

References

Achard, F. et al. 2002. Determination of deforestation rates of the world's humid tropical forests. *Science* 297: 999–1002.

ACIA. 2004. *Impacts of a warming Arctic: Arctic climate impact assessment*. Cambridge: Cambridge University Press.

Ballard, Chris. 2000. The fire next time: The conversion of the Huli apocalypse. *Ethnohistory* 47: 205–26.

Biersack, A. 1998. Sacrifice and regeneration among Ipilis: The view from Tipinini. In *Fluid ontologies: Myth, ritual and philosophy in the highlands of Papua New Guinea*, eds. L. R. Goldman and C. Ballard, 43–66. Westport, CT: Bergin & Garvey.

———. 2005. On the life and times of the Ipili imagination. In *The making of global and local modernities in Melanesia: Humiliation, transformation and the nature of cultural change*, eds. J. Robbins and H. Wardlow, 135–62. Burlington, VT: Ashgate.

Bourke, R. M. 1989. The influence of soil moisture on sweet potato yield in the Papua New Guinea highlands. *Mountain Research and Development* 9: 322–28.

Bourke, R. M., M. G. Allen, and J. G. Salisbury eds. 2001. *Food security in Papua New Guinea*. Canberra: ACIAR.

Brookfield, H. and B. Allen. 1989. High altitude occupation and environment. *Mountain Research and Development* 9: 201–09.

Brown, M. and J. M. Powell. 1974. Frost and drought in the highlands of Papua New Guinea. *Journal of Tropical Geography* 38: 1–6.

DeFries, R. S., L. Bounoua, and G. Collatz. 2002. Human modification of the landscape and surface climate in the next fifty years. *Global Change Biology* 8: 438–58.

Detwiler, R. P., and C. Hall. 1988. Tropical forests and the global carbon cycle. *Science* 239(4835): 42–47.

Ellis, D. M. 2003. Changing earth and sky: Movement, environmental variability, and responses to El Niño in the Pio-Tura region of Papua New Guinea. In *Weather, Climate, Culture*, eds. S. Strauss and B. Orlove, 161–80. Oxford: Berg.

Feddema, J., J. F. Johannes, K. W. Oleson, G. B. Bonan, L. O. Mearns, L. E. Buja, G. A. Meehl, and W. M. Washington. 2005. The importance of land-cover change in simulating future climates. *Science* 310(5754): 1674–78.

Filer, C., ed. 1999. *Dilemmas of development: The social and economic impact of the Porgera gold mine, 1989–1994*. Boroko, PNG: The National Research Institute.

Gibbs, P. 1975. *Ipili religion past and present*. MA Thesis, University of Sydney.

IPCC (Intergovernmental Panel on Climate Change). 2001. *Climate change 2001: Synthesis report. Summary for policymakers*. Geneva: IPCC Secretariat.

IPCC (Intergovernmental Panel on Climate Change). 2007. *Climate change 2007: The physical science basis. Summary for policymakers*. Geneva: IPCC Secretariat.

Jacka, J. K. 2005. Emplacement and millennial expectations in an era of development and globalization: Heaven and the appeal of Christianity among the Ipili. *American Anthropologist* 107: 643–53.

Lemonnier, P. 2001. Drought, famine and epidemic among the Ankave-Anga of Gulf Province in 1997–98. In *Food security in Papua New Guinea*, eds. R. M. Bourke, M. G. Allen, and J. G. Salisbury, 164–67. Canberra: ACIAR.

McAlpine, J. R., G. Keig, and R. Falls. 1983. *Climate of Papua New Guinea*. Canberra: CSIRO, Australian National University.

Pielke, R. A., Sr. 2005. Land use and climate change. *Science* 310: 1625–26.

Pielke, R. A., Sr., and L. Bravo de Guenni. 2004. How to evaluate vulnerability in changing environmental conditions. In *Vegetation, water, humans, and the climate: A new perspective on an interactive system*, eds. P. Kabat et al., 537–38. New York: Springer.

Pilkey, O., and L. Pilkey-Jarvis. 2007. *Useless arithmetic: Why environmental scientists can't predict the future*. New York: Columbia University Press.

Primack, R. and R. Corlett. 2005. *Tropical rain forests: An ecological and biogeographical comparison*. Oxford: Blackwell Publishing.

Salick, J. and A. Byg, eds. 2007. *Indigenous peoples and climate change*. Oxford: Tyndall Centre for Climate Change Research, Oxford University.

Sillitoe, P. 1996. *A place against time: Land and environment in the Papua New Guinea highlands*. Amsterdam: Harwood Academic Publishers.

Strauss, S., and B. Orlove. 2003. Up in the air: The anthropology of weather and climate. In *Weather, climate, culture*, eds. S. Strauss and B. Orlove, 3–14. Oxford: Berg.

Waddell, E. 1975. How the Enga cope with frost: Responses to climatic perturbations in the central highlands of New Guinea. *Human Ecology* 3: 249–73.

Wohlt, P. B. 1982. *An investigation of food shortages in Papua New Guinea: 24 March to 3 April 1981*. Boroko, PNG: Institute of Applied Social and Economic Research.

Chapter 11

Talking and Not Talking about Climate Change in Northwestern Alaska

Elizabeth Marino and Peter Schweitzer

Introduction

There is consensus among scientists that global climate change is occurring and that this change is in the direction of warming on a global scale (McBean et al. 2005, 994). We define *global climate change* here as changes in overall climate patterns over a given space and time. Global climate change, therefore, is a distinct phenomenon of global scale that has local effects.

A second phenomenon, however, has occurred as consciousness of climate change has emerged on a popular scale. As Lorenzoni and Pidgeon have noted, "Climate change has woven its way into the general consciousness worldwide" (2006, 75). Global climate change, as they explain it, did not come into worldwide consciousness through local experience, but rather through global public discourse. The second phenomenon therefore is the rise of climate change as discourse in the Foucauldian sense (Foucault 1972, 1977; Woolgar 1986), and as simply a new part of the lexicon the world over.

This chapter explores how the discourse of global climate change can complicate anthropological work that seeks to understand local change. This research was part of an interdisciplinary project investigating the relationship between fresh water, climate, and humans in the Arctic. Our work involved spending time in five communities in northwestern Alaska documenting, among other things, changes observed in the landscape and how those changes were perceived and understood on a local level. Surprising to us was the disconnect that existed between observations of local change and the generalizations of "climate change." While complex and detailed understanding of environment and change exist within local and culturally specific discourses, we found the term *climate change* to be detrimental and limiting to the anthropologist and consultant alike. The power of that term, we argue, alters patterns of speech on a local level. To those ends, we will explain the effects of global climate change in Alaska and the accompanying rise in global climate change research, some of which focuses on interviewing knowledgeable Inupiaq residents about those changes. We then show how the words "climate change" affected local discourse in Inupiaq villages in Alaska, and how *not talking* about climate change proved the best method for understanding local conceptions of change.

Climate Change in the Arctic

Scientists have shown that the climate in the Arctic is changing more rapidly than other world areas as a result of global climate change, an effect known as polar amplification. From 1954 to 2003, the mean annual atmospheric surface temperature in Alaska and Siberia has risen between 2 and 3 degrees Celsius, with warming particularly salient during winter and spring (McBean et al. 2005, 992). In response, snow and ice features have diminished and permafrost is melting as its boundaries move north, causing erosion and foundation problems for structures in Alaska (McBean et al. 2005, 997).

Confounding these amplified effects of climate change in the Arctic is a distinct lack of long-term data. As the Arctic Climate Impact Assessment makes clear, "The observational database for the Arctic is quite limited, with few long-term stations and a paucity of observations in general, making it difficult to distinguish with confidence between the signals of climate variability and change" (McBean et al. 2005, 22). While the Arctic lacks scientific longitudinal data, it has been observed by indigenous inhabitants for thousands of years. Scientists in the Arctic, therefore, are becoming interested in indigenous observations about climate and environmental change. Indeed "scientific examinations of nontraditional data sets collected by naturalists, hobbyists, or indigenous peoples have been instrumental in linking changes in biological communities to recent climatic changes" (Sagarin and Micheli 2001).

Inupiaq Perspectives

Inupiat and other indigenous groups have been highly successful at living at northern latitudes for millennia. Close observation, categorization and understanding of the local environment, and interactions among its ecological features are a central feature of indigenous knowledge in the Arctic (Burch 2006; Nelson 1899; Ray 1975). Among the Inupiaq people of northwestern Alaska, with whom we worked on this project, detailed knowledge of the local environment was evident and ever present. Conversation revolved around weather patterns, harvesting of species—both plant and animal—and areas traveled.

Extensive knowledge concerning the behavior, impacts, and feedback effects of plants, animals, and weather on one another abound in Arctic discourse. That chum salmon come after king salmon or that more snow and a later melt means a more prolific berry season are examples of such common knowledge among northern people who depend on their immediate environment for survival. Inupiaq place names on the Seward Peninsula like

Chikuchuilaqpiaq, "place where the river never freezes," or *IKmilaq*, "place without water," demonstrate a thorough awareness of conditions on the land and exemplify how longtime Arctic residents' local observations are encoded in the oral record (Marino 2005).

Inupiaq people today continue to observe and increasingly see a changing landscape that is affecting their subsistence and livelihood. For example, some Inupiaq people find hunting more dangerous and more difficult with thinning sheet ice in the winter (Tenenbaum 2005). In our research we found that traditional water sources are threatened by increasing and northward-moving beaver populations. Local experts report a previously unknown problem with beetles and other insects infesting and killing trees on the Seward Peninsula.

Figure 11.1: A hunter overlooking the frozen Norton Sound as it breaks up in the springtime. © Elizabeth Marino.

To date, the most significant and widely publicized impact in Alaska due, in part, to global climate change is the erosion of coastal area Inuit villages (GAO 2003), most notably the village of Shishmaref on Sarichef Island in the Chukchi Sea. Shishmaref was one of our case studies. There, the last thirty years of dramatic erosion (due in part to later freezing of the ocean ice) has triggered debate among local residents and federal and state governments about relocating the entire community. Failure to do so could lead to sudden flooding of the contemporary village and potentially loss of life.

INVASION OF THE PHOTO SNATCHERS

Arguably, indigenous peoples of the Arctic are some of the most severely affected populations worldwide of global climate change. With the rise in public discourse of climate change and the overwhelming desire to document the phenomenon, rural Alaska has been inundated with journalists, photographers, scientists, and politicians over the last twenty years in unprecedented numbers. These travelers to the north all seem eager to engage in a discussion and, even better, get a photo opportunity with the first victims of climate change.

During our four years of fieldwork in rural, primarily Inupiaq villages on the Seward Peninsula, the vast majority of summer visitors were either scientists or journalists studying or reporting on global climate change. While we were making dinner with Shishmaref resident Clifford Weyiouanna—who has already been in a Canadian documentary about climate change, and quoted and photographed for *People* and *Time* magazine—*two* television crews doing a story on climate change, one from Japan and one from Colorado, simultaneously filmed in his kitchen. Rita Buck of White Mountain has made the front page of *USA Today* discussing picking berries. A bed and breakfast owner in White Mountain has nearly exclusively relied on climate change scientists to fill the beds in the summer and keep her business going. Clearly the "discourse of global climate change" has affected indigenous Alaska.

These travelers to the Arctic often share their different views on climate change with local residents: what it is, what it is affecting, what it will affect in the future. Similarly, local residents are informed of a discourse on global climate change by the many forms of media they access. There exists a sharp contrast, however, between this global discourse, imported through individual actors and media outlets, and local knowledge based on daily observations of the environment.

THE POWER OF WORDS

What we wish to demonstrate are the two distinct discourses that arose during our field work. Both were seemingly about a changing climate, and yet the discourses were very different. In the summer of 2006 we conducted a number of formal and informal interviews. In some of the interviews we asked specifically about "climate change" or "global warming"; in others we intentionally asked only about change and local environment. In general, when asked about "climate change" or "global warming," consultants gave summaries about what they had heard from a scientist in the village or from media outlets.

A typical example from interviews follows:

> Whenever I hear about global warming it's that there is a layer of dust up there. When the sunlight hits the earth it can't escape and so it affects the weather conditions. . . . I try to consider that all the volcanic activity has something to do with it too. . . . I also considered it when I heard aerosol sprays

> were one of the causes of global warming. (J. B., personal communication with author, 2005)

Another woman, when asked to explain what global warming was, how it affected her, and who is responsible, said:

> I think it might be, they said it was our satellite field that has made a hole in our ozone layer and I'm pretty sure that it might be the government. As far as I understood that we have a satellite field that has a hole through the ozone layer and they turn on full satellite strength every thirteen or fourteen years to drain out some of the air pollution. But I shouldn't really be talking about that either. Or maybe if everyone could cut down on carbon monoxide. I shouldn't really be talking about that part. That's just something I read. (B. B., personal communication with author, 2005)

Notice how our consultant says that she should not be talking about this because it was "just something [she] read." This statement demonstrates how "climate change" or "global warming" as a general theory is not something she has experienced personally, but is a world of knowledge that exits outside of the local. Hesitancy to discuss something outside of personal experience is not unique to Inupiaq society, but it is a cultural norm within many Inupiaq groups. Anthropologists working among Inuit cultures throughout the circumpolar North (Briggs 1991; Morrow 1990; Omura 2005) note that often the goal of knowledge in these areas is the understanding of something personal and specific. This stands in contrast to scientific analysis, which ultimately seeks generalizations. While working with Canadian Inuit, Omura noted that generalization was culturally invalid and considered "childish":

> The ideal personality is someone who does not easily generalize about phenomena nor reduce complex phenomena into a simple principle without regard for the detailed context, but is sensitive to and gives careful consideration to the subtle details and contexts of phenomena in order to cope with them. (Omura 2005, 329)

If science has precedence and power within the discourse of global climate change to identify the goals and parameters of knowledge—including the high priority on generalization—then local knowledge becomes irrelevant. The discourse on global climate change, as it stands, is inherently bounded and limited by science and media. It exists in a theoretical framework that, for our consultants, triggered a preordained dialog with highly specific vocabulary. One, our consultant felt, she should not be talking about.

On the other hand, our conversations about change and the local environment elicited long, painstakingly detailed explanations. The same woman we mentioned above, when asked about ponds, said:

> I noticed that out in the tundra there were certain lakes that were really deep and they're all dried up now. Between here and Council. Between Dan's camp and Council there's some lakes that are absolutely dry and the only time they

> get wet is when it rains too much or in the springtime. (B. B., personal communication with author, 2005).

When we asked her to show the area where these tundra ponds were drying up, she located the exact tundra ponds on a map and showed detailed changes and drainages of the ponds.

When we asked about the reasons for a lack of fish one summer, a woman said, "Seems like it started when we took that kettle" (F. A., personal communication with author, 2005). The woman went on to explain a string of tragedies that ensued after taking an artifact from a turn-of-the-century miners' campsite upriver, including unusually rainy weather during the time the fish were running. This unexpected weather event meant the fish were unable to dry properly and the family had essentially no harvest for the summer.

Figure 11.2: A mother and daughter team unhook a dog salmon from a fishing line. © Elizabeth Marino.

By not using the words "climate change," "global warming," or the notion of climate in general, we instigated entirely different conversations with our consultants. Personal stories elucidate a changing environment on the Seward Peninsula; but local residents do not necessarily associate those stories with the global discourse of climate change.

These stories are not only rich in detail but also explain change over significant periods of time through multigenerational oral histories. "I was told

Figure 11.3: Fish drying on a rack at camp on the Fish River. Some families in White Mountain catch and dry upwards of 1,500 fish during the summer. © Elizabeth Marino.

by my grandmother that behind the washeteria, where we put the well, there was an old river there and now it's all dried up and only runs during the springtime" (B. B., personal communication with author, 2005). This is a different type of statement than a more general, "Rivers are drying up," but clearly shows long-term observational data.

Another complication in discourse arises when Inupiaq people in rural Alaska are asked to speak as "experts" of climate change—for political reasons. Wisniewski notes that when hunters or other local experts are asked to speak in the language of science about the environment, it can require breaking strict hunting taboos of talking to the future or being irresponsibly presumptive about a changing and sentient natural world. "While individuals do speak in general terms about climate change issues in order to meet certain community advocacy needs, assuming authority to speak about the environment can also put a hunter in a potentially compromising position" (Wisniewski and Marino 2007).

If we use the term *climate change* when wanting to know about change and the local environment, we can miss documenting local knowledge. How do we integrate scientific discourse with explanations that changing weather is due to a moved artifact and a consequence of "messing with" something that isn't one's own? Or the fact that an elder once explained a migrating caribou population by saying that the caribou did not like the sound wires make when they rub against one another, so they went farther away from civilization (cf. Cruikshank 2005 for similar kinds of explanation in the southern Yukon)? When anthropologists assume their studies are going to be about "climate change" we must ask ourselves, what place is there for these different perspectives?

The Discourse of Climate Change

Foucault has pointed out that systems of knowledge are essentially systems of discourse. "Discursive practices are characterized by the delimitation of a field of objects, the definition of a legitimate perspective for the agent of knowledge, and the fixing of norms for the elaboration of concepts and theories" (Foucault 1977).

We argue that the global discourse on climate change is bounded and limited; with a predetermined field of knowledge, agents of knowledge, norms of discourse, and acceptable concepts and theories. Its field of objects does not include stolen artifacts; its agents of knowledge do not include deceased Inupiaq grandmothers; the norms for theoretical explanation do not include personal experiences of traveling through specific river drainages. However, one cannot truly access local Inupiaq knowledge without taking these things into consideration. And so we reach an impasse.

We acknowledge that global climate change is occurring, with particularly visible effects in high northern latitudes. We also acknowledge that local input can greatly enhance Western scientific understandings of these changes. Anthropologists, in particular, however, must take great caution in how we develop research projects, how we frame anthropological questions, and how we elicit responses from local consultants.

There are places and occasions where local Inupiaq discourse about change and the scientific discourse about global climate change are converging, including the Inuit Circumpolar Council mentioned by Crate and Nuttall in the introduction to this volume. There is a danger, however, that much anthropological work in climate change will be reduced to a series of anecdotes about disenfranchised groups of people in the world. While politically moving, these anecdotes ignore the complex, sophisticated, and at times fundamentally un-"scientific" ways of organizing the world.

It has been an enduring legacy of anthropology to document and accept a multiplicity of worldviews and divergent explanations for the same phenomena. Perhaps anthropology's best contribution to understanding global climate change will come when, as anthropologists, we stop talking about it.

References

Briggs, J. L. 1991. Expecting the unexpected: Canadian Inuit training for an experimental lifestyle. *Ethos* 19(3): 259–87.

Burch, E. S. Jr. 2006. *Social life in Northwest Alaska: the structure of Inupiaq Eskimo nations*. Fairbanks: University of Alaska Press.

Cruikshank, J. 2005. *Do Glaciers Listen? Local Knowledge, Colonial Encounters, and Social Imagination*. Vancouver, UBC Press.

Foucault, M. 1972. *The archeology of knowledge: and the discourse on language*. Translated by A. M. Sheridan Smith. New York, Pantheon Books.

———. 1977. History of systems of thought. In *Language, counter-memory, practice: Selected essays and interviews by Michel Foucault*, ed. D. F. Bouchard, 199–204. Ithaca, NY: Cornell University.

Lorenzoni, I. and N. Pidgeon. 2006. Public views on climate change: European and USA perspectives. *Climate Change* 77: 73–95.

Marino, E. 2005. *Negotiating the languages of landscape: Place naming and language shift in an Inupiaq community*. Unpublished Master's thesis. University of Alaska Fairbanks.

McBean, G., G. Alekseev, D. Chen, E. Førland, J. Fyfe, P. Y. Groisman, R. King, H. Melling, R. Vose, and P. H. Whitfield. 2005. Arctic climate: Past and present in *ACIA, Arctic Climate Impact Assessment: scientific report*, 21–60. Cambridge: Cambridge University Press.

Morrow, P. 1990. Symbolic actions, indirect expressions: Limits to interpretations of Yupik society. *Etudes/Inuit/Studies* 14(1–2): 141–58.

Nelson, E. W. 1899. *The Eskimo about the Bering Strait*. Washington, DC: Smithsonian Institution Press.

Omura, K. 2005. Science against modern science: The socio-political construction of otherness in Inuit TEK (traditional ecological knowledge). In *Indigenous use and management of marine resources*, eds. N. Kishigami and J. Savelle, 323–44. Osaka, Japan: National Museum of Ethnology.

Ray, D. J. 1975. *The Eskimos of the Bering Strait*. Seattle: University of Washington Press.

Sagarin, R. and F. Micheli. 2001. Climate change in nontraditional data sets. *Science, New Series* 294(5543): 811.

Tenenbaum, D. J. 2005. Global warming: Arctic climate: The heat is on. *Environmental Health Perspectives* 113(2): A91.

United States General Accounting Office (GAO). 2003. *Alaska Native villages: most are affected by flooding and erosion, but few qualify for federal assistance*. GAO-04–142 report to Congressional committees. Washington, DC.

Wisniewski, J. and E. Marino. 2007. Being as knowing: knowledge as experience in a sentient world. Paper presented at the annual meeting of the Alaska Anthropological Association, March 14–17 in Fairbanks, Alaska.

Woolgar, S. 1986. On the alleged distinction between discourse and Praxis. *Social Studies of Science* 16(2): 309–17.

Chapter 12

Opal Waters, Rising Seas: How Sociocultural Inequality Reduces Resilience to Climate Change among Indigenous Australians

Donna Green

Introduction

When continental ice sheets melted about 15,000 years ago, rising seas inundated large regions of northern Australia, flooding valleys and creating vast, low-lying wetland areas. More recent transitions in some of these regions occurred as gradual siltation and levée formation reduced the movement of salt water to create freshwater swamps that were regularly filled by monsoon rains, a process thought to have occurred in Kakadu and Arnhem Land from about 4,000 to 1,500 years ago (Chappell 1988).

Archaeological records show how Australia's indigenous communities successfully adapted to a range of shifting landscapes. From the stone country of Kimberley, the sandstone escarpments of Arnhem Land, the rainforests of the Daintree, and the sandy palm-fringed islands of the Torres Strait, Aborigines and Islanders have effectively adapted to environmental change for thousands of years (Barham 1999; Roberts, Jones, and Smith 1990).

Indigenous Resilience to Climate Change

Given indigenous Australians' past ability to respond to environmental change, it is reasonable to assume that they would be some of the best placed among all Australians to cope with environmental impacts caused by anthropogenic climate change. In fact, the opposite is true due to at least two major factors. The first relates to the rate of environmental change. Projections of anthropogenic climate change indicate appreciable direct biophysical (and consequent secondary) impacts occurring over a timescale of decades. In contrast, prior environmental change—albeit of greater magnitude—occurred over millennia (Press et al. 1995). The recent climate projections suggest warming in central Australia of up to 6 degree celsius by 2070, with an increase in the absolute number of hot spells[1] to forty-three a year. Projections of precipitation change in the north indicate increasing extreme events:

monsoonal rain projected to increase by 23 percent, and the loss of nearly all precipitation in the dry season for the most affected regions (Green 2007).

The second factor relates to social and cultural resilience. Many of these communities are fighting a number of devastating social problems, the result of decades of profound government mismanagement and neglect (Arthur and Morphy 2005). The indicators are stark: indigenous Australians' life expectancy is twenty years lower than their nonindigenous counterparts (Ring and Brown 2002). Rates of suicide, diabetes, and other basic and treatable diseases are heightened and a daily reality for many outstation communities (McMichael 2006). Such widespread social ills have their roots in indigenous Australians' forced dispossession from their country and the past active suppression of their cultural practices (Rose 1996).

Traditional Knowledge and Environmental Justice

Over the past decade, I have focused my research on revealing the social and cultural dimensions of what many consider "one-dimensional environmental science" problems. This approach was guided by my academic background in natural and social sciences and several years working in the US with the emerging methods of community engagement. The environmental justice movement—which aims to expose the social and political dimensions of environmental pollution and directly involve those affected—has had a profound effect on my research in Australia.

In 2005, I returned from teaching in the US to explore whether and by whom participatory approaches to mitigate environmental problems, especially climate impacts, were being used in Australia. I found few researchers using an environmental justice framework, which may have helped to identify vulnerable social groups.

This concerned me for two reasons in relation to climate impact assessments. First, without explicit attention to the varying levels of resilience and coping ability among different cultures, it was unlikely that an assessment would identify the most vulnerable social groups. Secondly, the scientists developing the climate change projections were often unable to translate and communicate their findings to these people. Indeed, there was little recognition by government agencies that indigenous Australians might even be interested to know how climate change might impact their land, and therefore, their lives (AGO 2007).

Consequently, I chose to focus my initial research on how much climate change would affect remote indigenous communities, and what they wanted to do about it. The success of many Native Title land claims in the north of Australia suggested that it was likely communities that had maintained the strongest links to land would have the most intact traditional environmental knowledge (TEK).[2] This was important because I wanted to explore whether TEK passed down through oral histories could guide region-specific and culturally appropriate adaptation strategies. Additionally, I thought that it might be able to provide useful historical environmental observations for climate scientists—if the holders of the knowledge were willing to share it with them.

The first step was to understand how Traditional Owners[3] felt about the likely impacts of climate change, and to discover if they considered them a problem worth dealing with. In collaboration with colleagues at CSIRO,[4] I organized a workshop that, for the first time, brought together thirty elders and Traditional Owners from across the north with thirty researchers and scientists. This group included representatives from the most northwesterly Torres Strait Islands, Traditional Owners from the Kimberley in Western Australia, and Yolngu people from northeast Arnhem Land. The workshop aimed to identify and value both Western scientific and TEK forms of expert knowledge about environmental change in the north, and to acknowledge that both would be vital to design climate adaptation strategies that took Traditional Owners' priorities into account.

The workshop began with apprehension among the participants, due partly to the fact that many of these Traditional Owners would not normally meet and work together. It was also clear that for most of the elders, climate change was not an issue they had previously seen as a pressing challenge for their communities, given the more urgent problems confronting them such as chronic alcohol abuse and domestic violence.

However, as the scientists started to explain the latest temperature and sea-level rise projections, it became clear that climate change was likely to exacerbate the Traditional Owners' existing social and cultural problems, such as the potential for further dispossession from their land. Their initial apprehension and doubts about the importance of climate change were superseded by a unanimous demand for information to enable them to better adapt. Specifically, they requested localized "on country" workshops to allow further community discussions about likely climate impacts, and for TEK about weather and climate to be recorded.

The assembled group even agreed to prioritize urgent action in the Torres Strait, a decision influenced by a photo presentation made by one of the Islanders. The photos—taken just a month before the workshop during record king tides—showed Islanders wading down streets knee deep in water, damaged coastal infrastructure, and flooded rubbish tips (Green 2007). Although not suggesting a direct causal link between the record tides and climate change, the photos were a powerful reminder of how vulnerable these communities were and what damage extreme weather events (such as those likely to be more frequent due to climate change) could cause on the islands.

Climate Change Impacts on the Torres Strait Islands

People have lived for centuries on islands in the Torres Strait, lying in the shallow waters between mainland Australia and Papua New Guinea, close to the northern tip of the Great Barrier Reef (Figure 12.1).[5] The Islanders are a

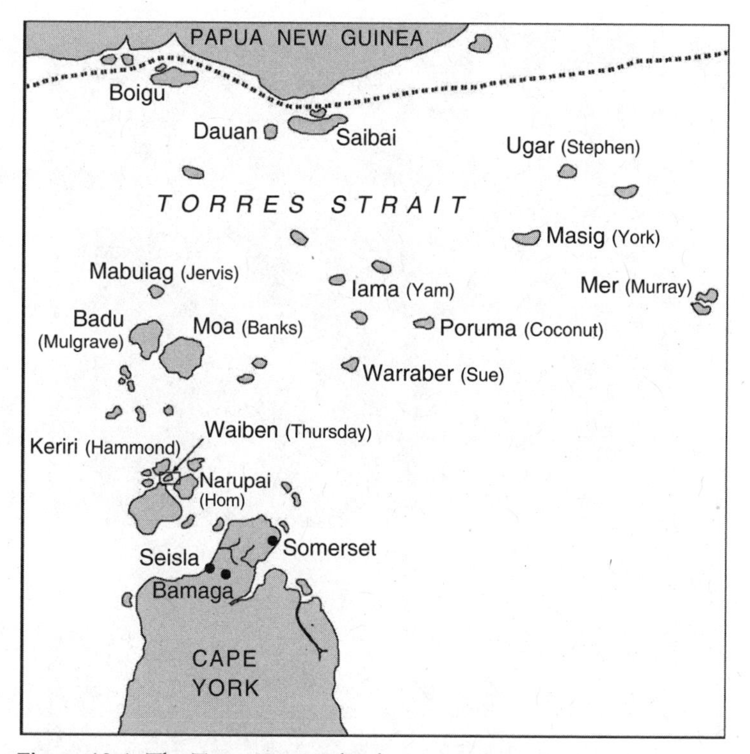

Figure 12.1: The Torres Strait Islands.

traditional seafaring people, who pride themselves on their intimate understanding of the seasonal shifts in the ocean and weather. Events such as the timing of the king tides are predictable for the Islanders. However, they had noticed that in recent years the waves occurring in these king tides seemed higher and more powerful (J. Warusam, personal communication with author, 2007). Consequently, on several of the islands, coastal tracks were being washed away and long-established graveyards and houses inundated. In addition to the psychological distress caused by the flooding, their remoteness makes repairing this damage extremely expensive, and Islanders lack access to the necessary resources to engage consultants to conduct assessments or to actually carry out maintenance work on the basic infrastructure.

The Islanders understand that the problem extends further than the initial flooding. They are concerned about indirect impacts of climate change, seeing how, for example, inundations could jeopardize public health caused by contamination of freshwater supplies or from the flooding of their landfill rubbish tips (Figures 12.2 and 12.3). A full suite of indirect impacts are harder to assess and quantify; however, they are crucial to consider in designing comprehensive adaptation strategies. During the workshop Islanders expressed their concerns that their TEK was no longer as reliable as it had once been in

Figure 12.2: Saibai village after inundation in 2006. © Rick Parmenter.

Figure 12.3: Damage to the sea wall on Saibai caused by storm surge in 2006. © Donna Green.

their living memory. They reported shifts in animal and plant behavior that did not accord with their past experience. These TEK observations showed that Islanders were acutely aware of changing temperatures and rainfall patterns, of shifting bird migrations and breeding seasons, and of changing abundance and distribution of particular species. For instance, a new species of mosquito had appeared on some islands, while perceived changes in marine habitats were thought to be disrupting Islanders' traditional subsistence hunting patterns. Given the profound cultural importance of totemic sea animals—such as turtles and dugong—for many Islanders, this issue takes on particular significance (DEH 2005; Marine and Coastal Committee 2005; Sharp 1993; Sutherland 1996).

Ailan Kastom, "Islander culture," refers to a distinctive Torres Strait Islander way of life, incorporating traditional elements of Islander belief systems combined with Christianity (Sharp 1993).[6] Ailan Kastom permeates all aspects of island life: it governs how Islanders take responsibility for and manage their land and sea country, how and by whom natural resources are harvested, and allocation of seasonal and age-specific restrictions on catching particular species (Johnson 1984; Mulrennan and Scott 1999). But at the heart of Ailan Kastom is the connection between the people and their land.[7]

Some of the key threats to the islands from climate change impacts have been initially identified in a "natural hazards" report commissioned by the state government level Island Coordinating Council (ARUP 2006). This report shows how some important cultural heritage sites—such as graveyards, monuments, and sacred sites—are vulnerable to storm surges on several of the islands (Figures 12.4 and 12.5). Although climate change impacts may only incrementally increase the total impact of a storm surge, such as through sea-level rise or increased cyclone intensity, given the extreme vulnerability of these sites any additional factor affecting the area inundated could have very serious consequences.

Figure 12.4: Sandbagged graves on Saibai. © Andrew Meares.

Figure 12.5: Flooded street on Masig. © Walter Mackie.

Crucial surface and ground water resources are also likely to be impacted by climate change, making resource management in the dry season increasingly difficult. In the past, many islands depended on fresh water lenses to provide drinking water, but overexploitation of this resource has caused problems and created the need for water desalination plants on many of the islands (Mulrennan 1992). Rainwater tanks and large lined dams are now used to trap and store water for use in the dry season. An increase in saltwater intrusion of fresh water supplies and reduced availability of fresh water is likely to add to difficulties of Islanders attempting to revive traditional gardening practices. Reduction in these practices has compounded health problems in recent years because of the lack of availability of affordable, fresh vegetables on the islands (Beadle 2005).

Public Recognition and Local Adaptation

Public awareness about climate change in Australia has grown enormously in the past few years, with most people now recognizing the threat it poses to iconic natural ecosystems such as the Great Barrier Reef and Kakadu. Yet relatively few Australians realize that the indigenous people who have lived in these regions for thousands of years are also threatened. This lack of public awareness is due to a complex combination of factors, one of which is the lack of information from remote communities. However, isolation appears not to be a limiting factor by itself, given the number of stories the mainstream media has carried about the need for Pacific Islanders to relocate.

In mid-2006 the major metropolitan newspapers began to report on the plight of several low-lying Torres Strait Islands (Michael 2007; Minchin 2006).[8] Shortly after, the threat was officially acknowledged by the Australian government, which signed off on the IPCC's Fourth Assessment Report.

For the first time, this report acknowledged the likelihood that around half of the 4,000 people living on the Islands would have to be relocated in the long-term (Hennessy et al. 2007).[9]

Understandably, the Islanders see relocation as an action of last resort, and are already working on adaptation strategies to delay, and ideally avoid, having to leave their ancestral homelands.[10] For example, on low-lying islands, such as Boigu and Saibai, all new houses are being built on stilts and nearly all the others have been raised so that they will not be affected by flooding. The Mer Island Council is negotiating to subdivide land further back from the shoreline and higher up to provide areas safer for new housing. Emergency management plans are being drawn up in consultation with government agencies, while on some islands, where the resources are available, sea walls are being reinforced and extended.[11]

Many of these local adaptation strategies are detailed on the *Sharing Knowledge* website, which is now run out of the Climate Change Research Centre at the University of New SouthWales.[12] The website provides climate change projections for most regions in northern Australia where remote indigenous communities live. It provides summaries of likely climate impacts in plain English, as well as in local languages.[13] Equally importantly, it serves as a clearinghouse for indigenous TEK on weather and climate. Additionally, through fulfilling requests for presentations to a range of local groups across the north (from school children to boards of management of national parks) the project strives to communicate locally relevant impact assessments to a range of audiences that otherwise have little or no access to such tailored information.

My experience in the Torres Strait, and increasingly in other parts of Australia, has shown that when indigenous people have access to information about how climate change is affecting their lives, often in tandem with seeing changes to their lands, they act. Funding to perform a limited number of climate impact workshops and begin TEK recording in the Torres Strait was made available by the federal government in mid-2007, soon after the IPCC report was released. But to date, no funding has been forthcoming to replicate this work in other regions of mainland Australia.

Acknowledgments

Funding for some of this research was kindly provided by the John D. and Catherine T. MacArthur Foundation, with helpful editorial comments provided by Liz Minchin.

Notes

1. A spell is a three-day period where the temperature is consistently over 35 degrees C.
2. Native title is the recognition in Australian law that some indigenous people continue to hold rights to their lands and waters, which come from their traditional laws and customs. Native title exists as a bundle of rights and interests in relation to land

and waters where the following conditions are met: the rights and interests are possessed under the traditional laws currently acknowledged and the traditional customs currently observed by the relevant indigenous people; those indigenous people have a "connection" with the area in question by those traditional laws and customs; and the rights and interests are recognised by the common law of Australia. Native title has its source in the body of law and custom acknowledged and observed by the claimant's ancestors when Australia was colonized by Europeans. Those laws and customs must have been acknowledged and observed in a "substantially uninterrupted" way from the time of settlement until now. For further discussion see the National Native Title Tribunal at http://www.nntt.gov.au.

3. A Traditional Owner is defined in accordance with aboriginal and Islander law and tradition. He or she has social, economic, and spiritual affiliations with, and responsibilities for, a specific region of land and or sea.
4. The Commonwealth Scientific and Industrial Research Organisation. The workshop was funded in partnership with the University of Melbourne.
5. Torres Strait Islanders are the smaller of the two recognized groups of indigenous Australians. The larger group are aboriginal Australians.
6. The London Missionary Society entered the Torres Strait in 1871.
7. As noted by Beckett (1987), "The predicament of Islanders on the mainland is that if their society can survive at all, it is only through the conscious perpetuation of island custom and the continual monitoring of its practice. The Strait does not have to worry about custom; the society of Islanders there remains axiomatic as long as they are in occupation of their ancestral islands."
8. An online multimedia presentation (Meares 2006) launched simultaneously with the newspaper stories was also well received, with international requests for its use including that it be shown at a side event on the impacts of climate change on indigenous people during the Sub-commission on Human Rights at the United Nations.
9. This report identified that the communities on the central coral cays and northwest islands communities most at risk from climate impacts.
10. The Torres Strait Islands have a particular significance for the protection of indigenous Australian culture and land rights more widely. In 1992, the High Court decided in favor of the traditional owners of Mer (or Murray) Island in what is known as the "Mabo case."
11. Individual Islanders are also doing their bit to reduce their contribution to the problem, with some Islanders beginning to install solar power systems and to encourage others not to waste electricity.
12. The SK website provides regional climate impact assessments and local TEK initially presented at the first SK workshop.
13. These are provided as requested—most frequently by local NGOs who are working with Elders and Traditional Owners on country.

REFERENCES

Australian Greenhouse Office. 2007. Projected impacts. www.greenhouse.gov.au/impacts/regions/te.html

Arthur, B. and F. Morphy. 2005. *Macquarie atlas of indigenous Australia: Culture and society through space and time*. Sydney: Macquarie Press.

ARUP. 2006. *Natural disaster risk management study*. Brisbane: ARUP.

Barham, A. 1999. The local environmental impact of prehistoric populations on Saibai Island, northern Torres Strait, Australia: Enigmatic evidence from Holocene swamp lithostratigraphic records. *Quaternary International* 59: 71–105.

Beadle, R. 2005. What's for dinner on Thursday? The impacts of supermarkets in the Torres Strait Islands. Report No. 23, School of Anthropology, Geography and Environmental Studies. Melbourne: University of Melbourne.

Beckett, J. Jeremy. 1987. *Torres Strait islanders: Custom and colonialism*. Sydney: Cambridge University Press.

Chappell, J. 1988. Geomorphic dynamics and evolution of tidal river and floodplain systems in northern Australia. In *Northern Australia: Progress and prospects*, vol. 2., eds. D. Wade-Marshall, and P. Loveday. Darwin: Northern Australia Research Unit.

Department of Environment and Heritage. 2005. *Sustainable harvest of marine turtles and dugongs in Australia—A national partnership approach*. Canberra: Commonwealth Government.

Green, D. 2007. Sharing knowledge. www.sharingknowledge.net.au

Hennessy, K., B. Fitzharris, B. Bates, N. Harvey, M. Howden, L. Hughes, J. Salinger, and R. Warrick. 2007. Australia and New Zealand. Climate change 2007: Impacts, adaptation and vulnerability. In *Contribution of Working Group II to the Fourth Assessment: Report of the Intergovernmental Panel on Climate Change,* eds. M. Parry, O. Canziani, J. Palutikof, C. Hanson, and P. van der Linden. Cambridge: Cambridge University Press.

Johnson, T. 1984. Marine conservation in relation to traditional lifestyles of tropical artisanal fishermen. *The Environmentalist* 4.

Marine and Coastal Committee. 2005. *Sustainable harvest of marine turtles and dugongs in Australia—A national partnership approach, Natural Resource Management Ministerial Council, Marine and Coastal Committee*. Canberra: Commonwealth Government.

McMichael, A. 2006. Climate change and risks to health in remote indigenous communities. Paper presented at the Sharing Knowledge workshop, March 30–31, 2006, Darwin, Australia.

Meares, A. 2006. Opal waters, rising seas. www.smh.com.au/multimedia/torres/index.html

Michael, P. 2007. Rising seas threat to Torres Strait islands. *Courier Mail,* 2 August, Brisbane.

Minchin, L. 2006. Going under. *Sydney Morning Herald*, 12 August, Sydney.

Mulrennan, M. 1992. *Coastal management: Challenges and changes in the Torres Strait Islands*. Darwin: Northern Australia Research Unit.

Mulrennan, M. and C. Scott. 1999. Land and sea tenure at Erub, Torres Strait: Property, sovereignty and the adjudication of cultural continuity. *Oceania* 70.

Press, T., D. Lea, A. Webb, and A. Graham. 1995. *Kakadu. Natural and cultural heritage and management*. Darwin: Northern Australia Research Unit.

Ring, I. and N. Brown. 2002. Indigenous health: Chronically inadequate responses to damning statistics. *Medical Journal of Australia* 177: 629–31.

Roberts, R., R. Jones, and M. Smith. 1990. Thermoluminescence dating of a 50,000-year-old human occupation site in northern Australia. *Nature* 345: 153–56.

Rose, D. B. 1996. *Nourishing terrains: Australian aboriginal views of landscape and wilderness*. Canberra: Australian Heritage Commission.

Sharp, N. 1993. *Stars of Tagai: The Torres Strait Islanders*. Canberra: Aboriginal Studies Press.

Sutherland, J. 1996. *Fisheries, aquaculture and aboriginal and Torres Strait Islander peoples: studies, policies and legislation*. Canberra: Environment Australia.

Chapter 13

The Glaciers of the Andes are Melting: Indigenous and Anthropological Knowledge Merge in Restoring Water Resources

Inge Bolin

The indigenous Quechua people of the high Peruvian Andes are worried as they look at their mountain peaks. Never in their lifetimes have they witnessed environmental changes of such drastic dimensions. One village elder expressed his concern by telling me: "Our *Apus* (sacred mountain deities) have always had sparkling white ponchos. Now some of their ponchos have brown stripes. Other peaks have shed their ponchos altogether" (Bolin 2001, 25). His feelings resonate throughout the hills and valleys where one often hears people say, "When all the snow is gone from the mountain tops, the end of the world as we know it is near, because there is no life without water" (Bolin 2003).

I first encountered the problem of melting glaciers in 1984–85 when I researched the organization of irrigation along the Vilcanota/Urubamba valleys (Bolin 1987, 1990, 1992, 1994). At that time Peruvian geologist Dr. Carlos Kalofatovich told me that the Chicon glacier above the Urubamba Valley had receded sixty meters in fifty years (personal communication). During the next two decades I continued to observe how glaciers melted in this and other adjacent regions. This process became much more visible starting in the mid-1990s, at which time my research and applied work among high-altitude pastoralists was focused on ritual activities (Bolin 1998), environmental issues (Bolin 1999, 2002; Bolin and Bolin 2006) and child rearing (Bolin 2006). Starting in 2004 when glacial retreat and water shortages had reached serious proportions, I shifted my research focus more directly to the problems of climate change, concentrating on melting glaciers, water shortages, and solutions that could improve the chances for the survival of the indigenous peoples and their cultures.

In this chapter I discuss glacial retreat in the high Andes with focus on the provinces of Quispicanchis and Urubamba in the Cusco region of southeast Peru (Figure 13.1). The people living in these areas deal with a rainy season that lasts from roughly October to the end of March, and a dry season from April to the end of September. During the dry season almost all the water that people and animals use throughout the Andes is derived from

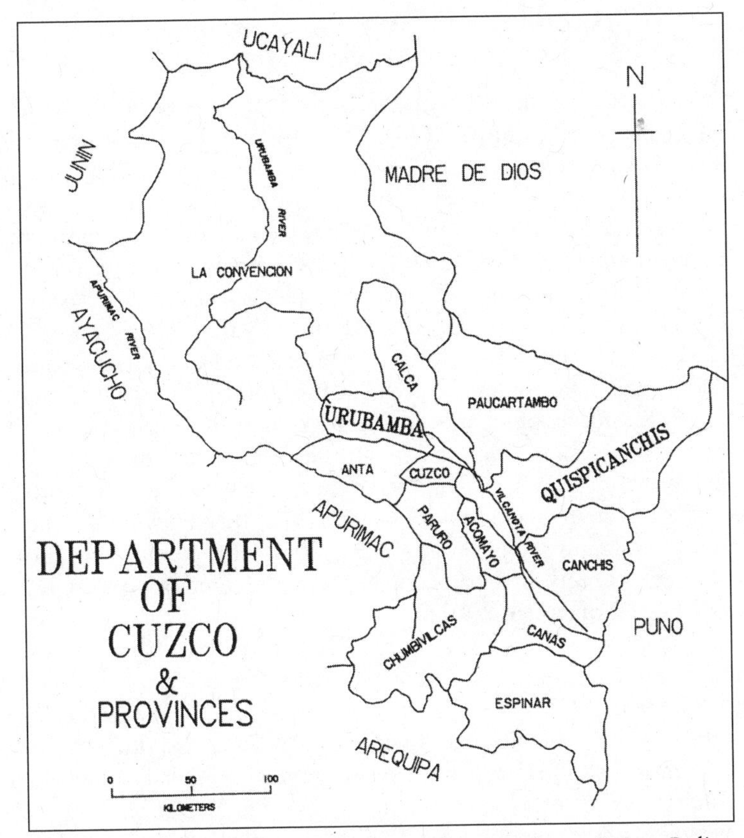

Figure 13.1: Map of the Cuzco region and its provinces. © Inge Bolin.

the glaciers in the mountains' high peaks. The indigenous people who reside along the hillsides of these provinces, between 3,000 and 5,000 meters above sea level, live primarily from working the land and pasturing their animals. I will describe how the melting process affects the natural environment, the livelihood of agriculturists and high-altitude pastoralists, and the impact it has on their culture and religious beliefs. The discussion will center on my role as a collaborative researcher and mediator between the local and the global by describing my interactions with the indigenous people as climate change intensified and we were forced to seek ways to mitigate the impending crisis caused by glacial retreat and water shortage. Together we started to consider local adaptive strategies and globally devised methods to preserve, capture, recycle, purify, and distribute water, and adjust irrigation and agricultural practices in ways that may allow the Andean people to remain in their homeland instead of migrating to overcrowded cities. As a collaborative researcher I have helped with such projects in the past; some are being implemented at the present time as discussed below; others will follow as funding becomes available.

Changes in Andean Glacial Topography and its Repercussions

What can science tell us about the changes that have occurred within the Andean glacial topography? At 7,250 kilometers in length, the Andean Cordillera is the longest mountain chain in the world. Within it, Peru's glaciers alone account for 70 percent of the tropical mountain glaciers of the planet (González 2003). Given their tropical latitudes, these glaciers are very close to the melting point and are therefore extremely sensitive to changes in the earth's temperature. Since climate change is greatest at high altitudes, it makes them prime indicators of global warming (Vásquez Ruesta et al. 2002). In the last twenty years the ice of the Peruvian Andes has been reduced by 20 percent, and this process is accelerating (González 2003). Renowned glaciologist Lonnie Thompson found that the Qori Kalis glacier, the largest glacier of the Quelccaya ice cap in the southern Andes mountains, which had been retreating an average of six meters per year between 1963 and 1978, has since retreated on average sixty meters a year (personal communication, 2007; see also Thompson et al. 2006). In a 2007 CBC News report, he and a team of scientists relayed evidence that Qori Kalis could be gone in five years (CBC News, February 16, 2007).

The Quelccaya ice cap, the world's largest tropical ice mass, covers 44 square kilometers and is located about 125 miles north-northeast of Lake Titicaca (Bowen 2005, 166). Thompson warns that it is the unprecedented rate of ice loss that concerns him most in the Andes and in other parts of the world where he and his crew have been working. Other leading scientists of the Intergovernmental Panel on Climate Change (IPCC) reported in 2007 that "warming of the climate system is unequivocal given increases in average air and ocean temperatures, widespread melting of snow and ice and rising sea levels" (CTV News, March 26, 2007).

The smaller glaciers that make up 80 percent of the glaciers within the Andean Cordillera may vanish in ten to fifteen years as predicted by Francou (2001) and Francou et al. (2003). But many of these small glaciers have already melted or are melting at a much faster rate than ever predicted. The forecast for rainfall, which is hoped to make up for at least part of the water loss from glaciers, is equally alarming. According to the fourth assessment report of the IPCC, the annual precipitation is likely to decrease in the southern Andes (Matthews 2007). More scientists now join indigenous peoples in their concern about glacial retreat and its consequences. During November and December of 2007 I discussed the issue of melting glaciers and water scarcity in the Andes with directors and scientists of environmental agencies in Cuzco, Peru, among them IMA (Instituto del Manejo de Agua), CONAM (Consejo Nacional del Ambiente), ANDES (Asociación para la Naturaleza y el Desarrollo Sostenible), Plan MERISS (International Irrigation), and Ayllus Ecológicos del Cusco. They unanimously agree that the situation is very serious and that steps must be taken at once to slow down the disastrous consequences of global warming.

Impact of Glacial Retreat on the Local Populations

Since virtually all the water available to Andean peasant farmers and pastoralists in the dry season comes from the snow and ice fields of their high mountains, the repercussions of melting glaciers are immense for local communities. Melting glaciers may provide added water in the short run, but they also cause rock falls, landslides, and floods. As snow masses diminish, however, mountain lakes and creeks shrink or disappear, and rivers no longer receive enough water from the glaciers to irrigate fields and meadows, which require 70 percent of the water supply, or to generate hydroelectric power. Water scarcity, combined with extreme weather conditions, result in bad or lost harvests. In addition, increasingly higher temperatures require that tender new plants be irrigated more frequently, though not enough irrigation water is available. To make matters worse, multinationals are building luxury hotels that use much of the precious water in the province of Urubamba, while international mining companies destroy the glaciers of many sacred mountains in the Andes and elsewhere to extract minerals, thereby poisoning water and land. The smaller mountain glaciers above the Urubamba Valley, which had receded slightly more than one meter in 1985, are now receding twelve meters each year (Tupayachi Herrera, personal communication, 2007); the last ones will soon be gone.

The unprecedented melting of the Andean glaciers is also posing serious drinking water problems for local inhabitants. Drinking water becomes scarce where springs dry up, and lower water levels in lakes and rivers are causing disease vectors from animal feces to increase. This requires that water be thoroughly boiled, but firewood is scarce. As a result, many people drink raw water, putting themselves at risk of contracting gastrointestinal ailments. These, in turn, require natural medicines that must be derived from plants that are increasingly scarce due to the uncertain water regime.

Furthermore, weather patterns in the Andes now tend to reverse, a trend also found in other climate-sensitive areas of the world. The absence of rain during part of the rainy season has interfered with the growing of food plants, while rains during the dry season have barely allowed for the freeze-drying of potatoes, the staple of the herders' diet. Hunger, combined with extreme temperatures, has caused much sickness, and led to new diseases (e.g., *Verruga peruana*) and new pests. Until a few years ago potatoes and other high-altitude tuber crops (*oqa*, *ulluku*, and *maswa*) were free from pests. Recently, however, some high-altitude communities have been forced to spray their potato crop, which has caused financial hardships for most families.

Views Expressed by Indigenous Andean People

Inhabitants of the Andean Cordilleras have been concerned about water scarcity for a long time. Myths and legends tell about courageous young people who dared to face severe obstacles to bring much-needed water from snowfields and high mountain lakes to villages in the valley. Water was sacred to pre-Columbian civilizations as it is to many indigenous people today. Pre-Columbian religions and the beliefs of today's indigenous societies have been based on the benevolence of Mother Earth, and the sacred Apus, those mountains whose snow and ice fields provide the life-giving waters, and on the mountain lakes that retain it. As the snowfields melt due to global climate change, these deities lose their powers. Eventually Andean religion may erode and these legends will become meaningless. Some indigenous people have wondered what they have done wrong to deserve the wrath of the gods who began to restrict the water that flows from their mountainsides. Although elders are often aware of the effects of El Niño that can cause havoc in the weather patterns, few know of the problems underlying global climate change and of those responsible for causing such a devastating process. Yet, the local knowledge of the Quechua people of the high Andes is invaluable to their survival. The slightest changes in the environment tell them when something goes wrong. Thus, for example, the people living along the hillsides above the Vilcanota and Urubamba valleys observed already in the mid-1980s (and perhaps earlier) that important medicinal plants became increasingly hard to find, and even where they persisted, their growth was stunted, usually because of water scarcity during at least part of the year. The Andean mountains that contain the most extreme range of landscape types, climates, and vegetation communities in the world are rapidly losing their biodiversity (Brack, Egg, and Noriega 2000). Since biodiversity is highest at high altitudes, indigenous knowledge in this area is paramount in our struggle to help preserve these plants (Gade 1975; Tupayachi Herrera 1997, 2005).

With the same degree of precision as with medicinal plants, the Quechua people's local knowledge tells about past weather patterns, either seen in their own lifetimes or learned through oral history. They know whether a change that is happening now has occurred before or in such an extreme form within living memory. When it comes to rainfall, for example, the Andean people refer to *veranillos,* which are dry periods that can last for three weeks and have occurred mainly during the height of the rainy season in January and February. Within recent years, however, *veranillos* have also appeared earlier. I experienced two *veranillos* within six weeks in November and December 2007. Abnormalities such as torrential rains, snow, and hail falling during the dry season have occurred increasingly within the last decade or two. These weather anomalies seriously affect the herds and crops, mainly the preservation of potatoes, the staple of high Andeans' diet. The freeze-drying process of potatoes can only take place between May and July when the days

Figure 13.2: Ukukus (bear men) bring ice, believed to be medicine, from the glacier to their villages in 1991. This age-old custom was abandoned in 2000 because of alarming glacial retreat. © Inge Bolin.

are sunny and the nights are frosty. Now this weather pattern can no longer be relied on. The rains that fall in the rainy season also tend to be stronger now, washing the potatoes out of their steep beds into the rivers. These situations have caused several years of hunger in high-altitude regions (Bolin 1999; see also Winterhalder 1994 on rainfall patterns). Also, during drought conditions high-altitude pastoralists often point to grasses that are of such weak texture that they break apart and even pulverize when grazed by llamas and alpacas. Melting glaciers and drying creeks and mountain lakes all add to the problems caused by drought.

Peruvian environmental organizations and village leaders are becoming increasingly concerned about the local and regional impacts of global climate change and in some cases have taken action. Attempts to slow down glacial retreat started several years ago along Peru's Sinakara mountain range. Here tens of thousands of pilgrims flock from high-altitude regions of Peru and Bolivia to celebrate at the sanctuary of Qoyllur Rit'i with ancient Andean and Christian rituals. Hundreds of *ukukus* who represent spiritual figures dressed in shaggy alpaca robes with masks of alpaca wool ascend to the glaciers under the full moon for initiation and other ceremonies. They leave a few drops of blood in the snow as a sacrifice to the mountain. In return, they used to chop off large chunks of ice and bring this potent medicine to the people in their villages (See Figure 13.2). Beginning in the year 2000, as the melting of glaciers became a frightening reality for many, the indigenous leaders who organized this great pilgrimage announced that the ancient custom of collecting ice from the glaciers must be abandoned. Since 2003 this law has been strictly enforced. Now each *ukuku* is allowed to fill only a tiny bottle with snow or water to bring to the valley.

Local Responses

Indigenous peoples live in close association with their land. In the high Andes the Quechua revere Pachamama or Mother Earth, the sacred mountains they call Apus, and lakes and meadows. These are omnipresent deities. Yet, in places where water became too scarce to make a living, families were forced to move to find a better environment in which to plant their crops or herd their animals. But few have been successful, relocating only to find similar issues with water or lack of land. Those who sold or abandoned their land and moved to the overcrowded cities were for the most part equally disappointed. Without extended families and *compadrazgo*, networks of fictive kinship ties, they found no support when they most needed it in an unknown environment.

Yet, Andean peasants and herders have been very resourceful throughout history in adapting to environmental changes. In order to defend themselves against the vagaries of the weather, for example, they have always used small parcels of land at different altitudes and within different microenvironments to ensure that at least part of a year's potato harvest can be saved. Now, with much greater changes in the weather pattern, with hotter summers and colder winters, with more variable precipitation, and with less or no water flowing from their mountain peaks, they contemplate growing drought-resistant species of food plants and think about methods of storing water. But the manifold effects of climate change, the activities that contribute to it, and governments that are not responding to the policy challenges are all too distant for most local people to comprehend. It is here that the role of an anthropologist or other professional as mediator between the local, national, and global levels becomes important.

Collaboration between Indigenous Peoples and Anthropologists

Just as many anthropologists have learned the strategies of survival in marginal environments from indigenous and other local peoples, indigenous peoples now need information from anthropologists about global climate change, the way it affects humanity, how future trends are detected and forecast, and new coping strategies. We all must understand, for example, that the disappearance of glaciers is not only felt locally, but also at the national and worldwide levels. Glacial retreat in the Andes causes mountain lakes and creeks to dry up, becoming unable to provide water to fill the rivers that make their way throughout the country to the dry, rainless coast or to the jungle regions. As aquifers also drop, even drinking water can become scarce during the dry season or whenever the rains do not arrive on time.

During more than two decades of research and applied work in the Andes, it became clear to me that migration to the cities or other parts of Peru is not the answer for people who want to get a better chance at survival. Andean people are attached to their land, lifeways, and religious beliefs, and it is here where efforts must begin. Since most of the world's leaders are doing little to curb climate change through implementing policies that restrict emissions from vehicles, factories, and billions of animals kept under atrocious conditions,[1] local people must become innovative and self-empowered to implement both short-term emergency projects to survive and projects that are sustainable in the long term. In most cases, indigenous inhabitants have a wealth of knowledge already available to them based on how their ancestors dealt with and adapted to weather extremes, like the scarcity of water.

In 1984–85, the villages along the hillsides above the Sacred Valley of the Incas in the province of Urubamba suffered from a serious water shortage that resulted in conflicts over the last few drops of water during the dry season. Given this emergency, the indigenous population asked for international cooperation to improve irrigation canals and reconstruct small Inca reservoirs. At various occasions the local people told me that their Inca ancestors knew the most stable regions along the mountainsides, where remnants of ancient canals and reservoirs could still be seen. Since many of these structures had for centuries been trampled on by animals, they were no longer functional.

The elders of the village of Yanahuara along the hillsides of the Sacred Valley of the Incas, where I studied the ways by which they organized their irrigation activities, approached me to assist them in writing a proposal to the international developers who were working in the Vilcanota Valley, 400 kilometers away. The elders requested that their broken ancient canals and reservoirs be repaired. Together with the local population, I wrote a proposal to get the necessary funding, which I took to the GTZ (Gesellschaft für Technische Zusammenarbeit, German International Development Corporation) in Germany who discussed the issues with their Peruvian counterparts.

The people of Yanahuara and adjacent regions rejoiced when in 1986 the international development agency Plan MERIS II (now Plan MERISS) in Cuzco, through which Peru and Germany cooperate, accepted our proposal to improve canals and reservoirs to provide enough water year round, and to also add complementary projects (e.g., a school building). Yet, within the last five years, with the glaciers along the Cordillera de Urubamba melting much faster and retreating at an average of twelve meters a year (Tupayachi Herrera, personal communication), water scarcity has again been sorely felt by the local people, especially during planting time. Within the last decade, climatic extremes here and elsewhere in the Andes have contributed to floods, catastrophic droughts, heat waves, and cold spells as never seen before.

Among other drastic events, in 1998 and 1999 harvests were destroyed by extreme weather conditions throughout large parts of the Andes (Bolin 1999). In 2005 an immense avalanche of snow and ice, estimated at about two hundred tons, tumbled from Mount Veronica, destroying everything in its path and finally obstructing the train tracks in the valley leading to Machu Picchu (Tupayachi Herrera, personal communication). The recent cold spell in May 2007 was more extreme than any previous one experienced by the Quechua people, killing some of the very old and very young. (See also Suarez 2008). Yet, as soon as this natural catastrophe was over, glacial melting continued as before.

With some peaks now free of ice and snow and others losing their glaciers at a rapid rate, major efforts are necessary to curb further destruction. Together with the volunteer organization Yachaq Runa, which I founded in 1992 in Cuzco, we have embarked on a program to help stop local environmental degradation and, hopefully, reverse it. The indigenous Quechua people along the hillsides of the Vilcanota and Urubamba Valleys have been eager to revert to Inca ways of managing the environment by planting native trees, recreating small forests on the hillsides and around their homesteads, and by planting bushes alongside irrigation canals to keep water evaporation low (see also Bolin 1987). Although Australian eucalyptus trees grow fastest and continue growing after being cut, they need much water, and their enormous roots destroy plants and buildings in their close proximity. Therefore reforestation with indigenous trees, such as Q'euña (*Polylepis incana*) and Quiswar (*Buddleja incana*), and indigenous shrubs, such as Tayanka (*Baccharis odorata*) and Chillka (*Baccharis latifolia*), is environmentally much more beneficial. These reforestation projects were started by the villagers of Chillihuani, in the province of Quispicanchis, in cooperation with the Yachaq Runa volunteer group and with international funding.[2] Reforestation in Challwaqocha and five other villages in the province of Urubamba is now underway. As soil and waterways are stabilized through reforestation, the simultaneous planting of the highly nutritious Maca tuber and other food plants is becoming more successful.

Since increasing water scarcity is already starting to affect the potential of hydroelectric plants, indigenous people in several of my study communities are happy as we cooperate to provide solar cookers, photovoltaic lights, and solar hot water to the health stations, schools, shower houses, and other facilities we help to build and equip. Yet, many more efforts are necessary to assure the survival of the Andean people should global climate change continue at the present rate. Unless precipitation patterns change to become more beneficial to agriculture and pastoralism, much more must be done to provide the amount and quality of water necessary for survival. We must consider primarily indigenous Andean knowledge—known in some areas, but forgotten in others—as waterways are restored, agricultural practices are adapted to prevailing climatic conditions, and new methods are devised to collect water and use it sparingly throughout the dry season. In cooperation with the indigenous population, and based largely on their ancestral knowledge system, we arrived at the following priorities: a) the reconstruction of ancient and building of new terraces, b) the use of conservation tillage, and, c) the rejuvenation of ancient irrigation systems.

Irrigation uses around 70 percent of the available water. Water is saved and erosion largely prevented when peasant farmers use terraces built by their forefathers and/or construct new ones. Where slopes are not terraced, fields must be leveled in such a way that water seeps to the root system without eroding the soil. The Incas did this masterfully, as seen on ancient fields. Secondly, to prevent runoff and erosion and keep soil and plants from drying out, contour drenches must be dug. Conservation tillage, used in pre-Columbian times and sometimes today, leaves the soil undisturbed and moist, as seeds are placed into narrow slits.

Above all, changes must be made in the way water is transported. The ancient Andeans used a variety of methods, including subterranean water channels, to bring water from the mountains to the dry, rainless coast. The ancient and venerated site of Tipon, 20 kilometers south of Cuzco, consists of a network of narrow irrigation canals that crisscross the region, providing the fields with small amounts of water throughout extended periods of time, causing no erosion and little evaporation. Modern drip irrigation systems are ideal, but still too expensive for most peasant communities (see also Schreier, Brown, and MacDonald 2006).

Conclusion

Anthropologists working with indigenous and other place-based peoples have a critical role in the issue of climate change, working as research collaborators and mediators between the local and the global. In this chapter I have shown how, in collaboration with the Andean people and in the context of Yachaq Runa, we continue to look into both old and new ways of collecting, using, and transporting as well as recycling and purifying water. Increasingly more reservoirs will have to be built to store water collected in the rainy season. Individuals and national and international nongovernmental organizations, those mentioned above and others, have been cooperating successfully with indigenous and other local people. Yet, much more help is required in the Andes and worldwide to guarantee survival. The situation is extremely serious (see also Hansen 2007) and it is unimaginable what will happen in the Andes and other parts of this planet if governments do not begin to act quickly: "With more than one-sixth of the Earth's population relying on glaciers and seasonal snow packs for their water supply, the consequences of these hydrological changes for future water availability are likely to be severe" (Barnett, Adam, and Lettenmaier 2005). For example, "Up to three billion people live from the food and energy produced by the Himalayan rivers"

(Schild 2007). In the Andes, "glaciers feed the rivers that feed the sprawling cities and shantytowns on Peru's bone-dry Pacific coast. Two-thirds of Peru's 27 million people live on the coast where just 1.8 percent of the nation's water supply is found"(CBS News 2007/02/11). The people of the high Andes have no choice but to remain in their mountains, keeping them moist, planting trees, digging trenches, collecting rain water, and hoping that the world's leaders will finally wake up and give all they have to help avert the worst disaster humankind ever had to face.

Notes

1. The suffering of billions of animals in animal factories is a disgrace to humanity and also a major contributor to our environmental dilemma. In a groundbreaking 2006 report, the UN declared that raising animals for food generates more greenhouse gases than all the cars and trucks in the world combined. Senior UN Food and Agriculture Organization official Henning Steinfeld reported that the meat industry is "one of the most significant contributors to today's most serious environment problems." Yet, this most significant issue is seldom, if ever, discussed at environmental conferences or elsewhere. Should we close our eyes to an issue that is at the very heart of global warming? Should we continue to waste 2,500 gallons of water required to produce one pound of beef?
2. Funding for this and other projects was provided by the Red Cross and Landkreis Böblingen, both of Germany, and by private donors; formerly also by Change for Children in Canada.

References

Barnett, T. P., J. C. Adam, and D. P. Lettenmaier. 2005. Potential impacts of a warming climate on water availability in snow-dominated regions. *Nature* 438: 303–09.

Bolin, I. 1987. The organization of irrigation in the Vilcanota Valley of Peru: Local autonomy, development and corporate group dynamics. PhD diss., University of Alberta.

———. 1990. Upsetting the power balance: Cooperation, competition and conflict along an Andean irrigation system. *Human Organization* 49(2): 140–48.

———. 1992. Achieving reciprocity: Anthropological research and development assistance. *Practicing Anthropology* 14(4): 12–15.

———. 1994. Levels of autonomy in the organization of irrigation in Peru. In: *Irrigation at high altitudes: The social organization of water control systems in the Andes*, eds. W. P. Mitchell and D. Guillet, 141–66, vol. 12, Society for Latin American Anthropology Publication Series.

———. 1998. *Rituals of respect: The secret of survival in the high Peruvian Andes*. Austin: University of Texas Press.

———. 1999. Survival in marginal lands: Climate change in the high Peruvian Andes. *Development and Cooperation* 5: 25–26.

———. 2001. When Apus are losing their white ponchos: Environmental dilemmas and restoration efforts in Peru. *Development and Cooperation* 6: 25–26.

———. 2002. Melting glaciers in the Andes and the future of the water supply. Paper presented at the meetings of IMA (Instituto de Manejo de Agua), October 31, Cusco, Peru.

———. 2003. Our Apus are dying: Glacial retreat and its consequences for life in the Andes. Paper presented at the annual meeting of the American Anthropological Association, Nov. 19–23, Chicago, Illinois.

Bolin, I. 2006. *Growing up in a culture of respect: Child rearing in Highland Peru*. Austin: The University of Texas Press.
Bolin, I. and G. Bolin. 2006. Solar solution for Andean people. *Development and Cooperation* 2: 74–75.
Bowen, M. 2005. *Thin ice: Unlocking the secrets of climate in the world's highest mountains*. New York: Henry Holt and Company.
Brack, E., A. Bravo, and A. Belen Noriega Bravo. 2000. Gestión Sostenible de los Ecosistemas de Montaña. In *Memoria del Taller Internacional de Ecosistemas de Montaña*. Cusco, Peru.
CBC News. 2007. Peru's glacier vanishing scientists warn. www.cbc.ca/technology/story/2007/02/16.
CBS News. 2007. Warming threatens double-trouble in Peru: Shrinking glaciers and a water shortage, by Leslie Josephs.
CTV News. 2007. Regional Climates may change radically. www.ctv.ca/servlet/ArticleNews/story/CTVNews/2007/03/26.
Francou, B. 2001. Small glaciers of the Andes may vanish in 10–15 years. http://unisci.com/stories/20011/0117013.htm.
Francou, B., M. Vuille, P. Wagnon, J. Mendoza, and J.-E. Sicart. 2003. Tropical climate change recorded by a glacier in the central Andes during the last decades of the twentieth century. Chacaltaya, Bolivia, 16oS. *Journal of Geophysical Research* 108(5): 4154.
Gade, D. 1975. *Plants, man and the land in the Vilcanota Valley of Peru*. The Hague: W. Junk.
González, G. 2003. Desaparecen Glaciares de Montaña. Tierramérica—Medio Ambiente y Desarrollo, August 21, 2003.
Hansen, J. 2007. Scientific reticence and sea level rise. *Environ.Res. Lett*, 2(2007).
Matthews, J. H. 2007. What you should know. *WWF Summary for Policymakers. IPCC Fourth Assessment Report—Climate Change 2007*.
Schild, A. 2007. Climate change, glaciers, and water resources in the Himalayan region. Speech given at the First Asia-Pacific Water Summit, December 3–4, Japan.
Schreier, H., S. Brown, and J. R. MacDonald. 2006. *Too little and too much: Water and development in a Himalayan watershed*. Vancouver: University of British Columbia.
Suarez, L. 2008. Climate change: Opportunity or threat in the central Andean region of Peru. *Mountain Forum Bulletin* (Jan.): 18–19.
Thompson, L. G., E. Mosley-Thompson, H. Brecher, M. Davis, B. León, D. Les, P. Nan-Lin, T. Mashiotta, and K. Mountain. 2006. Abrupt tropical climate change: Past and present. www.pnas.org/cgi/doi/10.1073.
Tupayachi Herrera, A. 1997. Diversidad arbórea en las microcuencas transversales al rio Urubamba en el valle sagrado de los Incas. *Opciones* 7: 41–46. Inandes UNSAAC, Cusco
———. 2005. Flora de la Cordillera de Vilcanota. *ARNALDOA* 12 (1–2): 126–44.
Vásquez Ruesta, P., with S. Isola Elias, J. Chang Olivas, and A. Tovar Narváez. 2002. *Cambio climático y sus efectos en las montañas sudamericanas*. Lima: Universidad Agraria La Molina.
Winterhalder, B. 1994. The ecological basis of water management in the central Andes: Rainfall and temperature in southern Peru. In *Irrigation at high altitudes: The social organization of water control systems in the Andes*, eds. W. P. Mitchell and D. Guillet, 21–67, vol.12, Society for Latin American Anthropology Publication Series.

Chapter 14

The Governance of Vulnerability: Climate Change and Agency in Tuvalu, South Pacific

Heather Lazrus

Introduction

During a visit with Tiliga at his home by the sandy edge of the lagoon on Nanumea Atoll, the retired doctor segued from our conversation about ways that people of the atoll prepare for disasters to the etymology of the atoll's name. Quiet waves lapped the shore as Tiliga related that *Nanugaomea*, as he says, it was once called, identifies the atoll in Tuvalu as "the place where things begin." Tiliga's interpretation reflects the strength of his pride in Nanumean traditions that he believes have spread to facilitate survival for other communities in the challenging environment of the Pacific island country of Tuvalu. On another occasion, following an especially informative interview about changes in the weather, sea, and land, Haulagi, a skilled pig farmer, asked me about the mechanisms of anthropogenic climate change. He had heard that industrial practices around the world are causing sea levels to rise in Nanumea. He questioned how distant activities could have such disastrous effects on the atolls of Tuvalu, where people have been living for so many generations according to the same traditions that Tiliga had told me of earlier.

Both of these anecdotes suggest important elements of how people perceive and cope with the environmental changes precipitated by climate change.On one hand, Tuvaluans past and present draw on traditionally available knowledge to understand, interpret, cope with, and respond to atmospheric and climatic disturbances. On the other hand, even in apparently remote places like Tuvalu, broader networks of knowledge inform understandings of and responses to climate change. Haulagi's triangulation between his own observations; historically referenced information about natural phenomena including wind direction, ocean currents, and precipitation patterns; and what he had heard about anthropogenic influences on the atmosphere belies uncomplicated distinctions between "traditional" and "local" versus "scientific" and "universal" ways of knowing.

Here, I refer to Tuvaluan and Nanumean understandings as being traditionally informed or locally based when implicit or explicit reference is made to the particulars of locally transmitted, observed, and deduced knowledge. This is not to reify a distinction between local understandings and transnational science that too is locally produced in laboratories, test sites, and conference venues. However, I do want to emphasize the local specificity of Tuvaluan and Nanumean knowledge and understandings because, while sometimes influenced by more transnational ideas, they are part of the unique way that people in Tuvalu will understand causes, consequences, and cures of climate change impacts. I do not want to promote traditional knowledge as a facile panacea or easy antidote to global ambitions insensitive to local specificities. Instead, I am attentive to Agrawal's argument that it is only "when we seek out bridges across the constructed chasm between traditional and scientific, that we will initiate productive dialogue to safeguard the interests of those who are disadvantaged" (1995, 433). With this is mind, how *can* anthropologists engage with people's understandings to promote productive dialogue, safeguard interests, and aid initiatives to mitigate the devastating effects of global climate change?

The rough sketch I did for Haulagi to illustrate my own understanding of how anthropogenic influences are changing the nature of the climate exemplifies the dissemination of knowledge in which anthropologists routinely engage. Our exchange of information prompted me to reflect on how "information concerning the troubles of a system [is] to be introduced into that system in ways that will help avoid, ameliorate, or correct those troubles rather than exacerbate them" (Rappaport 1993, 301). As this chapter demonstrates, I believe amelioration and correction are intrinsically linked to people's autonomy and sovereignty of the knowledge and governance of their vulnerability. I begin the chapter with a brief summary of Tuvalu and the impacts of global climate change. I then discuss my research in Tuvalu and identify three areas in which anthropology can play an important role in avoiding, ameliorating, and correcting problems precipitated by climate change: understanding how impacts are constructed, how agency is retained, and how governance can promote autonomy and sovereignty.

Tuvalu and Global Climate Change

> Long ago Eel and Flounder were great friends. One day they decided to find out which one of them was stronger by seeing who could carry the larger stone. As they competed, an argument broke out between them. Flounder became caught under the stone and his body became flat like the islands of Tuvalu. Eel was also wounded where Flounder hit him in the stomach and he vomited until his round body became long and thin like the coconut trees that are ubiquitous in Tuvalu. After this happened, the rock, which was colored black, white, and blue, broke into pieces that formed the dark night sky, the light sky of day, and the blue ocean that extends in every direction around the islands now known as Tuvalu.
>
> —Tuvaluan origin story

The Tuvaluan islands, a mix of atolls edged by small *motu* (islets that rise above sea level) and table reef islands, lie between 5°–10° south and 176° to 179° east. With distances of 125 to 150 kilometers between them, the islands' combined land area is just 26 square kilometers with an average elevation of less than three meters above sea level. Each island group is ecologically and culturally unique, and now each faces unique challenges from impacts of global climate change. The islands of Tuvalu are geologically very young and dynamic, subject to subsidence, reef growth, and eroding and accreting forces of the sea and weather (Baines, Beveridge, and Maragos 1974). In addition to complex geomorphology, limited technical research contributes to scientific debate about the specific impacts resulting from climate change (Hunter 2004). Nonetheless, there is broad agreement on general impacts that include sea-level rise, sea surface and subsurface temperature increases, ocean acidification and coral bleaching, coastal erosion, increased intensity but decreased frequency of rainfall, and increased frequency of extreme weather events, including drought (Mimura et al. 2007). Concerns summarized in a consultation report prepared for the National Adaptation Program of Action, discussed below, include salinization and pest infestation of subsistence crops, deterioration of reef fisheries, periodic shortages of potable water, increased occurrences of communicable diseases, and damage to infrastructure (Noa et al. no date, 9).

Figure 14.1: At the highest tides, seawater rushes past this broken seawall on Nanumea making the ground too saline for most vegetation. © Heather Lazrus.

Figure 14.2: Children play on the edge of a motu in the Nanumea lagoon. The coastline is constantly shifting, shaped by lagoon currents. © Heather Lazrus.

The scientific consensus expects that the effects of climate change will dramatically transform the Tuvalu islands, settled by Polynesian seafarers from elsewhere in the Western Pacific as recently as eight hundred and as long ago as two thousand years ago (Chambers and Chambers 2001; Munro 1982). Land resources are limited due to sparse alkaline coral sand soils covered only by a thin layer of accumulated vegetable deposit. Freshwater that "floats" in places on the heavier saltwater permeating the porous coral soil, known as the Gyhben-Herzberg lens, is also unevenly distributed and limited (McLean et al. 1986). Narratives, stories, and legends tell of how the challenges of daily atoll life are endured and droughts and storms are survived. Such stories provide mythical metaphors for explaining and coping with disruptive geological and environmental processes (Cronin and Cashman 2007).

In the nineteenth and twentieth centuries, European interest in the islands waxed and waned. Whaling and copra production generated short-lived interests; Protestant missionaries gained footholds on each island; and American forces established defenses on Funafuti, Nukufetau, and Nanumea at the frontline of WWII (McQuarrie 1994). In 1889, forty years after sustained European contact, Tuvalu, then named the Ellice Islands, came under British rule. Independence came almost a century later, in 1978. The national government is a parliamentary democracy with a prime minister and representatives from each island except Niulakita, which is settled by Nukulaelae and together are considered a single political entity. Tuvalu's economy is now dominated by migration, remittances, aid, and bureaucracy (known as a MIRAB economy) and has proven to be surprisingly resilient, contrary to expectations (Knapman, Ponton, and Hunt 2002). The growing importance of a cash-based economy drives heavy urbanization in the capital on Funafuti Atoll where the international airport, wharf, and government buildings offer economic opportunities. Funafuti's burgeoning economy and increasing population density differentiate the cadence of life in the capital from the outer islands.

Interrelated material, economic, and political processes all affect how impacts of climate change will be manifested and managed. A very important example is the intensive excavation of reef barriers and quarrying of sandy shores for building materials, which has exacerbated coastal erosion and disrupted the integrity of the freshwater lenses. These activities were initiated first by US forces stationed on three of Tuvalu's atolls at the allied frontline of WWII and subsequently by islanders for local building requirements (Webb 2005).

A Shifting Tide: Approaches to Climate Change Impacts

My interviews with Tuvaluans who have spent a lifetime observing and interacting with the environment, centered on their observations of changes in the sky, sea, and land together with their knowledge of traditional responses to extreme weather events such as storms, droughts, and floods. To begin fieldwork, I traveled to Tuvalu's capital Funafuti by boat from Fiji. Unexpectedly rough seas and high winds extended our journey, typically a three- or four-day passage, to an entire week at sea. Several unseasonable cyclones were twisting across the Pacific Ocean between us and Tuvalu, generating impassable high winds and rough seas and forcing us to wait at anchor in sheltered waters protected by Fiji's outer islands. Tropical cyclones may develop close to Tuvalu, but it is unusual for them to remain in the area once they grow into storms. However, memories of Cyclone Bebe that decimated Funafuti in 1972 remained vivid among my fellow travelers, all of whom were Tuvaluan. In spite of that intimate experience with the potentially deleterious forces of Pacific weather, I was overwhelmed by people's energy and vitality. In very real terms, I understood the need to rethink a so-called top-down rubric of vulnerability. The dominant paradigms of the ways in which people are vulnerable too often deny local agency and meaning making, aspects that were evident to me in peoples' response to the present storm and memories of past storms.

Early approaches in disaster research tended to be technocratic and othering. At the same time that Western society and nature were considered separable and isolated from each other, non-Western societies were considered different, primitive, and located within nature. Such approaches described disruptive events as one-dimensional, physical (e.g., geological, biological, technological) calamities befalling internally homogenous populations. More holistic understandings of disasters, accommodating the cultural, social, and physical, are more representative and offer useful approaches to climate change research (Gaillard 2007; Glantz 1996; Hewitt 1983; Lewis 1999; Oliver-Smith 1986; Schipper and Pelling 2006). These approaches begin to reorganize earlier notions of nature and culture, and question the "naturalness" of natural hazards (O'Keefe et al. 1976). Like other disasters, climate impacts can be abrupt or subtle, social or physical, and "unavoidable in the context of a historically produced pattern of 'vulnerability'" (Oliver-Smith and Hoffman 2002, 3). Although still a debated concept, one definition of vulnerability refers to "a person's or group's capacity to anticipate, cope with, resist and recover from the impact of a hazard" (Wisner et al. 2004, 11). Accordingly, climate change will have different impacts on different people and groups within a society. Oliver-Smith offers a more dynamic interpretation of vulnerability that forefronts individuals and communities' interdependence as "the conceptual nexus that links the relationship people have with their environment to social forces and institutions and the cultural values that sustain or contest them" (2004, 10).

Tuvalu and the Local Experience of Climate Change

In the following sections I offer illustrations from Tuvalu of how climate change impacts are socially constructed, how the assessment of vulnerability is related to local agency, and how management of vulnerability and responses to climate impacts occur across multiple scales of governance. My intention is to offer some examples of how anthropology can be attentive to the complexity often obfuscated in approaches to vulnerability. Vulnerability often arises among people and places with histories of domination. In turn, vulnerable people and places are denied access to power, resources, and decision-making processes, often on the basis of their vulnerability, dangerously perpetuating their domination. Hence, the governance of vulnerability need not only redress marginalization as a cause of social vulnerability, but should also rely on understandings of vulnerability that are locally identified according to threats to the culturally defined "normal order of things" (Oliver-Smith 2007). Governance of vulnerability refers broadly to the mechanisms and strategies that manage risk, mediate vulnerability, and promote capacities for resilience and adaptation.

You Are Not Okay with One Hand in the Freezer and the Other Hand on the Stove: Construction of Environmental Impacts

The National Tidal Facility, housed in the Australian Bureau of Meteorology, determined that sea level had fallen in the area around their monitoring station on Funafuti Atoll over the decade leading up to 2000 (Eschenbach 2004). The Tuvalu Meteorological Office, in their analysis of the same data set, obtained contradictory results indicating a sea-level rise of several millimeters over the same time frame. An assistant secretary in a government department explained this discrepancy to me as the result of different approaches to averaging the data points: the two agencies had treated numerical extremes in the data differently. This produced their contradictory outcomes. The disparity goes beyond divergent analyses to how problems are perceived and valued according to internal sets of logic and meaning. In other words, it is not just that the mathematics were different, but that environmental processes, including hazards, are perceived and understood within the particular cultural context of the relationship between people and their natural world.

To explain the Meteorological Office's approach to the data, the assistant secretary elaborated: "We are not interested in the averages of the numbers; we are interested in the extremes. If you have one hand on a hot stove element and one hand in a running freezer, you might average out to be body temperature, but you are *not* okay!" Something Tuvaluans may well know from experience and observation is that, because of coastal morphology, a rise in sea level of just a few millimeters can lead to severe coastal erosion at alarming rates (Hunter 2004). On islands that narrow to the width of a North American road in many places, even minor coastal erosion can be catastrophic. In Nanumea, people pointed out where erosion and accretion is accelerating and some expressed concern that the tips of the islands ringing the atoll were shifting. These areas, known in Nanumea as the *mata*, or eyes of the land, are associated with particular families who trace their lineage and related rights to positions of power within the community. Coastal erosion—due to sea-level rise, other impacts of climate change, and perhaps exacerbated by practices such as coastal extraction—has significant social and cultural implications in terms of loss, displacement, and organized hierarchies of power.

We Know How We Are Vulnerable: Agency through Vulnerability Assessment

As the incident described above illustrates and as Oliver-Smith notes, the "perception of risk and vulnerability, and even impact, is clearly mediated through linguistic and cultural grids, accounting for greater variability in assessments and understandings of disaster" (2004, 17). Culturally meaningful and locally apposite assessments must inform managed responses to climate change. A better appreciation of what constitutes disaster, how people are susceptible to it, and more effective means of responding demand the real participation of those most at risk (Hewitt 1997, 338). Allowing traditionally available knowledge and local understanding to define the problem goes a long way towards accomplishing this, while simply tolerating traditional knowledge as alternative discourse tends to reinforce the dominant paradigms of disaster risk reduction (Bankoff 2004, 35).

One instance that illustrates this point occurred toward the end of my fieldwork, while visiting an officer in the Tuvalu Department for the Environment. A fax arrived from the regional office of an international NGO in Suva, Fiji, announcing that Tuvalu was one of two locations chosen to pilot a vulnerability assessment that would eventually be applied across the Pacific region. The officer's response endorsed the imperative for local agency in assessing and responding to vulnerability: "Why do we need them to come here, with their ideas from other places, and tell us what is wrong with us here? We know our islands and our people. We know how we are vulnerable."

The Place Where Things Begin: Grounding Governance in Nanumea

The governance of vulnerability to climate change must necessarily be linked across international, regional, national, and local levels. Much has been done to demonstrate that vulnerable people and places around the world are excluded from the decision-making, power, and resources involved in governance of risk and disaster. Long-lasting adaptation to climate change needs to be framed by local perceptions, and spearheaded by local governance efforts via robust local institutions. Though they have played a diminishing role in island governance, first under colonial British administration and now a postcolonial democratic government, local practices that link disaster preparedness with political processes have sustained life in Tuvalu for generations and remain important to daily and political life.

One example of international action on climate change that not only bridges scales of governance but may go so far as to privilege local understandings and decision-making processes is a mechanism established under the United Nations Framework Convention on Climate Change. National Adaptation Programs of Actions (NAPA) are being implemented in several Least Developed Countries. Although it is an international initiative that is coordinated by a national government body (in Tuvalu it is the Department for the Environment), the guidelines for preparation of the NAPA seek to ensure that the perspectives and priorities of those affected are incorporated into planning and decision-making at national and international levels. Planning of the NAPA in Tuvalu is currently underway and it is too early to know how political and institutional pressures will influence it. Encouragingly, one development officer on Nanumea who is employed by the national government told me that the NAPA represents a real opportunity for locally identified adaptation priorities to be addressed by locally defined projects that are nationally coordinated and internationally sponsored.

Placing the Local in Global Climate Change

Global climate change enters a world patterned by different abilities to cope with and recover from its impacts. Anthropology has inherited a mandate from voices like Agrawal and Rappaport, among countless others, to be sensitive to these patterns and to approach climate impacts with a deep empathy to the cultures and conditions of those affected. Arriving in Funafuti by plane is an experience that inspires reflection on the tenacity of people to inhabit these slim coral crescents at the center of the planet's largest ocean. When I first spotted the *motu* from the boat after our precarious, seven-day journey from Fiji, I marveled at what now appeared to be invincible fortresses of land rising above the indigo waves of that deep ocean water. The strong spirits of some people living in threatened environments are also like fortresses, and I am reminded of their steadfast sincerity, dignity, and vision.

As much as it is social, political, and physical, vulnerability is also a matter of representation to which questions of agency are central: Who is doing the representing, under what conditions, and for what purposes? Tuvalu is often portrayed by outsiders as a "canary in the mine"; a microcosm of our planet that is the world's first casualty of global climate change. The impacts of climate change in Tuvalu demand immediate attention and action. Those fatalistic representations may be contradicted by capabilities woven into the brightly colored fabric of daily life, buoyant traditions, and robust leadership.

References

Adger, W. N. 2006. Vulnerability. *Global Environmental Change* 16: 268–81.

Agrawal, A. 1995. Dismantling the divide between indigenous and scientific knowledge. *Development and Change* 26: 413–39.

Baines, G. B. K., P. J. Beveridge, and J. E. Maragos. 1974. Storms and island building at Funafuti Atoll, Ellice Islands. *Proceedings of the Second International Coral Reef Symposium* 2: 485–96. Brisbane, Australia.

Bankoff, G. 2004. The historical geography of disaster: 'Vulnerability' and 'local knowledge' in western discourse. In *Mapping vulnerability: Disasters, development, and people*, eds. G. Bankoff, G. Frerks, and D. Hilhorst, 25–36. London: Earthscan.

Chambers, K. and A. Chambers. 2001. *Unity of heart: Culture and change in a Polynesian Atoll society*. Prospect Heights: Waveland Press.

Church, J. A., N. J. White, and J. R. Hunter. 2006. Sea-level rise at tropical Pacific and Indian Ocean islands. *Global and Planetary Change* 53: 155–68.

Cronin, S. and K. Cashman. 2007. Volcanic oral traditions in hazard assessment and mitigation. In *Living under the shadow: The archaeological, cultural and environmental impact of volcanic eruptions*, eds. J. Grattan and R. Torrence. Tucson: University of Arizona Press: Tucson.

Eschenbach, W. 2004. Tuvalu not experiencing increased sea level rise. *Energy & Environment* 15(3): 527–43.

Gaillard, J. C. 2007. Resilience of traditional societies in facing natural hazards. *Disaster Prevention and Management* 16(4) 522–44.

Glantz, M. 1996. *Currents of change: El Nino's impacts on climate and society*. Cambridge: Cambridge University Press.

Hewitt, K., ed. 1983. *Interpretations of calamity from the viewpoint of human ecology.* London: Allen & Unwin.

———. 1997. *Regions at risk: A geographical introduction to disasters.* Essex: Longman.

Hunter, J. R. 2004. Comment on 'Tuvalu not experiencing increased sea level rise'. *Energy and Environment* 15(5): 927–32.

Knapman, B., M. Ponton, and C. Hunt. 2002. *Tuvalu 2002 economic and public sector review.* Manila: Asian Development Bank.

Lewis, J. 1999. *Development in disaster-prone places: Studies of vulnerability.* London: Intermediate Technology Publications.

Mclean, R. F., P. F. Holthus, P. L. Hosking, C. D. Woodruffe, D. V. Hawke. 1986. *Tuvalu land and resources survey: Nanumea.* Department of Geography, University of Auckland: Auckland.

McQuarrie, P. 1994. *Strategic atolls: Tuvalu and the second world war.* Christchurch: University of Canterbury McMillian Brown Center for Pacific Studies, and Suva: University of the South Pacific Institute of Pacific Studies,

Mimura, N., L. Nurse, R. F. McLean, J. Agard, L. Briguglio, P. Lefale, R. Payet and G. Sem. 2007. Small islands. In *Climate change 2007: Impacts, adaptation and vulnerability. Contribution of Working Group II to the Fourth Assessment Report of the Intergovernmental Panel on Climate Change*, eds. M. L. Parry, O. F. Canziani, J. P. Palutikof, P. J. van der Linden and C. E. Hanson, 687–716. Cambridge: Cambridge University Press.

Munro, D. 1982. The lagoon islands: A history of Tuvalu 1820–1908. PhD diss., Sydney: Macquarie University.

Noa, P., N. Maisake, P. Latasi, S. Silu, and P. Faavae. No date. National Adaptation Programme of Action (NAPA) multi-disciplinary team consultation to all islands of Tuvalu except Niulakita. Department of the Environment, Tuvalu.

O'Keefe, P., K. Westgate, and B. Wisner. 1976. Taking the naturalness out of natural disasters. *Nature* 260: 566–67.

Oliver-Smith, A. 1986. *The martyred city: Death and rebirth in the Andes.* Albuquerque: University of New Mexico Press.

———. 2004. Theorizing vulnerability in a globalized world. In *Mapping vulnerability: Disasters, development, and people*, eds. G. Bankoff, G. Frerks, and D. Hilhorst, 10–24. London: Earthscan.

———. 2007. Social vulnerability: The nexus of disaster management. Presentation at the annual meeting for the Natural Hazards Center, July 8–11, Boulder, Colorado.

Oliver-Smith, A. and S. M. Hoffman. 2002. Why anthropologists should study disasters. In *Catastrophe and culture: The anthropology of disaster*, eds. S. M. Hoffman and A. Oliver-Smith, 3–22. Santa Fe, NM: School of American Research Press.

Rappaport, R. 1993. Distinguished lecture in general anthropology: The anthropology of trouble. *American Anthropologist* 95(2): 295–303.

Schipper, L. and M. Pelling. 2006. Disaster, risk, climate change and international development: Scope and challenges for integration. *Disasters* 30(1): 19–38.

Webb, A. 2005. *Tuvalu technical and country mission report: Assessment of aggregate supply, pond and lagoon water quality and causeway construction on Funafuti and Vaitupu Atolls.* EU EDF 8/9—SOPAC Project Report 36: Reducing vulnerability of Pacific ACP States. Suva, Fiji: SOPAC.

Wisner, B., P. Blakie, T. Cannon, and I. Davis. 2004. *At risk: Natural hazards, people's vulnerability and disasters.* New York: Routledge.

Chapter 15

From Local to Global: Perceptions and Realities of Environmental Change Among Kalahari San

Robert K. Hitchcock

Introduction

As Simms and Reid (2005, 2) have argued, the continent of Africa is particularly vulnerable to the impacts of climate change. This vulnerability exists because the majority of the continent is tropical and subtropical, a significant percentage of Africa's population is dependent on natural resources (soils, grazing, wildlife, timber, and other wild plants), poverty levels are relatively high, and there are significant challenges facing systems of governance.

The San (Bushmen, Basarwa) of the Kalahari Desert region of southern Africa are all too familiar with environmental and socioeconomic challenges. For generations they have experienced serious droughts; floods; cold spells; hot spells; and outbreaks of human, livestock, and wildlife diseases that have affected their livelihoods (Campbell 1979; Campbell and Child 1971; Hitchcock 1979, 2002; Vierich 1981, 1982). Like many other peoples, San employed a variety of often-ingenious strategies to cope with environmental change, ranging from diversifying their subsistence bases to depending on other groups or, in some cases, governments or international agencies for food and support. San informants note that some of them have slipped further into poverty as a result of economic and environmental changes that have occurred over time in southern Africa.

From 1990 to 2004 I have studied the familiarity of San peoples with issues surrounding global climate change and local environmental situations. Interviews with approximately 250 San from eight ethnic groups in the Kalahari Desert region in Botswana included questions about perceptions of the state of the Kalahari, adaptive strategies, and short-term and long-term environmental change.

Responses to Environmental Change

The Republic of Botswana has more San than any other southern African country, and approximately 30 percent of its remote rural area is made up

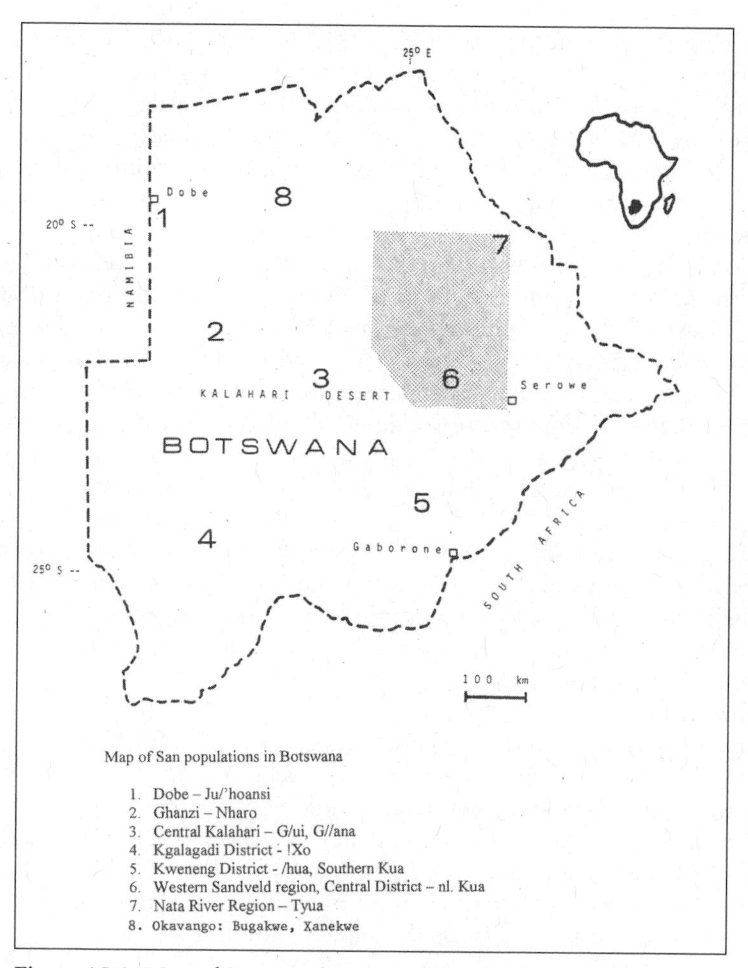

Figure 15.1: Map of San populations in Botswana.

of San, also known in Botswana as Basarwa (Hitchcock 1978; Saugestad 2001). Many San are poor and heavily dependent on natural resources, agriculture, livestock production, and small-scale income-generating projects (Hitchcock et al. 2006; Suzman 2001a, 2001b). Employment opportunities in the rural areas are limited and wages tend to be low. The Kalahari Desert, a vast, undulating "sand-sea" that covers much of southern Africa, possesses many of the attributes of semiarid savanna ecosystems. Some analysts prefer the term *thirstland* to *desert*, as water is the main variable driving biological productivity.

Anthropologists have worked closely with San groups for over fifty years (Hitchcock 2004; Schapera 1930, 1939) and have always been particularly interested in ecological issues (Katz, Biesele, and St. Denis 1997; Lee 1979;

Marshall 1976; Silberbauer 1965, 1981; Tanaka 1980; Thomas 2006; Wiessner 1977, 1982). A number of anthropologists have worked with San informants (or "consultants") to document the varied ways in which people use and manage natural resources; adjust themselves to changing social, economic, and environmental conditions; employ cultural techniques for resolving conflicts; and cope with environmental and development constraints.

According to both observations and interviews, San have experienced hunger and privation as a result of climatic events and outbreaks of pests (e.g., locusts) and human diseases (e.g., malaria) (Hitchcock 2002; Hitchcock, Ebert, and Morgan 1989; Vierich 1981). In the twentieth century, outbreaks of foot-and-mouth disease caused particular difficulties when government agencies imposed restrictions on the movement of milk, meat, and other domestic animal (and sometimes wild animal) products. In 1995, after the outbreak of lungsickness among cattle (Contagious Bovine Pleuropneumonia, CBPP), the government of Botswana opted to destroy the entire herd of livestock in the North West District (Ngamiland) (Hitchcock 2002). People in rural areas were compensated to some extent for the loss of their animals; they used the cash they received to buy food. Many rural people increased their foraging activities, collecting wild foods and medicines for both domestic use and sale.

The rural economy in Botswana is very varied and complex, and rural people in Botswana (and elsewhere) tend to spread their risk, diversifying their subsistence and income sources. The relative importance of natural resources and in-kind income is greater among poorer households, and the poorer one is, the greater the dependency on other people or the state for in-kind transfers (e.g., food, goods, and services) (Hay 1988). Remittances have always been important to rural households in Botswana but have become less so, at least as far as the mines in South Africa were concerned, since the 1980s.

Nearly all of the people to whom my research collaborators and I spoke recognized that environmental change was occurring, but there was some disagreement about the sources of that change. Some people said that the changes were due to natural oscillations in the ecosystem that had been occurring "since time immemorial." Others said that current environmental conditions are worse than those they remembered from their childhoods and from stories of their parents and grandparents. They said that the current environmental and socioeconomic conditions were much more difficult to deal with, because there were more people on the land and their options were more limited (for example, they could no longer opt for migration to a new area since there were other people there already). San who engage in farming said that they were using some of the strategies that they had used in the past to cope with environmental stress—staggering their planting, planting drought-resistant crops, diversifying the locations where they placed their fields, and introducing water control techniques—but these strategies were no longer as effective as they used to be.

People in the Kalahari gave a wide variety of reasons for why and how the local environment was changing. Some said that the Kalahari was becoming drier and hotter. Others said that the winters were cooler and that this was affecting people and animals. Still others said that it is not so much hotter now as it is more variable—there is more uncertainty, they say, in the environment. Many of the environmental factors they believed to be indicators of seasonal shifts (for example, the growth of certain species of plants or the appearance of certain insects in the trees) are no longer as obvious as they used to be. Not all people agree that this is due to global warming; in fact, some said that parts of the Kalahari were actually getting cooler, and that this cooling trend was affecting their well-being. Nearly all of the people who were interviewed agreed that the biggest challenge that they faced was increased environmental variability and risk.

The most important resource for all populations in our study area is water. The average person needs approximately 2,500 milliliters of drinking water per person per day. The Basic Water Requirement (BWR) is 50 liters per person per day, which includes sufficient water for people for drinking, sanitation, cooking, and bathing. One of the problems facing human, animal, bird, and other populations in the Kalahari Desert is that much of the landscape lacks surface water. For a long time it was assumed that all groundwater in the region was fossil water and that subterranean aquifers received no recharge whatsoever from rainfall. Rainfall water, it was believed, was held in the upper few meters of the soil, from which it evaporated in a relatively short period of time. It is now believed that some of the water contained in geological layers is fossil water, but that there may also be some recharge occurring (Mazor 1982; Thomas and Shaw 1991).

People obtain water in several ways in the Kalahari. There are permanent (year-round) sources—for example, wells in calcrete areas (calcium carbonates that form under conditions that alternate between wet and dry) tap groundwater trapped in the upper layers of the sand. Seasonal sources include pools that form during and after rains in the wet season and holes in trees in which water accumulates, for example, in *marula* (*Sclerocarya caffra*) or *mongongo* (*Schinziophyton rautanenii*) trees.

A major problem for people in the Kalahari, according to San informants, has been what they see as a significant decline in the water table. They blame this decline on the increased number of water points, especially boreholes, that have been drilled in the Kalahari over the past several decades, beginning especially during the drought of 1961–1965 and as a result of the Tribal Grazing Land Policy after 1975 (Hitchcock 1979; Hitchcock, Ebert, and Morgan 1989; Sanford 1977; Richard Morgan, John Holm, personal communications, 1988). They also blame reduced rainfall, although rainfall data do not tend to support such a claim.

A critical problem was access to water in the dry season. One means of coping was through the aggregation of one's moisture source, a strategy employed by a number of San groups who cached ostrich eggs filled with water in places they planned to visit during the dry season. San groups also dug wells or deepened areas in the sides of pans (low-lying playa-like features on the landscape) in order to increase the length of time that water was available after the rainy season (Cashdan 1984; Hitchcock 1982). Another strategy employed by San was to diversify the types of plants that they exploit for food and moisture. Some groups, such as the Tyua of the Nata River region, broadened the numbers and types of plant species that they exploited in the face of competition for resources with people and animals.

In some parts of the Kalahari, cattle and other livestock have contributed to local habitat deterioration and a decline in wild foods, although there is considerable debate over the extent of their role in overgrazing (see Behnke, Scoones, and Kerven 1993; Brandenburgh 1991; Perkins 1991; Sporton and Thomas 2002). According to informants, the numbers of wild animals, especially grazers such as wildebeest and zebra, have declined in areas dense in cattle. One consequence is that foraging success rates in such areas have gone down and people have had to resort to alternative strategies to earn a living, such as shifting to more logistically organized systems of resource procurement, engaging in livestock-related labor to earn income,

Figure 15.2: A Kua family in the east-central Kalahari Desert, an area in which a large Botswana government–World Bank livestock ranching program contributed to significant environmental change. Photo by Robert Hitchcock.

and depending on goods provided by relatives, livestock owners, or, in a number of cases, the government of Botswana, which began mounting drought relief feeding and labor-based relief and development programs in rural areas (Hitchcock, Ebert, and Morgan 1989; Holm and Morgan 1985).

When asked what adaptive strategies San and other people employed in dealing with environmental constraints, they responded in a variety of ways. Some said that they had no special ways of dealing with environmental change; they "just lived." Overall the responses can be summarized as follows: People said that they a) diversified their subsistence and income generation strategies; b) sold off personal assets such as livestock or plows; c) borrowed goods or cash from other people; d) tried out new and innovative technologies; e) left areas where they were experiencing environmental and economic difficulties (i.e., they employed mobility and migration strategies); f) shifted from food production to foraging; g) specialized in the collection of wild plants for medicine; h) engaged in ritual activities aimed at influencing the environment (e.g., rain making) and influencing the actions of other people; i) worked for other people who paid them cash or in kind (e.g., in the form of food); and j) depending on the state, nongovernment organizations, or faith-based institutions (churches) for food and other kinds of assistance. A number of people noted that those who had been resettled or relocated away from their ancestral areas—places with which they were familiar—were especially vulnerable.

Botswana Government Strategies for Coping with Economic Stress

As early as the late nineteenth century, Bechuanaland Protectorate officials attempted to avert environmental problems using various technological and management solutions (Sir Seretse Khama, Alec Campbell, Brian Egner, Chris Sharp, personal communications, 1976, 1978, 1979). Drought-relief strategies have ranged from the distribution of food and goods to borehole drilling and transporting livestock feed into grazing-deficient areas (Campbell 1979; Hay 1988; Hinchey 1979; Holm and Cohen 1988; Holm and Morgan 1985; Sanford 1977). There have been a number of different kinds of subsidy programs in Botswana, including agricultural subsidies for seed, fuel, fertilizer, feed, vaccines, and boreholes (given to groups), as well as write-off of loans such as those provided by the National Development Bank. Botswana is well known for its drought relief, labor-based relief, and development programs, which have ensured that most people have been able to sustain themselves in spite of environmental and economic downturns.

Beneficiaries of Botswana's livelihood support programs include a) supplementary feeding programs for school-going children, b) aid to pregnant and lactating mothers, c) aid to destitutes (defined as people who cannot support themselves; see Republic of Botswana 1990), d) aid to medically selected (specially targeted) malnourished children, and e) aid to remote area dwellers (RADs). In the case of labor-based relief and development projects, individuals and groups engage in activities such as labor-intensive road construction, hand-stamping (pounding) of sorghum and other grains, and destumping of agricultural fields, which aids in planting.

Remote area dwellers are a special case in Botswana (Wily 1979), and the government has specific programs targeting them. In 1985, 90 percent of households surveyed by Hay (1988) were receiving some kind of food ration, and 43 percent were receiving between one hundred and three hundred kilograms of food per annum. Approximately 60 percent of all of the people in the country were receiving food aid at the peak of the feeding program in 1985–86. This program was oriented heavily toward feeding children. Remote area dwellers were also provided a ration intended to cover all their food requirements. The government made no attempt to target (e.g., by means testing) within this category of Remote Area Dwellers, as all of them were seen as being seriously affected by drought and as having few alternative sources of income. Food rations made up more or less 100 percent of the total food consumed by remote area dwellers. An examination of the amounts and types of commodities made available to people, combined with comments of informants, including government officials, suggests that there may have been some micronutrient deficiencies because of the composition of the ration, for example, of vitamins A and C. The usual sources of these micronutrients are from products such as wild bush foods that were unlikely to be available to the people because of drought (Helga Vierich, personal communication, 1979).

The Botswana government feeding program in the past was almost entirely dependent on food aid from donors (Hay 1988; Holm and Cohen 1988). According to analysts, the program was implemented and executed smoothly, although there were some initial logistical difficulties, and there was some leakage (i.e., some of those who were supposed to get specified amounts of food did not get it, or got only a portion of what they were supposed to receive). It was particularly successful in addressing the entitlement failure for the poorest households that were most vulnerable to drought (Hay 1988; Holm and Cohen 1988; Holm and Morgan 1985).

In 2002, the Botswana government revised its National Policy on Destitutes (see Republic of Botswana 2002) to alter foodstuffs provided to people, ensure that adult beneficiaries received 1,750 kilocalories per day, and provide some cash. The food ration now includes maize meal, sugar, sorghum meal, bread flour, vegetables, greens, oil, pulses, salt, meat, tea, and milk (Republic of Botswana 2002).

As some of the San and government officials in Botswana who were interviewed pointed out, the government and international donors must look at an annually selected basket of goods, this basket of goods must be balanced nutritionally, and the goods basket must be adequate in terms of the minimum daily requirements of individuals and follow World Health Organization (WHO) standards. The problem for many San, they say, is that they are living below the poverty line and they are unable to get sufficient support from the current basket of goods supplied by the government.

Coping with the Impacts of Environmental Change

The San and their neighbors in the Kalahari expressed a great deal of concern about environmental change and often ask researchers, government officials, and others many questions about what global climate change is and whether it will continue to affect them. The main questions they raised include: a) What do the government and the international community propose to do about global climate change? b) What impacts will global climate change have on them, given that many of them are poor and especially vulnerable to environmental perturbations? and c) What can they do themselves to mitigate the impacts of global and local environmental change?

A significant number of San informants said that they felt that their government and the international community should do more to reduce the effects of global and local climate change. A few people suggested that the government and international donors should help develop energy-saving technologies that were not based on fossil fuels. It was pointed out that some San communities and nongovernment organizations working with them have opted for wind-powered borehole pumps. Others are seeking alternative technologies that can be used to render water safer for drinking.

Several San informants suggested that the governments of the countries where they lived should be more environmentally responsible, and they said that they were aware that they themselves needed to vote for political candidates who were more environmentally oriented. Over two dozen San communities in the Kalahari had opted to establish community-based natural resource management programs in an attempt to diversify their income sources and enhance local-level conservation. Some of these community-based natural resource management programs incorporated the planting of trees and the creation of shelter belts aimed at reducing soil erosion. A few San communities engaged in the construction of campsites for tourists in order to try to influence where they camped, and they supplied firewood to these places so that people would not overexploit local trees and shrubs.

There were those who asserted that people should return to more traditional kinds of activities, such as the use of small-scale burning practices to increase grazing quality for wild and domestic animals. One of the concerns expressed by a number of San who were interviewed was that there would be both losers and winners when it comes to environmental change, and that they, as people who were largely poor, would lose out to people with more capital, many of whom were not San themselves. A number of San would agree with Thomas and Twyman (2005) who argue that equity and justice are issues of crucial concern when it comes to climate change impacts among societies that are relatively heavily dependent on natural resources.

Some elderly San informants and ones involved in traditional healing suggested that people needed to engage more in ritual activities aimed at improving the health of the land and the people living on the land (see some of the discussions, for example, in the volume *Healing Makes Our Hearts Happy*, Katz, Biesele, and St. Denis 1997). Three people said that some modern-day healers were more powerful now than they used to be because they specialized in rain making and weather prediction. These healers, they noted, were marketing their services not just to San but to people throughout southern Africa.

Half a dozen people said that they felt that current environmental problems exist because many people were no longer following traditional customs of reciprocity and collective burden sharing. The custom of sharing of goods and services, they said, was breaking down in part as a result of the social, demographic, and natural changes occurring in the Kalahari. A number of people said that the changes that they were seeing in the Kalahari were simply part of a pattern of a series of wet and dry periods. There was no long-term trend in climatic change; they argued, rather, the change was regular, occurring every two decades or so. Some elderly people said that they remembered at least two times in their lives when there were very dry periods and very wet periods.

Approximately a third of the people I interviewed said that they had heard about global climate change and global warming on the radio. Several young San noted that they had read the 2001 report of the Intergovernmental Panel on Climate Change on the internet. A number of people had seen reports on climate change on the websites of the International Institute for Environment and Development (www.iied.org) and the United Nations.

Several San said that governments and international finance and development agencies such as the World Bank and the United Nations needed to be much more proactive than they were in terms of addressing poverty alleviation. Poverty, they said, was a major constraint and one that limited the options that they had available to them in trying to adjust to what they considered to be new natural and social environmental realities.

It is interesting to note that at least some of the San who were interviewed said that the problems they were experiencing were due to external factors rather than to the actions of local people. They pointed to government policies that were aimed at grouping people together into settlements, which resulted in large population densities and therefore greater competition for resources. A common local perception of environmental change was that it was a product of larger forces over which local people had little control.

Quite a few people asked and attempted to answer the question, what can they do themselves to mitigate the effects of global and local climate change?

Numerous suggestions were made, including working more closely with their neighbors to solve environmental problems (for example, putting social pressure on people who were keeping too many livestock on the range or who were hunting too many animals on their citizens' hunting licenses). Quite a few people said that diversifying their income and subsistence strategies would have beneficial effects, not least of which would be to reduce pressure on natural resources and thus contribute, they hoped, to greater conservation. "Returning to the old ways" of egalitarianism, communal land and resource tenure, and extensive systems of reciprocity and exchange were recommendations heard frequently in the course of discussions on environmental change in the Kalahari. A useful strategy, several people argued, would be to incorporate lessons about how to cope with climate change in the curricula of primary schools.

Perhaps the biggest lesson that could be communicated to the young, one San woman said, was that people needed to be more equitable in the ways that they treated the environment and dealt with other people. Global climate change, she said, was something that would be difficult to deal with, but if everyone worked together, the old and the young, San and non-San, Africa and the rest of the world, they could cope with the challenges they faced, drawing on both scientific and local knowledge about ways to deal with natural disasters, social conflict, and environmental variability.

Acknowledgments

Support of the research upon which this chapter is based was provided by the US National Science Foundation, the US Agency for International Development, the Norwegian Agency for Development Cooperation, the Ford Foundation, the Working Group of Indigenous Minorities in Southern Africa, the Kuru Family of Organizations, the International Work Group for Indigenous Affairs, and various nongovernment and community-based organizations in Botswana, Namibia, South Africa, and Zimbabwe. I wish to thank Susan Crate and Mark Nuttall who provided useful feedback and information on climate change impacts. Judy Miller and George Silberbauer reviewed the chapter and pointed out areas that needed clarification and improvement. I also wish to thank Alec Campbell, Paul Devitt, Megan Biesele, Jackie Solway, Jim Ebert, Melinda Kelly, Axel Thoma, Magdalena Brormann, Alice Mogwe, Gakemodimo Mosi, Masego Nkelekang, Chris Sharp, Derek Hudson, Roy Sesana, Joram/Useb, Aron Crowell, John Cooke, Helga Vierich, John Holm, Richard Morgan, Larry Robbins, Richard Lee, Ted Scudder, Stephen Sanford, Festus Dikgale, Rod Brandenburgh, and Janet Hermans for their insights and advice. I cannot express sufficiently my indebtedness to all of the San and other people in the Kalahari who provided information on their adaptations and strategies for coping with environmental, economic, social, and political stress and variability.

References

Behnke, R., I. Scoones, and C. Kerven, eds. 1993. *Range ecology at disequilibrium: New models of natural variability and pastoral adaptation in African savannas*. London: Overseas Development Institute.

Brandenburgh, R. L. 1991. An assessment of pastoral impacts on Basarwa subsistence: An evolutionary ecological analysis in the Kalahari. MA Thesis, University of Nebraska, Lincoln.

Campbell, A. C. 1979. The 1960's drought in Botswana. In *Proceedings of the symposium on drought in Botswana*, ed. M. T. Hinchey, 98–109. Gaborone: Botswana Society in collaboration with Clark University Press.

Campbell, A. C. and G. Child. 1971. The impact of man on the environment of Botswana. *Botswana Notes and Records* 3: 91–110.

Cashdan, E. 1984. The effects of food production on mobility in the central Kalahari. In *From hunters to farmers: The causes and consequences of food production in Africa*, eds. J. D. Clark and S. A Brandt, 311–27. Berkeley: University of California Press.

Hay, R. W. 1988. Famine incomes and employment: Has Botswana anything to teach Africa? *World Development* 6: 1112–25.

Hinchey, M. T., ed. 1979. *Symposium on drought in Botswana*. Gaborone, Botswana: Botswana Society and Hanover, NH: University of New England Press.

Hitchcock, R. K. 1978. Kalahari cattle posts. Two volumes. Gaborone, Botswana: Ministry of Local Government and Lands.

———. 1979. The traditional response to drought in Botswana. In *Proceedings of the symposium on drought in Botswana*, ed. M. T. Hinchey, 91–97. Hanover, NH: Clark University Press and Gaborone, Botswana: Botswana Society.

———. 1982. The Ethnoarcheology of sedentism: Mobility strategies and site structure among foraging and food producing populations in the eastern Kalahari Desert, Botswana. PhD dissertation, University of New Mexico, Albuquerque.

———. 2002. Coping with uncertainty: Adaptive responses to drought and livestock disease in the Kalahari. In *Sustainable livelihoods in Kalahari environments: Contributions to global debates*, eds. D. Sporton and D. Thomas, 161–92. Oxford: Oxford University Press.

———. 2004. Mobility, sedentism, and intensification: Organizational responses to environmental and social change among the San of Southern Africa. *In Processual Archaeology: Exploring Analytical Strategies, Frames of Reference, and Culture Process*, ed. Amber Johnson, 95–133. Westport, CT: Praeger.

Hitchcock, R. K., J. I. Ebert, and R. G. Morgan. 1989. Drought, drought relief, and dependency among the Basarwa of Botswana. In *African food systems in crisis*, Part One: *Microperspectives*, eds. R. Huss-Ashmore and S.H. Katz, 03–36. New York: Gordon and Breach Science Publishers.

Hitchcock, R. K., M. Biesele, and K. Ikeya, eds. 2006. Updating the san: Image and reality of an African people in the twenty-first century. *Senri Ethnological Studies* 70. Osaka, Japan: Senri Ethnological Studies.

Holm, J. and M. Cohen. 1988. Enchancing equlity in the midst of drought: The Botswana approach. *Journal of Social Development in Africa* 3(1): 31–38.

Holm, J. D. and R. G. Morgan. 1985. Coping with drought in Botswana: An African success. *Journal of Modern African Studies* 23(3): 463–82.

IPCC (Intergovernmental Panel on Climate Change). 2001. *Climate change 2001*. Cambridge: Cambridge University Press.

Katz, R., M. Biesele, and Verna St. Denis. 1997. Healing makes our hearts happy: Spirituality and cultural transformation among the Kalahari Ju/'hoansi. Rochester, VT: Inner Traditions International.

Lee, R. B. 1979. *The !Kung San: Men, women, and work in a foraging society*. Cambridge: Cambridge University Press.

Marshall, L. 1976. *The !Kung of Nyae Nyae*. Cambridge, MA: Harvard University Press.

Mazor, E. 1982. Rain recharge in the Kalahari: A note on some approaches to the problem. *Journal of Hydrology* 55: 137–44.

Perkins, J. 1991. The impact of borehole-dependent cattle grazing on the environment and society of the eastern Kalahari Sandveld, Central District, Botswana. PhD Dissertation, University of Sheffield.

Republic of Botswana. 1990. Tourism Policy. Government Paper No. 2 of 1990. Gaborone, Botswana: Government Printer.

Sanford, S. 1977. *Dealing with drought and livestock in Botswana*. Gaborone, Botswana: Government Printer.

Saugestad, S. 2001.The inconvenient indigenous: Remote area development in Botswana, Donor Assistance, and the First People of the Kalahari. Uppsala, Sweden: Nordic Africa Institute.

Schapera, I. 1930. The Khoisan peoples of South Africa: Bushmen and Hottentots. London: Routledge and Kegan Paul.

———. 1939. A Survey of the Bushman Question. *Race Relations* 6(2): 68–83.

Silberbauer, G. B. 1981. *Hunter and habitat in the central Kalahari Desert*. New York: Cambridge University Press.

Simms, A. and H. Reid. 2005. *Africa—Up in smoke? The second report from the Working Group on Climate Change and Development*. London: New Economics Foundation and International Institute for Environment and Development.

Sporton, D. and D. Thomas, eds. 2002. *Sustainable livelihoods in Kalahari environments: Contributions to global debates*. Oxford: Oxford University Press.

Suzman, J. 2001a. An introduction to the regional assessment of the status of the San in Southern Africa. Windhoek, Namibia: Legal Assistance Center.

———. 2001b. An assessment of the status of San in Namibia. Windhoek, Namibia: Legal Assistance Center.

Tanaka, J. 1980. *The San: Hunter-gatherers of the Kalahari. A study in ecological anthropology*. Tokyo: University of Tokyo Press.

Thomas, D. S. G. and P. A. Shaw. 1991. *The Kalahari environment*. Cambridge: Cambridge University Press.

Thomas, D. S. G. and C. Twyman. 2005. Equity and justice in climatic change: Adaptations amongst natural resource dependent societies. *Global Environmental Change* 15: 115–24.

Thomas, E. M. 2006. *The old way: A story of the first people*. New York: Farrar, Straus, Giroux.

Vierich, H. I. D. 1981. The Kua of the southeastern Kalahari: A study in the socio ecology of dependency. PhD dissertation, University of Toronto.

———. 1982. Adaptive flexibility in a multi ethnic setting: The Basarwa of the southern Kalahari. In *Politics and history in band societies*, eds. E. Leacock and R. Lee, 213–22. Cambridge: Cambridge University Press.

Wily, E. A. 1979. Official policy towards San (Bushmen) hunter gatherers in modern Botswana: 1966–1978. Gaborone, Botswana: National Institute of Development and Cultural Research.

PART 3
Anthropological Actions

Chapter 16

Consuming Ourselves to Death: The Anthropology of Consumer Culture and Climate Change

Richard Wilk

Defining "The Cause" of Global Climate Change

The way different experts define the causes of global climate change tells a great deal about their training, world view, and the limitations of the partitioned knowledges we have inherited from the nineteenth-century division of the world into physical, natural, and social sciences. Ask an atmospheric scientist, and they will tell you that the cause is greenhouse gases, cloud cover, and the balance of the carbon cycle. An ocean specialist is likely to talk about thermohaline circulation, currents, glaciers, and sea levels. You get the same myopia from economists, who focus on the industries that spew carbon dioxide and other gases into the atmosphere, and from political scientists who talk about the failure of regulatory regimes. A tropical ecologist is likely to give you a lecture about deforestation and the burning of Amazonia.

At the other extreme are consumer activists and the "simple living" movement, which tell us that climate change is the result of Western society's extravagant lifestyles that consume too many of the world's resources, and other world areas emulating the West. From this perspective, consumption is "using up" the resources of the world, leaving nothing but a despoiled wasteland for the next generations. While this is an easy way of defining the problem, with a clear moral vision defining "good guys" and "bad guys," it is in many ways as myopic and inaccurate as a technical focus on smokestack emissions or rainforest destruction. It may work as an agitation tactic to wake people up and get them to see the urgency and magnitude of the problem of climate change, and warn them that the solution will impact all of our lives. However, in some important ways, this image of "using things up" misrepresents the problem and can make consumers think that just using less of everything is the answer.

If the world was simply a shopping bag of groceries that humans could either open and eat, or store for later, then the idea of "using up" would make sense. Instead human impacts on the planet are far more varied than simply using up stocks of resources. Most nonrenewable resources—like copper, iron, and coal—are still abundant and are in no danger of running out. In fact, cost, not availability, is usually the obstacle to obtaining them. Additionally, the ocean is an almost inexhaustible supply of gold and other valuable minerals, but the cost of recovering them is very high. The irony is that *renewable* resources like timber and fish are the ones most in danger of being destroyed by human overexploitation. The most immediate ecological dangers of pollution, extinction, and climate change are due more to waste, poor regulation, and unregulated emission than to the using up of resources.

The vast engines of the economy, especially in rich, developed countries, *are* ultimately driven by the relatively luxurious lifestyles of what is called the "consumer economy," a way of life based on moving and transforming huge amounts of materials and energy. Averaged on a per person basis, for example, citizens of the US use about sixty times as much material each year as citizens of poor countries like Mozambique, and ten times as much as people in middle-income countries like Turkey or Mexico. On the other hand, because of their huge populations, large countries like China and India, even though they consume far less on a per capita basis, are now rivaling the national levels of consumption and carbon emission of rich countries with far smaller populations. Said another way, if 1.3 billion Chinese were to start producing waste and emitting CO_2 at the same level as North Americans, the rate of climate change would increase dramatically. The Chinese, however, see this is an issue of development and justice, and ask why they should stay in poverty in order to compensate for the past greed and wealth of North Americans and Europeans (Patterson and Grubb 1992). Even though rich countries want to focus on the present and the future, the legacy of the past injustice and inequality will not go away in debates over sustainability. There is always a moral and political dimension to issues of consumption, and because moral standards vary from place to place, we also have to think about different cultural ideals about justice, comfort, needs, and the future. Anthropology is uniquely situated as a discipline to address just these issues, all of which have historically been part of our research program.

But defining the cause of climate change as overconsumption and prescribing a kind of global belt tightening tends to put the burden of *systemic* change on individuals. Many of the ways North American consumers live are beyond their individual control: suburbs are built in a way that requires people to own cars, since there is no public transportation. Businesses are actually the largest consumers of many kinds of goods like paper and cardboard and so even if every private citizen were to recycle every single piece of household paper waste, this would still account for less than 10 percent of the paper used in the country. Many consumption decisions are not made by consumers at all, but by governments, regulatory agencies, and businesses.

To give just one example, the automobile industry in the US has not produced and sold large numbers of electric vehicles to a mass public market since 1914, despite abundant evidence for demand. Archaeologist Michael Schiffer (2003) has persuasively argued that the death of the electric car was a cultural event, related to the marketing of electric cars as "feminine," rather than an economic or technical imperative. The industry's antipathy to the electric car has continued to the recent past, when General Motors, forced to produce an electric vehicle by the state of California, eventually killed the program and destroyed the vehicles despite strong public demand (Sartotius and Zundel 2005).[1]

My point is that it is only easy to condemn overconsumption in general terms. On a large scale there is no question that many of the environmental problems the world faces are due to high levels of material and energy that are used, wasted, and disposed of. But the devil is in the details. A simple message to "consume less" may be a satisfying moral message, but in different contexts it can carry class, cultural, and religious overtones. Most religions of the world tell us to be charitable, because wealth is corrupting and materialism distracts us from spiritual and ethical matters (Belk 1983; Miller 2005). But on the other hand, poor people are often accused of wasting their money and resources, consuming improperly or immorally on clothes, drink, or drugs. The moral critique of consumption is a difficult terrain, in which we have to tread carefully to make sure we are not passing along class or religious prejudices, rather than thinking about environmental ethics and issues (Horowitz 1988; Miller 2001; Wilk 2001).

In practice it is very hard to separate "good" consumption from bad; why should visiting museums to view fine art be inherently better than going to amusement parks? Is a collection of old master paintings less materialistic than a box of comics, or a garage full of motorcycles? Why should a fast food hamburger be censured, while a plate of fresh foie gras in a gourmet restaurant is praised? Recent studies show that eating centrally prepared, highly processed food is actually more energy-efficient than home cooking the same number of meals from fresh ingredients purchased individually in many stores (Sonesson et al. 2005a, 2005b). When it comes to organic, ethical, local, and energy-efficient, what yardstick do we use to measure the negative aspects of consumption, and when do we acknowledge the importance of aesthetics and pleasure? Even tracking the total energy cost or CO_2 emissions of a single product turns out to be a complex task, full of uncertainties and arbitrary choices. For the problem of climate change, some kinds of consumption are pretty irrelevant—what is most important is how much fossil fuel was burned to produce, move, and dispose of an item, and how much fossil fuel an item (like a lawnmower) consumes over its life.

So far most of the information that has been provided to the consumer public about these kinds of issues has been oriented towards helping individuals be "smart consumers' who make the right choices. But what do consumers do as public citizens, as politically engaged members of society, and as anthropologists who want to address the major pressing issues that face humanity during our lifetime?

At the general, strategic level, can anthropologists contribute to understanding the complex dynamics of consumer culture, so that they can play a part in the modeling and prediction that has become so crucial to the public policy battles going on at national and international levels? In other words, can anthropologists use what they know about society in general to understand consumer culture in a way that is precise and concrete enough to have an influence on government policy? The second way anthropologists can work is the tactical level of applied anthropology, on specific projects aimed at changing specific kinds of consumption—can anthropologists, for example, find ways to get people to weather-strip their houses or compost their organic trash?

Anthropology and Climate Change Strategy

One way to recast the problem of consumer culture and climate change in anthropological terms is to phrase it this way: *What makes human wants and needs grow?* How do things that were once distant luxuries—say, hot water—become basic necessities that people expect on demand for civilized life (see Illich 1977)? Air conditioning in personal automobiles, an expensive and uncommon option just twenty years ago, is now standard—even on cars sold north of the Arctic Circle in Norway! Why do Western consumers expect their standard of living to keep rising?[2]

Anthropology offers the scope and sweep of time to step back and offer a bigger picture of how the human species got itself into its present dilemma of rapid growth in greenhouse gas emissions. Archaeology shows that the growth of human wants and needs is not a new thing. The Neolithic period, for example, saw human societies all over the world becoming used to a wide assortment of new goods and possessions, from pottery to personal jewelry. Once people built a way of life around these goods, they seem to have been very reluctant to give them up and go back to being mobile hunters and gatherers. But compared to the pace of change today, the consumer culture of most early civilizations was relatively stable, and new wants and needs grew very slowly. The Nile Valley of Egypt and Sudan hosted a civilization that seems to have provided a stable, modest life with the same basic domestic consumer culture for millions of people over at least three thousand years. There were only short periods of instability. A small urban elite class lived in relative luxury, with bigger houses, servants, and many kinds of exotic arts and crafts, but even their repertoire was, by our standards, remarkably conservative and slow to change. The daily material culture and rhythm of everyday life did not substantially change until cheap manufactured goods and machines from Europe arrived in the late nineteenth century.

These historical examples contrast sharply with today's furious pace of change in Chinese and Indian consumer societies or with the constant stream of new goods that flow into supermarkets, electronic stores, clothing stores, and car dealerships in the US and other rich countries. Marketing and selling, as Kalman Applbaum (2003) persuasively argues, are the dominant ideology, the unquestioned daily work, of Western consumer society. Disposing of no-longer-wanted goods has itself become a major enterprise and environmental problem. North Americans are buying bigger and bigger houses (and renting huge amounts of extra space) just to have the room to keep all of their possessions.

Why have human beings become so insatiable? Social science, to date, has provided only fleeting and partial answers to this query. Here anthropology can potentially play an important role. Economic anthropologists have convincingly argued that the modern capitalist economic system, itself an enormous engine of growth requiring constant expansion in consumption, is itself a cultural artifact (Miller 1997; Yang 2000). Humans are operators and participants whose behavior is not simply determined by advertisers. Because anthropology has a systemic view of human society and does not isolate our evolutionary heritage from our technology, economy, consumption, morality, and religion, and which commands the full sweep of the human past, it alone has the potential to assemble a comprehensive approach. But since the time of Leslie White's now outdated evolutionary model of human energy use, anthropologists have not used their magnificent cross-cultural and long-term data on human societies to think synthetically about the problems of growth and consumption.

This is not to say that anthropologists have not recognized consumption's key importance to the problem of climate change. In the last decade, an interdisciplinary approach to understanding consumer culture has been slowly growing under the rubric "sustainable consumption" (Murphy and Cohen 2001). However, anthropologists have not had a notable presence in any of the discussions about how public policy can respond to the challenge of making consumer culture more sustainable. Willett Kempton, a cognitive anthropologist trained at Berkeley, and I were the only anthropologists who gave papers at the 1995 National Research Council conference on Environmentally Significant Consumption. Since then, sociologists, social psychologists, and ecological economists have taken the lead in thinking about the problems of high consumption in North America (e.g., Jackson 2006; Princen 2005; Princen, Maniates, and Conca 2002).

The idea of making consumer culture more sustainable through government policy is much more widely discussed in Europe than in the US, and there have been many congresses and high-level policy groups on the issue (OECD 1997). There are already dramatic results: for example, German legislation requiring companies to take responsibility for recovering and recycling the consumer products they sell (papers in Reisch and Ropke 2005). Yet anthropologists have again been almost invisible in these discussions, their place taken by social scientists who lack cross-cultural experience and have a global perspective.

At least anthropologists have begun to take consumer culture seriously as a research topic in the last two decades.[3] The number of studies and their breadth is truly impressive (e.g., Miller 1995), though some find the anthropological approach fragmented, overly symbolic, and poorly contextualized with political economy (Carrier and Heyman 1997). Anthropologists have also been lumped together with cultural studies scholars, whose approach to consumer culture is often to celebrate its richness and variety, the opportunities it offers for expression, agency, and developing identity, rather than critiquing it.

Nevertheless, for a globally contextualized picture of how consumer culture is becoming established in new territories around the world, or for a detailed understanding of the social and moral meaning of daily grocery shopping in rich countries, anthropologists offer rich and comprehensive case studies and sophisticated theory. So far, what anthropologists have not done is to sit down together to compare and synthesize results, or put them into the kind of language or format that could be useful to policy makers. Why not?

Fundamentally, anthropologists are methodological individualists. We are not trained in collaborative research, and we are not socialized to work together—instead, we compete for publications, jobs, and visibility. Our collaborations tend to be fleeting affairs, at most on a single research project or a publication, more often confined to a few days of conference discussion or committee meetings. Contemporary anthropology departments rarely offer the time or the resources for faculty members to actually do research together, much less synthesize and discuss their work with an eye towards policy. These are not the kinds of work that the academic reward structure supports.

In contrast, the disciplines that have made effective contributions to policy have developed appropriate institutions that bring numbers of scholars together for an extended period of time in specialized research institutes, think tanks, and policy centers, often with direct foundation support. There have been a few applied anthropology "shops" associated with major anthropology departments over the years, but they have usually been oriented toward field research and contract supported, with limited policy capabilities.[4] About the closest thing we have today are the two Sloane Foundation–supported centers on American working families at UCLA and the University of Michigan, both of which have done a great deal of innovative research and are beginning to produce policy-relevant publications. But otherwise there are no Brookings Institutes, World Resources Institutes, or Russell-Sage Foundations to support groups of anthropologists to translate research into useful advice and policy, and to act as a public voice for the discipline. The result is that anthropologists are left to their own devices to try to bring the relevant results of their individual research to the attention of a policy community, a task akin to whispering in a room full of people screaming through bullhorns.

Tactical Responses to the Problems of Consumption

Global climate change has brought the issue of consumption forward into public and academic attention. But it was the so-called first energy crisis in the early 1970s and the OPEC oil boycott that gave the world its first taste of what a world with energy scarcity and higher prices would be like. This initiated a first round of scientific research on conservation, energy efficiency, and alternative energy, funded by the US federal government and electric utilities. For a time in the early and mid-1980s, several anthropologists became involved in what was a flowering of applied social science in the field of energy consumption.

Willett Kempton was perhaps the most prominent and prolific researcher, and his innovative ethnographic work on home hot water use, the way people used and understood thermostats, and folk knowledge about energy costs was widely read and cited among energy researchers (Kempton 1986; Kempton, Darley, and Stern 1992; Kempton, Feuermann, and McGarity 1992). He had influence on the eventual shape of the DOE Energy Star labeling program. His experimental work on the way people could be encouraged to lower their electricity use by showing their neighbors' consumption on their monthly bill was pathbreaking. He has also done important research on how Americans think about and understand global climate change, which attracted attention from congressional legislators (Kempton 1997). Steve Rayner is another anthropologist who had a major role in energy research, through his leadership position at the Oak Ridge National Laboratories and then the Pacific Northwest National Laboratory, where he was chief scientist (see Rayner and Malone 1998).

During this time, Harold Wilhite and I conducted an ethnographic study contrasting people who were and were not investing in home energy-saving improvements, funded by the University of California Energy Research Group. We were the only anthropologists who applied for funding. The funding organization was initially quite skeptical, but, once they saw our results, became quite supportive. We found a receptive audience for our research in the energy community, and our technical publications continue to be cited (Wilhite and Wilk 1987; Wilk and Wilhite 1984, 1985). But even with grant funding, neither of us could get secure university positions in anthropology, and we both had to leave California. Wilhite moved to Norway, where he continued to do innovative research on culture and energy consumption (Wilhite 1996; Wilhite and Ling 1995). I left academia for a position as a rural sociologist with the US Agency for International Development.

If anyone knows how to change consumer behavior, it is advertisers and marketers. Consumer research is the applied science of consumption, taught in marketing programs in hundreds of business schools in the US and elsewhere. During the 1980s business schools became the sole haven for anthropologists interested in the applied study of consumption. Eric Arnould, John Sherry, Barbara Olsen, and Grant McCracken, among others, effectively introduced cultural anthropology to marketing programs, which had previously been dominated by social psychologists, survey researchers, and demographers. They produced a huge volume of new and creative research, textbooks, and collaborations (e.g., Arnould, Prince, and Zinkhan 2003; Sherry 1995; Sunderland 2007).

Anthropology has become both an accepted part and a fashionable trend in mainstream consumer research and marketing. Anthropologists also helped bring a critical perspective towards consumer culture into business schools, which have questioned the impact of consumption on culture, gender, class, individual identity, and the environment, in ways that were previously quite alien in a pro-business environment. This movement matured in 2005 with the "Transformative Consumer Research" initiative and conferences sponsored by the Association for Consumer Research, both of which explicitly seek to turn the tools of marketing towards socially positive ends.[5] In 2006, the University of Auckland founded an "anticonsumption" institute within its business school, a radical initiative unmatched by anything imagined so far by anthropologists or sociologists.

Given the potential for an applied anthropology of consumption, however, the total response of our discipline has been late, random, and feeble. Anthropologists have started to study the environmental movements that have spawned simple living, bioregionalism, farmers' markets, Community Supported Agriculture (CSAs), local currencies, and a host of other initiatives. However, anthropology plays catch-up with popular movements for ethical consumption, vegetarianism, and global equity, and against the World Trade Organization, genetically modified organisms, and the industrialization of the food system. These movements are growing by leaps and bounds in the US and throughout the world, and many of these groups have also formed effective international networks. Not only are anthropologists slow to recognize the importance of consumption activism and study it as an important phenomenon in the world, they also have lagged in putting anthropology to work as an applied tactical tool in furthering activist goals.

THE TEACHING MISSION

Our students live and breathe consumer culture, and just the way fish living in water never really see the water they live in, our students pass through their four years at a university without ever learning anything systematic about the consumer culture that forms the very fabric of their daily lives. In some ways this is a strange paradox—the most conspicuous aspect of modern life and the part that students have direct daily experience with is largely passed over during their studies. A liberal arts curriculum is supposed to equip them for intellectual engagement with the world they live in. This is a great tragedy, for it misses an important chance to show students how their daily life engages them with global issues, complex moral choices, and cultural complexity.

This huge gap also presents an enormous opportunity for anthropology to step in as the one discipline that can integrate a topic that is otherwise fragmented and scattered across the entire university curriculum from nutrition to economics, history to physics. Teaching about consumption is also a great opportunity to put ethnography right into the classroom, and to get the university connected with its surrounding community through various kinds of service learning and community engagement in active learning.

Among other consumption-related classes, I teach a freshman course on "global consumer culture" to 120 students every other year.[6] I give students a variety of assignments that require them to go out in the community to talk to people in supermarkets, shops, restaurants, and food banks, and I also ask them to inventory their own food, bottled water, and clothing consumption. A smaller group does service-learning projects with our Bloomington community Center for Sustainable Living.[7] A significant number of students every semester find that thinking about their own consumption is challenging, transforming, and exciting, and I am always gratified when it recruits new anthropology majors, but I am equally happy to see others heading off towards history, psychology, media studies, international studies, and other useful majors.

Teaching this class keeps reminding me that teaching university students is itself one of the most important kinds of applied anthropology. Each public speaking opportunity, every lecture is an opportunity to bring anthropology to bear on problems of consumption and to spread the message that sustainability is not an issue that can wait for the next generation. Students want to see the university itself as a laboratory for a more sustainable way of life. Anthropologist Peggy Barlett (see her chapter in this section) is among the pioneers of the "Greening Campus" movement—first working effectively to get sustainability ingrained into the mission of her own Emory University, and then seeing how to extend the message to other institutions around the country (this volume, and Barlett and Chase 2004). Any anthropology class can take up fair trade coffee, bottled water, or another global or equity issue as an applied project with direct relevance to daily campus life.

Conclusion

"Sustainability is a term like truth or beauty," says Fred Kirschenmann, a senior fellow at the Leopold Center for Sustainable Agriculture at Iowa State University. "We struggle but never get there."[8] But we have no choice but to join this struggle, because the world cannot survive any more with business as usual. Since we are all consumers, we have power to change our own participation in the system as users, though many of our decisions will be negative ones—we will decide not to eat certain things, not to buy particular products, or to travel by train instead of by air, for example. But this is only a small and relatively passive part of our possible role in building a sustainable future that is going to depend on new kinds of social, cultural, and economic systems. The challenge of the next generation is inventing those systems and getting them deployed in the world while there is still time. Anthropology has tremendous potential to play a productive role in the transformation to a more sustainable economy, but only if we are willing to enter new fields of study, improve our communication skills, and think of ourselves as participants in change rather than just critics. So far we have mainly exploited our methodological skills as specialists in ethnography and participant observation to gain entry into areas that are already full of active and engaged social scientists from rival disciplines. Now that we have an entry, it is time to use our knowledge of social change and process, our synthesis of biology and culture, our command of global issues, and our holistic understanding of the economy to make a greater contribution and increase our voice. It would help a great deal in this enterprise if we could find a way to temper methodological individualism, and create new models of comparison and collaboration, so that we could represent anthropology as more than a quarreling and fractious gaggle of scholars.

Notes

1. This incident was the topic of the documentary film "Who Killed the Electric Car," http://www.whokilledtheelectriccar.com/.
2. Surprisingly few philosophers and theorists have written extensively about this very fundamental question, which Adam Smith considered fundamental to economics, but which later economists fumbled and then dropped.
3. Although Eric Arnould and I submitted a paper on growing consumer culture in developing countries to *American Anthropologist* in 1982, it was rejected on the grounds that "consumption was not an anthropological topic."
4. One of the longest lived is the Bureau of Applied Research in Anthropology at the University of Arizona, which has concentrated on border-studies issues, and there have been similar institutions focused on international development at Harvard and SUNY Binghampton.
5. The TCR website is at http://www.acrwebsite.org/fop/index.asp?itemID=325.
6. The syllabus from the last time I taught the course can be found at http://www.indiana.edu/~wanthro/e104_05.htm
7. http://www.bloomington.in.us/~csl/
8. http://pelennor.leopold.iastate.edu/about/moreaboutfred/kirschenmann.htm

References

Applbaum, K. 2003. *The marketing era: From professional practice to global provisioning*. London: Routledge.

Arnould, E. J., Price, L., and G. M. Zinkhan. 2003. *Consumers*. Columbus. OH: McGraw-Hill.

Barlett, P. and G. Chase. 2004. *Sustainability on campus: Stories and strategies for change*. Cambridge, MA: MIT Press.

Belk, R. 1983.Worldly possessions: Issues and criticisms. *Advances in Consumer Research* 10: 514–19.
Carrier, J. and J. Heyman. 1997. Consumption and political economy. *Journal of the Royal Anthropological Institute* 3(2): 355–73.
Horowitz, D. 1988. *The morality of spending*. Baltimore: Johns Hopkins University Press.
Illich, I. 1977. *Toward a history of needs*. New York: Pantheon.
Jackson, T. 2006. *The earthscan reader on sustainable consumption*. London: Earthscan Publications Ltd.
Kempton, W. 1986. Two theories of home heat control. *Cognitive Science* 10: 75–90.
———. 1997. How the public views climate change. *Environment* 39(9): 12–21.
Kempton, W., J. Darley, and P. Stern. 1992. Psychology and energy conservation. *American Psychologist* 47(10): 1213–23.
Kempton, W., D. Feuermann, and A. McGarity. 1992. "I always turn it on super": User decisions about when and how to operate room air conditioners. *Energy and Buildings* 18: 177–91.
McCracken, G. 1988. *Culture and consumption: New approaches to the symbolic character of consumer goods and activities*. Bloomington: Indiana University Press.
Miller, D. 1995. Consumption and commodities. *Annual Review of Anthropology* 24: 141–61.
———. 1997. *Capitalism: An ethnographic approach*. Oxford: Berg.
———. 2001. The poverty of morality. *Journal of Consumer Culture* 1(2): 225–43.
Miller, V. J. 2005. *Consuming religion: Christian faith and practice in a consumer culture*. London: Continuum International Publishing Group.
Murphy, J. and M. J. Cohen. 2001. *Exploring sustainable consumption: Environmental policy and the social sciences*. London: Pergamon.
Murthy, J. 2006. *Environmental sustainability: A consumption approach*. New York: Routledge.
OECD. 1997. *Sustainable consumption and production*. Paris: Organization for Economic Co-operation and Development.
Paterson, M. and M. Grubb. 1992. The international politics of climate change. *International Affairs* 68(2): 293–310.
Princen, T. 2005. *The logic of sufficiency*. Cambridge: MIT Press.
Princen, T., M. Maniates, and K. Conca, eds. 2002. *Confronting consumption*. Cambridge MA: MIT Press.
Rayner, S., and E. Malone. 1998. *Human choice and climate change*. Colombus, OH: Battelle Press.
Redclift, M. 1996. *Wasted: Counting the costs of global consumption*. London: Earthscan Publications.
Reisch, L. A., and I. Ropke. 2005. *The ecological economics of consumption*. London: Edward Elgar Publishing.
Sartorius, C., and S. Zundel. 2005. *Time strategies, innovation, and environmental policy*. London: Edward Elgar Publishing.
Schiffer, M. B. 2003. *Taking charge: The electric automobile in America*. Washington, DC: Smithsonian.
Scitovsky, T. 1992. *The joyless economy*. New York: Oxford University Press.
Sherry, J. F. 1995. *Contemporary marketing and consumer behavior: An anthropological sourcebook*. New York: Sage Publications.
Sonesson, U., B. Mattsson, T. Nybrant, and T. Ohlsson. 2005a. Industrial processing versus home cooking: An environmental comparison between three ways to prepare a meal. *Ambio* 34(June): 414–21.
———. 2005b. Home transport and wastage: Environmentally relevant household activities in the life cycle of food. *Ambio* 34(June): 371–75.
Sunderland, P. 2007. *Doing anthropology in consumer research*. Walnut Creek, CA: Left Coast Press.

Stern, P. C., T. Dietz, V. W. Ruttan, R. H. Socolow, J. L. Sweeney, et al. 1997. *Environmentally significant consumption: Research directions*. Washington, DC: National Academy Press.

Wilhite, H. 1996. *The dynamics of changing Japanese energy consumption patterns and their implications for sustainable consumption*, American Council for an Energy Efficient Economy Summer Study, Human Dimensions of Energy Consumption, ACEEE, Washington, DC.

Wilhite, H. and R. Ling. 1995. Measured energy savings from a more informative energy bill. *Energy and Buildings* 22: 145–55.

Wilhite, H. and R. Wilk. 1987. A method for self-recording household energy use behavior. *Energy and Buildings* 10: 73–79.

Wilk, R. 2001. Consuming morality. *Journal of Consumer Culture* 1(2): 245–60.

Wilk, R., and H. Wilhite. 1984. Household energy decision making in Santa Cruz County, California. In *Families and energy: Coping with uncertainty*, eds. B. Morrison and W. Kempton, 449–59. East Lansing: Michigan State University.

———. 1985. Why don't people weatherstrip their homes? An ethnographic solution. *Energy* 10(5): 621–31.

Yang, M. 2000. Putting global capitalism in its place: Economic hybridity, Bataille, and ritual expenditure. *Current Anthropology* 41: 477–509.

Chapter 17

Global Change Policymaking from Inside the Beltway: Engaging Anthropology[1]

Shirley J. Fiske

In November 2006, the US crossed a threshold to a policy orientation towards action on carbon accumulations in the atmosphere, signaling a strong congressional interest in regulation of carbon emissions and greenhouse gases (GHGs). The US is now at the cusp of a national policy focused on what we can *do* about global climate change, in addition to documenting the likelihood and dimensions of its occurrence.

The intent of this book is to explore anthropologists' roles with respect to global climate change and to contribute to the dialog regarding our roles as witnesses, advocates, communicators, and activists. The Acting and Communicating section explores "what anthropologists are doing and can/need to do to reach out to wider audiences from the local to the global, including affecting policy."

My starting point is the intersection of national policy and opportunities I see for the engagement of anthropology, given the expected shift in US policy from research and assessment to mitigation and regulatory policy. Anthropologists have contributed a great deal to the witnessing of climate change, as described in Roncoli, Crane, and Orlove's chapter in this volume. Anthropologists have also contributed to understanding the development and use of climate and seasonal forecasts (Roncoli 2006) and undertaken critical analyses of how the science and modeling of climate change in the US is constructed and used (Lahsen 2005a, 2005b). I am suggesting that anthropologists now need to make more explicit links with the key policy concepts in national legislation and Kyoto that are shaping the implementation of tradeable emissions and carbon markets globally.

This chapter starts with the description of the salient policy provisions being debated in Congress, and the parallel and related phenomena of carbon markets. I try to raise useful policy questions that social science can address in a timely fashion, and suggest how concepts of sequestration and offsets might broaden participation in these markets beyond global financial institutions and political institutions of nation-states. I highlight examples of anthropologists who have made the link between local communities and policy instruments such as carbon offsets and sequestration and the questions they raise.

By way of background, I have been both a participant in and observer of global climate change policy for twenty-five years at the national level, in both the executive and legislative branches of government.[2] From my perspective, the nation's global change policy has been firmly rooted in research and assessment via the US Global Change Research Program (USGCRP) (P.L. 101–606), the primary authorization for the involvement of the federal government in climate change. The de facto and de jure national policy in the last twenty years has been to identify the dynamics and effects of global climate change, played out through an interagency and international effort to fund research, outreach, and assessment.

A Sea Change in Dialog

With the change in congressional leadership brought about by the 2006 elections, an energized and highly focused dialog about global climate change and greenhouse gas policies has been unleashed in Washington. The change may not be visible outside the Beltway, but it is portentous. With new leadership in the House and Senate and new chairs of key committees, Democratic legislators released an abundance of new proposals for limiting carbon dioxide and other greenhouse gases, signaling their concern about the effects of climate change and the lack of US leadership globally. In the first month of the 110th Congress, seven major bills were introduced in the Senate and House; House Speaker Pelosi announced a new committee (a relatively rare occurrence) to investigate climate change, and Senate leadership promised to make it a legislative priority. At the close of the session in December 2007, eighteen greenhouse gas emission limits proposals had been introduced, compared to two in the previous session, and one in the previous decade. This vigorous interest in and number of concrete proposals for regulating greenhouse gases to slow carbon accumulation in the atmosphere had not occurred in the last fifteen years under the Global Change Research Act of 1990.

The policy questions have migrated to options to mitigate or control carbon emissions. The debate in the second half of the 110th Congress will become a competition among variants of cap and trade proposals. If passed and enacted, these bills will shape how the US will deal with climate change domestically and internationally in the coming decades.

Anthropologists have a well-established track record in witnessing climate change impacts across the globe, but have been more hesitant to be actors in the areas of national policy making. Now that the debate has moved to implementing carbon controls and how that might work, there are opportunities to bring anthropologists' participation, collaboration in the field, and critical analyses to the specific programs and activities that are being discussed.

The current Washington debate reframes the traditional scientific debate (whether global climate change exists or not) with the focus on what to do about it—namely, regulating carbon emissions through taxes or a cap and trade system. Following is a brief overview of the basic concepts embedded in legislative proposals for a national cap and trade system to limit greenhouse gases.

A Cap and Trade Field Day

Cap and trade systems generally rely on market oriented emission reduction schemes, along the lines of the trading provisions of the current acid rain program.[3] While there are many greenhouse gas emission limits proposals, I will focus my discussion on cap and trade systems, because of the intersection of carbon market with the cap and trade provisions of the Kyoto protocol[4] and because a cap and trade system is the most likely type of regulatory system to be enacted.

Cap and trade systems set limits on specific emissions, such as carbon dioxide or methane, in the case of greenhouse gases. The regulated entities—e.g., oil producers, refineries, electricity generating plants, or carbon-intensive industries—would be required to meet these targets, either as "upstream" producers or "downstream" emitters, by reducing their own emissions or purchasing credits (buying or trading) from other regulated entities that have reduced emissions below their individual allotment. Allocations of allowances (sometimes called credits or permits) would generally be by government authorization or auction, depending on the legislation.

The provisions in the seven major bills for greenhouse gases vary across a number of factors: breadth of emissions limitations (economy-wide vs. specific entities such as petroleum or natural gas refineries); specific emissions limits (year of implementation, specific scientific standards vs. relative reductions); variety of allocation proposals (allowances, auctions, safety valves); and provisions for early reduction credits and bonus credits. Many of the bills include a public auction of allowances, returning the revenues to the public treasury, variously called a Climate Change Credit Corporation, Climate Action Trust Fund, or Climate Reinvestment Fund. Revenues would be used for a number of purposes, including as carbon-reduction technologies to be shared with developing countries, rewards for early reductions of greenhouse gases, or assistance for low-income households and dislocated workers (Parker and Yacobucci 2007); the importance here is the interest in authorizing a mechanism to allocate carbon revenues to public purposes.

Other provisions of note are the inclusion of research/studies of the impact of climate change on the world's poor and provisions to encourage domestic agriculture and forestry offsets. In this vein, several of the bills include provisions for an offset program to reward sequestration, the process of capturing carbon dioxide in such a way that it is sequestered, or prevented from entering the atmosphere. S. 2191, America's Climate Security Act of 2007, for example, would establish a domestic offset program to sequester GHGs in agriculture and forests.

I review these provisions in the hope that anthropologists will consider taking a closer look at the domestic policy debate—writing op eds, contacting congressional members, organizing professional associations, and commenting or testifying on legislative provisions that clearly affect both national policy and various stakeholder groups or constituencies. I would expect that many of these provisions could be of interest to anthropologists because each raises questions surrounding access to and use of carbon auction public funds, distributional effects of carbon taxes, environmental justice concerns, and equity among stakeholder groups in implementing them.

Proponents of a strong global change policy argue that the US must take an aggressive position on climate change to signal national intent for the upcoming post-Kyoto treaty negotiations (it expires in 2012). Only via strong US leadership and policy, it is argued, can the US hope to influence China and India (who are not signatories to the Kyoto Protocol), whose rates of growth in carbon emissions now surpass the US (Mufson 2007), to join the treaty and reduce carbon emissions.

Deconstructing Cap and Trade

The domestic discourse about greenhouse gas emission reductions, cap and trade, carbon taxes, allocations of credits, carbon reduction technologies, etc., is only peripherally about the effects on communities, families, and people—things that are traditionally of interest to anthropologists. Global climate change is seen as a technology problem that can be overcome through market forces and the creation of a carbon proxy currency, including a system where incentives nicely cushion regulation of industry, utilities, agriculture, and manufacturing sectors.

Figure 17.1, widely circulated among staff and members of Congress, elegantly symbolizes the techno-environmental priorities and dimensions of the cap and trade dialog by mapping the effects of each legislative proposal on metric tons of carbon emitted over time. The implicit assumption is that the greater the reduction of metric tons of carbon dioxide, the better for global stabilization, the environment, and therefore the better legislation. Global climate change is seen as a carbon accumulation problem for the US, not necessarily an environmental justice or human rights issue, which are largely secondary in the debate inside the Beltway.

The discourse on legislative proposals is driven by the hearings process. In gathering public comment on the cap and trade bills since 2005, congressional committees carefully selected witnesses (which have to be agreed to by both majority and minority staff) that continually represented a standard set of sectors and interest groups that it viewed as most instrumental and most affected: the authoritative voice of data and analysis of energy use and

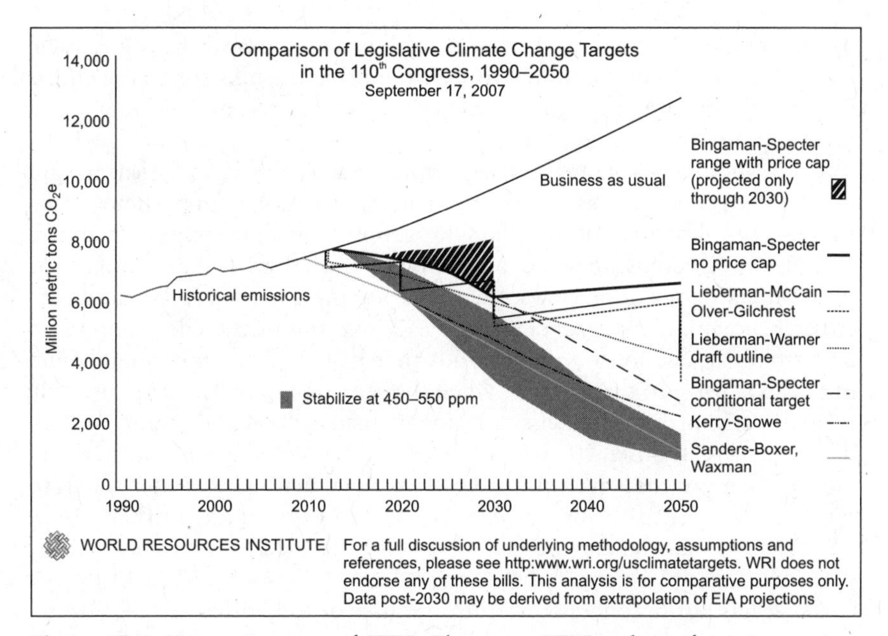

Figure 17.1: Figure courtesy of WRI. Please see WRI website for most current estimates, continuously updated based on legislative developments. http://images.wri.org/usclimatetargets chart1 big.gif (Larsen and Heilmyr 2007).

modeling, the Energy Information Administration (part of the Department of Energy); the executive director of the National Commission on Energy Policy, a bipartisan, energy-specific think tank funded by the Hewlett Foundation; witnesses from what were seen as environmentally linked advocacy groups, the Natural Resources Defense Council, or the World Resources Institute; high-profile think tanks and foundations with substantial programs in the field, such as the Pew Center on Global Climate Change; a representative of financial and energy interests on Wall Street, an economist with an energy consulting or capital markets firm; and a CEO from an energy holding company, a utility, or oil and gas firm (or the American Petroleum Institute, the oil industry association).

Committee staff rarely turned to expert scientists from the National Academy of Sciences or universities on this policy trajectory (because it was not about science anymore; it is about mitigation of carbon accumulation). When a committee wanted specific testimony on economic costs and impacts of the bills, they invariably turned to economists who describe them in macroeconomic terms. The point is that neither anthropology nor other social sciences, which can give finely tuned characterizations of impacts, are among the opinion leaders on global climate change emissions policy.

The momentum carrying the debate in 2007 is clearly with the interest groups and national associations that have a grip on the US economy—oil, utilities, automobile manufacturers, or associations that are heavily backed by industry. It is clear that the beneficiaries in cap and trade legislation will be industry and manufacturing, to the extent that the bills soften the effects of regulation through a trading and early credit system, and even more if a price cap on carbon is enacted.

Anthropologist Myanna Lahsen has provided critical analytical insight into US climate politics and the process of policy making. Of particular relevance to this discussion is her characterization of the buffeting of the public by "media-driven campaigns" and "political interests." She identifies the two competing scientific views: "One side attempts to collect data and construct prudential, risk-reducing policies while the other side attempts to undermine any action on the grounds that it would be unreasonable and overly expensive to do anything when the scientific models are at best only probabilistic products integrating inconclusive data and significant indeterminacies" (Lahsen 2005b, 142). Her analysis also suggests "inequities of power" that are challenges in the US—she documents the greater access to power and influence of "high-profile climate dissidents" (Lahsen 2005b, 143), and suggests that the playing field needs to be leveled for more democratic decision-making (see also Lahsen in press).

What has not become part of the discourse or fully discussed on Capitol Hill are discussions of the regressive nature of cap and trade as currently structured. The increased costs of producing electricity, gasoline, and other goods and services are expected to trickle down disproportionately to lower- and fixed-income families and individuals. For one bill, S. 1766, gasoline prices are projected to rise by 22 cents per gallon by 2030, and electricity prices would increase by 19 percent in the same time period. The average household income would decrease by $435 by 2030. The GDP would continue to increase, but would be slower by 0.5 percent, according to a recent analysis by the US Environmental Protection Agency (2008). Proponents for emissions limits (those aligned with the first of Lahsen's scientific interpretations) make every effort to minimize these allegedly minor adjustments, as if there will be very little effect on America's families or the economy, so that these price vectors do not create political backlash.

On the other hand, those who believe that global warming "is the biggest hoax ever perpetrated on mankind," as Senator James Inhofe (R-OK) was fond of saying on the Senate floor, marshal data that show that greenhouse gas emissions limits will bring economic calamity for the United States. These are the second of Lahsen's competing scientific views, embodiment of the "anti-greenhouse" coalition (Lahsen 2005b, 144). Testifying before the US Senate Committee on Environment and Public Works, Kevin Book, a senior vice president at FBR Capital Markets Corporation, stated, "Efforts to internalize the cost of greenhouse gas emissions could seriously disrupt one or several economic sectors, particularly power generation, heavy industry and fossil energy production." He went on to say what many economists have argued, that a carbon tax would be more efficient than a cap and trade policy (Senate Committee on Environment and Public Works 2007).

While serious discourse on impacts on families, marginal populations, or those most at risk due to climate change has been limited to macroeconomic analyses, some in Congress have been concerned about this lacuna. A number of legislators have added provisions for assistance for low-income households to help address increasing costs. But actual testimony on behalf of low-income consumers and/or marginal populations—subsistence hunters in Alaska, tribal groups in the Southwest, bayou residents in Louisiana, families in Pacific island states—was not invited nor heard at congressional hearings. While it may have come through the front doors of the personal offices of the members, it was not evident in formal hearings. In general, social and environmental justice concerns are not front-burner issues in the cap and trade policy discussions; rather, the front-burner issue is ability to stem emissions at a specific scientific target of 450–550 parts per million carbon equivalents, and secondly, political expediency (meaning trades and compromises) in order to get legislation through Congress. The human dimensions of global climate change are obscured in the greenhouse gas policy discourse.

As debate continues from 2008 forward, and as different committees review the variety of cap and trade bills under Democratic leadership, more attention may be paid to social justice and locality-based issues. The Senate Committee on Environment and Public Works, for example, held a hearing on "their" cap and trade bill in 2007 (S. 309) and invited testimony on the Chesapeake Bay region, signaling an interest in real effects on vulnerable populations, not solely a macroeconomic view of cap and trade impacts.

Growing Market for Carbon Sinks

As Washington debates policy and Kyoto heads to its next round of negotiations, financial institutions are creating capital through carbon markets, a phenomenon closely tied to the Kyoto treaty and US legislative proposals. Carbon futures are being traded on the Chicago stock exchange in the US and the European Union (EU), where auctions of carbon offsets are being sold.

The EU began regulation of carbon dioxide emissions in January 2005, as energy, metals, minerals, and pulp and paper came under limits—and the future may include additional sectors such as transportation. The EU adopted a directive that allows companies to buy Certified Emissions Reductions (CERs) via the Kyoto Protocol's Clean Development Mechanism (CDM) emission allowances, creating a strong market for CDMs. When a company buys a CER, the firm gets an EU emission reductions unit in exchange for surrendering the CER to the country government—which the country can use to offset its Kyoto reduction targets.

In the US, even though emissions reductions are not yet in effect, the first carbon trading corporation, the Chicago Climate Exchange (CCX), where companies can trade credits for greenhouse gas reductions among themselves, was launched in 2003.[5] CCX auctions carbon reduction credits in exchanges between the US and developing countries for the Chicago Climate Futures Exchange (CCFE)—a derivatives exchange, a wholly-owned subsidiary of CCX (www.chicagoclimatex.com). Another firm, Point Carbon, an analytic services provider including "market intelligence" of carbon markets, generated $26 million in a new investment (global venture capital) through a new share issue in Fall 2007 (www.pointcarbon.com).

Revenue is being generated. The beneficiaries of the carbon markets are publicly traded corporations, privately held corporations, and venture capital and hedge funds, in general. The question now becomes whether the economic engine that accompanies carbon reduction has any benefits for the victims of the carbon transgressions of developing and developed countries. Theoretically, yes; but there are multiple policy, organizational, and social institutional hurdles to overcome, as will be discussed below.

At the time of writing, there is a 50–50 expectation inside the Beltway that Congress will pass some kind of cap and trade legislation, probably after the 2008 election. The Senate has moved forward with a bill ready to go to the floor (S.1291) but the House of Representatives does not yet have a bill, leading to speculation that Representative John Dingell (D-MI), the House of Representatives' top energy leader, is not anxious to embrace emissions limits legislation in order to protect Michigan's powerful automobile manufacturers.

A cap and trade bill, if enacted, would put a premium on reducing carbon emissions and the technologies, hard or soft, to achieve those reductions. Obviously there would be dramatic growth in carbon markets. One of the soft technologies being talked about is carbon sequestration through biomass production or soil or ocean sequestration—increasing forests (afforestation or reforestation), grasslands, tropical forests, or land cover that sequester carbon through vegetative growth or reduce its transfer to the atmosphere in the case of soil or ocean sequestration. For example, CCX's offsets program is broadly based, including agricultural methane, landfill methane, agricultural soil carbon, forestry, renewable energy, coal mine methane, and rangeland soil carbon. These carbon sink assets (offsets) are part of the carbon market.[6] US firms would be able to buy credits from developing countries and businesses. There are transaction costs since the credits need to be applied for, registered, certified, and monitored, incurred by the project implementers.

CARBON CYCLING, OFFSETS, AND ANTHROPOLOGICAL INSIGHTS

Anthropologists have helped make the link between the policy options being discussed and the constituencies potentially affected, scaling up understandings of sequestration projects. In particular, anthropologists Carla Roncoli, Carlos Perez, and a number of colleagues, as part of the Carbon from Communities project funded by the US National Aeronautical and Space

Administration (NASA), offered a bottom-up perspective on the institutional, social, and political challenges and constraints to pastoralist and agriculturalist participation in programs that sequester carbon in drylands Africa (Perez et al. 2007a; Roncoli et al. 2003). [7] It should be noted that there are other community-based carbon sequestration projects being implemented, typically forest-based systems in Latin America or South Asia. However, anthropologists have not been involved or studied those cases (e.g., Asquith, Vargas Rios, and Smith 2002).

The carbon offsets program, the CDM, under Article 12 of the Kyoto Protocol to the UN Framework Convention on Climate Change (UNFCC), permits Annex 1 countries (largely developed countries) to obtain additional credits (CERs) from GHG reduction projects in a developing country that they can use to meet their emission caps. The process is overseen by a CDM Executive Board that registers and validates projects, issues CERs, and manages a series of panels and working groups. A critical component of the process is the requirement that CERs issued under the CDM represent only reductions in excess of those that would have occurred in the absence of the project. Almost all of the current CDMs are emissions reductions projects (renewable electricity projects), or afforestation and reforestation; but CERs may be expanded to other sequestration projects in the future: the details of the CDM under Articles 3.3 and 3.4 of the Kyoto Protocol, which identify activities in addition to afforestation and reforestation, have yet to be decided upon by the UNFCC. Mechanisms for participation in carbon trading are included in most of the GHG cap and trade bills under consideration by the US Congress.

Participation in carbon markets through the provision of offsets is beginning to receive critical evaluation, particularly in Africa (Perez et al. 2007b). In the context of multidisciplinary research, such as the above-mentioned Carbon from Communities project, anthropologists are elucidating the relationship between social and cultural realities on the ground, land management interventions, carbon sequestration, carbon offset policies, and ultimately markets. Roncoli et al. have examined the role of institutional factors in the adoption of carbon sequestration practices among farmers using sustainable tilling practices and pastoralists using rotational grazing practices on common rangelands in Mali[8] (Roncoli et al. 2003, 5). They also examine the potential for existing institutions at the local level to mesh with supralocal institutions that can support collective/aggregate land use decisions and represent communities in carbon trading negotiations.

Participation in carbon trading would mean creating supralocal institutions, aggregating or consolidating households, village associations, communities, or producers associations such as the herders' association. It would mean that units need to have or choose a representative unit that is recognized as legitimate by internationally recognized corporations or the national government—they could be brokers or brokering institutions for local communities, which in any case would create additional costs to producers. Roncoli et al. speculate that at the local level "the Commune and the herders' associations are both likely to play key roles." At the national level, community-based groups would likely join in broader coalitions. "New federative bonds are also developing among community-based groups, enabling them to have a voice in national level policy making" (Roncoli et al. 2003, 17–18). She points out that being organized into a federation would allow communes and associations to accumulate enough carbon credits to engage in trading and negotiation.

One of the core findings is that carbon offset projects generally rely on a number of assumptions that are problematic for pastoralists in north-central Mali and farmers in southern Mali. The assumptions of clearly established land tenure, dedicated to a single use, within clearly demarcated administrative or territorial units do not make sense given the seasonally fluctuating, multiple use of common property (Roncoli et al. 2007).

Carbon sequestration projects are possible, operationally, but will need serious rethinking for them to be viable, equitable, and useful for communities to see the benefit of carbon markets (Perez et al. 2007b; Roncoli et al. 2003). To this end, the work of human geographer Petra Tschakert is particularly instructive (2004, 2007). Her work shows that carbon sequestration projects are likely to be adopted by, *and more able to be adopted by*, the better-off farmers with larger asset bases, raising concerns about equity among participants, a concern that Roncoli et al. (2003) have also noted with respect to farming families in southern Mali. Tschakert directly addresses the question of whether poverty reduction through environmental services provision such as carbon sequestration is an option for smallholders in the Sahel (2007). Through her case study in Senegal, she concludes that access to the projects is likely to be inequitable, and she questions whether benefits will reach poor smallholders. Additionally, she notes that in regions where there is a great deal of uncertainty and risk (as in Sahelian regions), factors that constrain participation of the poor must be addressed—such as the need for more flexible management and payment mechanisms, including local credit schemes and cost-sharing mechanisms. She also suggests the need for strengthening institutional structures including "cooperative institutions facilitate the bundling and bargaining of contracts" so that a group of organized smallholders' services can compete against those offered by a corporation or privately held contract (Tschakert 2007, 84).

CONCLUSION

Anthropologists often bemoan their discipline's "short reach" into policy (Haenn and Casagrade 2007, 100). With the change in dialog about global climate change, this is an opportune time to focus more squarely on policy and policy options for greenhouse gas emissions and to broaden our reach by becoming conversant in the lingua franca and discourse of global climate change policy, tradable emissions, and cap and trade, among other things.

I have suggested contributions to policy from multiple levels in this chapter. At the macro level, Lahsen raises epistemological questions about the fundamental production of science and knowledge; and reveals political dimensions, such as access to power, behind the climate science debate. At the micro level, Perez, Roncoli, and Tschakert provide highly relevant, focused, and grounded research on the viability and feasibility of carbon sequestration projects. An important contribution of anthropology here is to articulate the *institutional* constraints and contexts of policy options.

A model for far-reaching impacts on policy comes from elsewhere in natural resource anthropology, namely, marine fisheries and fisheries management, where anthropologists have informed and energized paradigm and policy shifts in common property resources, raised and addressed questions of equity and distributional impacts, investigated and championed the viability of comanagement options, and advised on day-to-day management decisions (allocations, closures, effort) in marine fisheries, both domestically and globally (see McCay and Acheson 1987; McGoodwin 1990; National Academy of Sciences 1999; Pinkerton and Weinstein 1995). In this domain, actions of anthropologists occurred at multiple levels, scales, and localities over a thirty-five-year period of research and policy engagement. The work of anthropologists spanned many roles, from manager, evaluator, researcher (Colburn, Abbott-Jamieson, and Clay 2006), and funder of research and extension efforts (Fiske 1990), to advocate and coalition-builder (see McGuire 2005, 87–118).

The policy arena is one of both research on the policy process itself as a topic and of action (Haenn and Casagrade 2007). Within the arena of action, there are well-recognized roles for anthropologists to play, from listening and gathering community environmental knowledge, to translating, advocating, researching, providing expert witness, building coalitions, and engaging in activism and politics (Ervin 2000). Political engagement also includes taking direct roles in crafting legislation (creating policy) or managing and funding research (helping to form the research agenda).

In a melding of these two types of roles, I have posed the question of whether and how it might be possible to scale up our research, critical knowledge, and activism to enable resource-poor communities to participate in one of the prominent policies being discussed—carbon sequestration. Here I have focused on carbon markets as an environmental justice issue, suggesting that local communities deserve to benefit from global carbon markets when undertaking carbon reduction activities. In a larger sense, I am raising the question about how we can govern common property resources (the earth's atmosphere) on a global scale (Dietz, Ostrom, and Stern 2003), but also, very importantly, questions of equity and environmental justice in the allocation of benefits and losses under domestic and international global climate change regimes. Anthropologists, with their finely nuanced understandings of cultures, social institutions, meanings, assessments of power and vulnerability, and familiarity and advocacy for marginalized peoples, are well positioned to identify institutional constraints or distributional impacts (two elements that I have discussed in this chapter), unintended consequences, inequities in access, or regressive economic impacts on sensitive populations. I am posing the larger policy challenge for anthropologists: "How can global climate change policies be designed to work to benefit local communities in the developing world or domestically—or work to their detriment?" I would like to see anthropological insights scale up in a way that links them with current policy discourse on global climate change policies now being implemented worldwide.

Lastly, a note on the voice of anthropology in climate change policy: anthropologists in the past have focused their attention on the multiple ways that climate and changing conditions affect local peoples and on climate assessments and climate forecasts such as El Niño and other climate anomalies,[9] and have made important contributions to the formulation of research pathways for the human dimensions of global change (National Academy of Science 1999). I suggest that anthropologists now change direction and become actors in the policy process. I realize that individual action has its limits, whether on Capitol Hill or in the executive branch, and earlier in the chapter suggested public interest actions (testimony, op eds). I also realize the inherent difficulties that anthropology has in acting collectively or with a unified voice on an issue through the associations of which we are members (that is, cultural anthropologists). But at this point it is worth considering using the collaboration that has become a hallmark of our research to build relationships with other organizations, associations, think tanks, and foundations, who have a stake in this issue—or create new ones. Organizations such as the Pew Center for Climate Change and Physicians for Social Responsibility have made climate change a human rights issue and a priority for action. Working relationships with association heads, foundations and think tanks, legislators and staff, executive branch program officers, UN officers, foundation analysts, and NGO directors can help accumulate constituency and jump start anthropology's engagement with the larger policy community as the dialog of global climate change moves forward.

Acknowledgments

I would like to thank the anonymous reviewer and Susan Crate and Mark Nuttall, the editors of this book, for their thought-provoking questions and editing; and I would especially like to thank Carla Roncoli for her generosity of spirit in sharing information and insightful comments on the draft manuscript.

Notes

1. "Inside the beltway" is an American political term that refers to a beltway (interstate freeway) that encircles Washington, DC and all the activities that are important primarily within the offices of the federal government, its contractors, lobbyists, Congress, and the media that cover them. It can be used to denote something that is of very limited importance to the general public, but by the same token, something that is important insider information.
2. As an anthropologist at the National Oceanic and Atmospheric Administration (NOAA) in the 1980s and 1990s, the scientists who were documenting the increase in carbon dioxide and hypothesizing ocean-atmosphere interactions were my colleagues. After the US Global Change Research Program (USGCRP) was authorized in 1990, I worked closely with the NOAA Office of Global Programs (OGP) to help develop a human dimensions research agenda. I was concerned that any national program should have a robust human dimensions research agenda that included anthropology. As the NOAA representative on the interagency group that formed the human dimensions of global climate change agenda, I worked closely with my interagency colleagues to put together a research agenda and program for funding research. I helped publicize the opportunities and fund anthropologists either through the NSF or NOAA HDGC programs, through peer-reviewed programs. The original scope of the NOAA HDGC program was a broad-based social science research agenda, seeking knowledge about GCC as it was occurring in contemporary human-scapes across the globe, but also seeking to understand what we could learn about human adaptation to climate change from an archaeological perspective. Subsequently, the OGP focused more closely on climate and climate forecasts in the hopes of providing more finely tuned and useful forecasts nationally and internationally. More recently, as a staff member working in the personal office of the Senator Akaka (D-HI), I advised the senator on key global change legislation, helped draft legislation dealing with cap and trade systems, and staffed the senator for his responsibilities on the Senate Committee on Energy and Natural Resources, one of the primary committees whose jurisdiction included oversight, funding, and regulatory frameworks for GCC.
3. Acid rain reductions programs, generally considered among the most successful models in cleaning up air pollution, were authorized in the 1990 Clean Air Act amendments.
4. Other types of climate change legislation not utilizing cap and trade systems include a) climate change research and studies; b) technology to reduce greenhouse gases (using tax credits, grants, foreign assistance); c) international agreements, urging the administration to reengage in international climate change negotiations; d) preparing communities to adapt to climate change; and e) greenhouse gas reporting and registry bills.
5. The Chicago Climate Exchange (CCX) began with a grant in 2000 from the Joyce Foundation, as part of a series of grants made by the foundation to "catalyze, support and reinforce ideas, concepts or institutions of lasting intergenerational significance." An initial grant of $347,000 was made to the Kellogg Graduate School of Management at Northwestern University to provide technical support to Dr. Richard Sandor and colleagues to examine whether a cap and trade market was feasible in the US. Dr. Sandor is chairman of Climate Exchange Plc.
6. Carbon offsets are a very unusual commodity. "Its substance is intangible, the absence of something. Some pollution would have existed, somewhere, sometimes, the seller says, but it now won't." The market for carbon offsets grew by 80% in 2006 alone (Farenthold and Mufson 2007).
7. The Carbon from Communities project was implemented by two of the US Agency for International Development's Collaborative Research Support Programs (CRSP),

namely the Soil Management CRSP and the Sustainable Agriculture and Natural Resource Management CRSP.

8. Namely, a) a contour-tillage system to retain water and nutrients and expand vegetative cover; and b) an open-range rotational grazing system. The southern Mali tillage system can be adopted by individual farmers, as contrasted to rotational grazing, which takes a coordinating and decision-making mechanism or group. The constraints for tilling are availability of labor, animal traction equipment, and technical support services. Farmers grow mostly cotton, but also grains—sorghum, millet, and maize. Also grown are peanut, bambara nut, cowpea, sesame, soja, calabash, and vegetables. Farming lies across the path of traditional transhumant herding, whereby livestock—cattle, sheep, and goats—are moved between drylands and wet land pasture.
9. Anthropologists have helped federal and state agencies and ministries refine their understandings of how seasonal climate forecasts are used (and sometimes misused) and how standard forecasts must be adjusted to be useful for Brazilian and Costa Rican agricultural communities, and Peruvian fishermen, during El Niño events (Broad 2000; Nelson and Finan 2000; Otterstrom 2000). Similarly in Mali, the seasonal/temporal distribution of rainfall, rather than the overall quantity, was needed (Roncoli and Magistro 2000).

REFERENCES

Asquith, N. M., M. T. Vargas Rios, and J. Smith. 2002. Can forest-protection carbon projects improve rural livelihoods? Analysis of the Noel Kempff Mercado Climate Action Project, Bolivia. *Mitigation and Adaptation Strategies for Global Change* 7(4): 323–37.

Broad, K. 2000. El Niño and the anthropological opportunity. *Practicing Anthropology* 22(4): 20–23.

Colburn, L. L., S. Abbott-Jamieson, and P. M. Clay. 2006. Anthropology applications in the management of federally managed fisheries: Context, institution history, and prospectus. *Human Organization* 65(3): 231–39.

Dietz, T., E. Ostrom, and P. Stern. 2003. The struggle to govern the commons. *Science* 302: 1907–12.

Ervin, A. M. 2000. *Applied anthropology: Tools and perspectives for contemporary practice.* Boston: Allyn and Bacon.

Farenthold, D. and S. Mufson. 2007. Cost of saving the climate meets real-world hurdles. *Washington Post,* August 16.

Fiske, S. J., guest ed. 1990. Anthropology and marine extension. Special issue of *Practicing Anthropology* 12(4).

Haenn, N. and D. Casagrande. 2007. Citizens, experts, and anthropologists: Finding paths in environmental policy. *Human Organization* 66(2): 99–102.

Lahsen, M. 2005a. Seductive simulations: Uncertainty distribution around climate models. *Social Studies of Science* 35: 895–922.

———. 2005b. Technocracy, democracy, and U.S. climate politics: The need for demarcations. *Science, Technology & Human Values* 30(1): 137–69.

———. In press. Experiences of modernity in the greenhouse: A cultural analysis of a physicist "trio" supporting the backlash against global warming. *Global Environmental Change.*

Larsen, J. and R. Heilmyr. 2007. Comparison of legislative climate change targets in the 110th Congress. World Resources Institute. http://images.wri.org/usclimatetargets chart1 big.gif.

McCay, B. J. and J. M. Acheson. 1987. *The question of the commons: The culture and economy of communal resources.* Tucson: The University of Arizona Press.

McGoodwin, J. R. 1990. *Crisis in the world's fisheries: People, problems and policies.* Stanford: Stanford University Press.

McGuire, T. R. 2005. The domain of the environment. In *Applied anthropology: Domains of application,* eds. Satish Kedia and John van Willigen, 87–118. Westport, CT: Praeger.

Mufson, S. 2007. In battle for U.S. carbon caps, eyes and efforts focus on China. *Washington Post,* June 6.

National Academy of Sciences. 1999. *The human dimensions of global environmental change: Research pathways for the next decade.* Commission on Behavioral and Social Sciences and Education, National Research Council. Washington, DC: The National Academies Press.

———. 1999. *Sharing the fish: Toward a national policy on individual fishing quotas.* Ocean Studies Board, National Research Council. Washington, DC: National Academy Press.

Nelson, D. R. and T. Finan. 2000. The emergence of a climate anthropology in northeast Brazil. *Practicing Anthropology* 22(4): 6–10.

Otterstrom, S. M. 2000. Variation in coping with El Niño droughts in northern Costa Rica. *Practicing Anthropology* 22(4): 15–19.

Parker, L. and B. D. Yacobucci. 2007. Greenhouse gas reduction: Cap and trade bills in the 110th Congress. Washington, DC: Congressional Research Service report RL33846, updated July 13.

Perez, C. A., C. Roncoli, C. Neely, and J. L. Steiner. 2007a. Can carbon sequestration markets benefit low income producers in semi arid Africa? Potentials and challenges. *Agricultural Systems* 94(1): 2–12.

———. guest eds. 2007b. Making carbon sequestration work for Africa's rural poor: Opportunities and constraints. Special issue, *Agricultural Systems* 94(1).

Pinkerton, E. 1989. *Cooperative management of local fisheries: New directions for improved management and community development.* Vancouver: University of British Columbia Press.

Pinkerton, E. and M. Weinstein. 1995. Fisheries that work. Sustainability through community-based management. Vancouver, Canada: The David Suzuki Foundation.

Roncoli, C., guest ed. 2000. Anthropology and climate change: Challenges and contributions. *Practicing Anthropology* 22(4): 2–28.

———. 2006. Ethnographic and participatory approaches to research on farmers' responses to climate predictions. *Climate Research* 33: 81–99.

Roncoli, C., K. Moore, A. Berth, S. Ciss, C. Neely, and C. Perez. 2003. An analysis of institutional supports for community-based land management systems with carbon sequestration potential in Mali. Paper presented at invited workshop on Reconciling Rural Poverty Reduction and Resource Conservation: Identifying Relationships and Remedies. May 2–3, Cornell University, Ithaca, NY.

Roncoli, C. and J. Magistro. 2000. Global science, local practice: Anthropological dimensions of climate variability. *Practicing Anthropology* 22(4): 2–5.

Roncoli, Carla, Christine Jost, Carlos Perez, Keith Moore, Adama Ballo, Salmana Cissé, and Karim Ouattara. 2007. Carbon sequestration from common property resources: Lessons from community-based sustainable pasture management in north-central Mali. *Agricultural Systems* 94: 97–109.

Senate Committee on Environment and Public Works. 2007. Testimony submitted legislative hearing on S. 2191, November 15. http://www.epw.senate.gov/.

Tschakert, P. 2004. The costs of soil carbon sequestration: an economic analysis for small-scale farming systems in Senegal. *Agricultural Systems* 81: 227–53.

———. 2007. Environmental services and poverty reduction: Options for smallholders in the Sahel. *Agricultural Systems* 94: 75–86.

US Environmental Protection Agency. 2008. Analysis of the Low Carbon Economy Act of 2007, S. 1766 in 110th Congress. Washington, DC: US Environmental Protection Agency. January 15, 2008.

Chapter 18

Living in a World of Movement: Human Resilience to Environmental Instability in Greenland

Mark Nuttall

Over the last decade, scientific research arguing that current climate change is more pronounced in the Arctic than in any other region has been enhanced by compelling observations from indigenous peoples (ACIA 2005; Huntington and Fox 2005; Krupnik and Jolly 2002; Nuttall et al. 2005). As well as revealing the limits of science in understanding the experience and lived worlds of the peoples who dwell in high latitude regions, the significance of these observations of a rapidly shifting environment goes far beyond the northern reaches of the Earth and enriches our understanding of living on a planet undergoing constant change. As elsewhere in the circumpolar North, Greenland's residents are reporting that they are noticing changes in the weather and climate of their Arctic homeland. Certainly, on recent travels in Greenland, I hear people say with increasing frequency "*Sila kiagukkalattuinnarpoq*"—"the weather is getting warmer and warmer." Hunters in communities along the northwest coast talk of having to travel further in search of seals and fish, of the sea ice forming later and breaking up earlier, and of not being able to live as they once did. Politicians and Inuit activists reinforce these comments about local difficulties and environmental risk with authoritative sound bites about how people can no longer hunt or fish in the ways they have always been used to because the climate is changing—or worse, that people are getting lost when out traveling because of climate change, or that they are falling through thin ice and drowning because of climate change.

I have always been a little cautious about how to respond to such statements. For one thing, I have never felt entirely comfortable with the mere chronicling of indigenous observations of climate change and their incorporation into scientific assessments, particularly when such observations are removed from their lived, everyday social and cultural context and offered as supporting evidence for scientific research on climate change without much critical interrogation. Too often this is done within a politicized research context that plays on the well-rehearsed argument about the modernist dichotomy between traditional knowledge and scientific knowledge.

Indigenous and local observations deserve serious attention, especially when we seek to understand them with reference to the everyday life, and social and cultural meanings of local people. But such attention raises questions to do with epistemology and challenges us to ask how and why people know what they know (in the same way as science is scrutinized), and to demonstrate how we can distinguish between an observation about the weather and a claim that indigenous knowledge provides evidence of climate change. Climate change is becoming an explanatory account for almost everything that seems unusual in the weather, the environment, or in people's actions and encounters with the natural world. Political correctness and the changing dynamics of research in many parts of the circumpolar North, it also seems to me, prevent many researchers working on the human dimensions of Arctic climate change from questioning why death on the ice is all too quickly explained away as a result of climate change, when it could simply be a tragic accident or because a person's inexperience is to blame (both cases could be weather related, i.e., in terms of an inability to read the weather or a lack of skill, but not necessarily because the climate is changing).

In these kinds of encounters with the use and claim of indigenous and local knowledge about climate change, anthropologists find themselves asking ethical and intellectual questions about the nature of anthropological action and the need for analytical sophistication. We need to listen carefully when people tell us that the weather does not appear to be normal for the season, or that the habits of birds, seals, and caribou are changing. Yet we need to think about what these statements actually mean in terms of their rhetorical and metaphorical senses too. In one sense, there are taken-for-granted assumptions about what is real and true and observable about the world; while in another sense, we know we need to remain attendant to understanding the multiplicities of meanings deriving from circumstances that are particular to social, cultural, and political interpretations and contexts. This is not to reduce things to a distinction between realist and constructivist ideas about environmental problems and concerns, but rather that, in arguing for anthropological action when it comes to climate change, anthropologists should not lose sight of some of the fundamental perspectives that mark out anthropology as a distinctive discipline with a claim for contributing to climate change science, policy, and discourse. Anthropology reminds us that our task is both epistemological and ontological in how we grapple with understanding what people know about the world, how they move within it, how they relate to it, how they think and feel about it, and what they say about it.

My point of departure in this chapter is with understanding the complexity of responses to climate change as an intermingling of concern, a range of possibilities, irony, and contradiction. I show that the various ways climate change is perceived, and how urgent or otherwise climate change is felt to be, depends on how individuals or communities are positioned. I do so by drawing on several years of intermittent anthropological research in various parts of Greenland, working in both small villages and larger towns but also following political discourses about the environment and resource development. Indigenous and local perspectives on climate change, for example, are not only reports from the front line of climate change about the real nature and character of environmental problems and risks—they are suggestive of something else beyond mere description of changing ecosystems. In particular, they are entanglements of moral and emotional feelings and meanings, social and cultural claims, and political processes. This is illustrated by the difference between various Inuit perceptions about climate change and diverse views about its nature as a global crisis.

In 2005, Sheila Watt-Cloutier, the former international chair of the Inuit Circumpolar Council (ICC, and then still known as the Inuit Circumpolar Conference), submitted a 167-page petition to the Inter-American Commission on Human Rights on behalf of all Inuit of the Arctic regions of the United States and Canada. The petition dealt specifically with the violation of Inuit human rights caused by greenhouse gas emissions from the United States. In it, Watt-Cloutier argued that climate change is harming every aspect of Inuit life and culture and drew attention to the intimate relations between Inuit and the Arctic environment:

> Like many indigenous peoples, the Inuit are the product of the physical environment in which they live. The Inuit have fine-tuned tools, techniques and knowledge over thousands of years to adapt to the arctic environment. They have developed an intimate relationship to their surroundings, using their understandings of the arctic environment to develop a complex culture that has enabled them to survive on scarce resources. The culture, economy and identity of the Inuit as an indigenous people depend upon the ice and snow. (Watt-Cloutier 2005, 1)

The petition goes on to describe how this delicate balance between Inuit and the environment is now threatened by climate change and how Inuit are struggling to adapt. A careful reading of the petition, however, reveals that it is about more than just Inuit concerns with the impacts of climate change. Its recasting of climate change as a human rights issue, not just an environmental one, draws attention to the position of Inuit as indigenous people within nation-states, and in particular to broader aspects of indigenous rights.

The petition was not submitted on behalf of Greenland Inuit, who have achieved a greater degree of self-government than any other Inuit population since Home Rule was introduced by Denmark in 1979. However, ICC Greenland has also been involved with documenting Inuit observations of climate change, and the list makes familiar reading (ICC Greenland 2006). But Alaskan and Canadian Inuit perspectives on climate change are not necessarily a pan-Inuit view, and as the Greenlandic context reveals to us, we must be careful in distinguishing between an Inuit NGO view and an Inuit government view, but also not fail to acknowledge the diversity of views about climate change within families, households, and communities (Nuttall 2008). In April 2008, Josef Motzfeldt, a member of parliament in the Greenland Home Rule government, and a former minister of foreign affairs, told the audience at the Trans-Atlantic Climate Conference in Torshavn, the capital of the Faroe Islands:

> While reduction of ice cover may have a negative impact on some hunting activities, it may open up new opportunities for other activities in our society, like fisheries. A new generation of hunters and fishermen, building on their ancestor's skills, knowledge and cultural socialisation, adds to this by learning how to cope with the changes. In the way we look at climate changes we have a saying that "nothing is so bad that it is not good for something else." (Motzfeldt 2008)

Motzfeldt also went on to remark in Torshavn that "climate change has already opened up new areas for the exploitation of mineral resources as the ice cap is retreating." His comments reflect a Greenlandic government view that contradicts Watt-Cloutier's argument that Inuit economic activities and identity—indeed the very cultural survival of Inuit as a people—are inextricably linked to the continued presence of snow and ice. In a sense, Alaskan and Canadian Inuit are arguing for the right to continue to be cold, whereas Greenland is literally warming to the idea of less snow and ice. It is almost tempting to place this neatly within Cotgrove's (1982) catastrophist and cornucopian analysis of divergent views of nature. Yet these views do offer two contrasting—and interesting—cases: one that suggests that Inuit cultural survival is not possible without the familiar winters of sub-zero temperatures and the icescapes that make life possible, and another that suggests that the continued presence of snow and ice hinders the Inuit right to political, economic, and social development. Watt-Cloutier argues that Inuit are struggling to adapt to climate change; Motztfeldt argues that Inuit will adapt.

The Icelandic *Saga of Eirik the Red* famously relates how, in the tenth century, Eirik named the land to the west of Iceland "Greenland" because people would be attracted to go there if it had a favorable name. Greenland's Home Rule politicians and business leaders of the new millennium, eager to attract energy multinationals, mining companies, and aluminium producers, are similarly extolling the virtues of their country as a green land, or at least a land that is getting greener. And this is framed within a Greenlandic political discourse of nation building and development that gives a positive spin to the prevailing global discourse of climate change as a cataclysmic force about to devastate human existence. It is also distinctive in that it differs considerably from many other indigenous perspectives on Arctic climate change as a social and environmental crisis.

The Greening Land

The regional texture to climate change means changing environments are perceived and experienced differently. The disappearance of sea ice in northern Greenland may well hasten the end of traditional Inuit hunting lifestyles, yet at the southern tip of the island sheep farmers shake their heads in wonder as they dig potatoes from the ground and pluck their first harvests of broccoli, cauliflower, and cabbage from increasingly larger plots of cultivated soil. Also in south Greenland, researchers at the agricultural research station at Upernaviarsuk near the fishing town of Qaqortoq speak enthusiastically about a future of productive vegetable farms and viable forests of imported pine, spruce, larch, and firs. Their imaginative construction of this part of Greenland is that it is an agricultural frontier where temperature is a limiting factor for human survival. A slight warming of two or three degrees can have tremendous significance for those dreaming of lush forests and fertile soil. A warmer climate, as an elderly sheep farmer put it to me in spring 2008, means that the younger generation has more options for the future. Elsewhere in the circumpolar North, particularly in Canada and Alaska, indigenous politicians and activists portray indigenous peoples as victims of climate change (as well illustrated by Sheila Watt-Cloutier's petition), but the official Greenlandic response to climate change diverges from this prevalent view. For politicians in the Home Rule government, hopeful of greater political and economic independence from Denmark, climate change means more than possible self-sufficiency in vegetable production. A warmer climate brings opportunities for opening up this self-governing North Atlantic territory to mining and hydrocarbon development. Greenland Inuit are not a people imperiled by a shrinking northern icescape—climate change is revealing a bigger, greener land, and Greenlanders may be on the verge of greater political independence from Denmark because of it.

The melting of the Greenland ice sheet is a much-reported and potentially catastrophic impact of climate change. Covering 1.7 million square kilometers, with an average thickness of 1,600 meters and total volume of some 3 million cubic kilometers, Greenland's inland ice consists of a northern and southern dome, with maximum surface elevations of approximately 3,200 and 2,850 meters respectively. It is a frozen archive of the climate of the past, with each frozen snow layer retaining memories of what conditions were like with each annual snowfall. The ice core record is incredibly detailed, extending back through the present interglacial period, through to the last ice age (when temperatures on the ice were 20° C colder than at present), and into the preceding interglacial when the sea level was some five meters higher than at present (Thomas 2005). What this ice archive shows is that the climate has experienced remarkably abrupt and severe changes. Scientists have drilled ice cores that reveal how, between these periods, there were dozens of abrupt temperature warmings and coolings. During the glacial period, for example, there were twenty-six abrupt temperature increases of about 7°–10° C. These glacial warm periods, named *Dansgaard-Oeschger events* after the two scientists first observing them, may be random, chaotic, and unpredictable.

Globally, this ice record reinforces our knowledge of how human development, especially over the last 11,500 years (the Holocene), has taken place against an environmental backdrop of climatic and geological instability. Sudden and dramatic climatic shifts and extreme biophysical events have always ensured that nature is in flux, and not in static balance. Alarm about the melting of the Greenland inland ice, however, arises from scientific scenarios that suggest the scale and nature of climate change in the coming decades may be greater than previous changes in the earth's history. Within the context of global climate change, the Arctic and sub-Arctic regions of the circumpolar North will experience a greater degree of change than countries in the tropics (e.g., ACIA 2005; Weller 2000). The melting of the entire Greenland ice sheet is projected to raise global sea levels by seven meters over the next two or three centuries.

As the inland ice melts, a new Greenland is emerging. Geographically, mountains, headlands, and islands are appearing and cartographers are beginning to revise maps and charts—for example, the retreat of the Sermeq Avannarleq glacier near Ilulissat in Disko Bay has recently created a new island, which has been named Qarsunga (Always Pale Island). But this process of topographical reshaping is coinciding with the emergence of a new Greenlandic nation that is redefining people's relationships to place, to the environment, and to one another. Politically, Greenlanders say they are ready for the challenge of greater autonomy—indeed a warmer climate is seen by some as a positive transformation helping Greenland to become a modern nation. Yet this responsibility will also challenge the Home Rule authorities as they face the magnitude of possible environmental change and its local effects. Accordingly, research efforts should not only focus on local observations and community perceptions of change, but, perhaps more importantly, on identifying the nature of human agency and resilience, assessing community vulnerability, and understanding community responses to past and current change.

Recent work suggests that Arctic communities are facing greater change, and that they need to be prepared for the unpredictability of the weather and an increase in extreme climatic events. They are told to brace themselves for a future of living precariously on thin ice, and researchers and indigenous leaders report to the wider world that the peoples of the Arctic are becoming strangers in their own lands (ACIA 2005). But change is nothing new and one way to know how to be resilient to it, as far as my work in Greenland seems to suggest, is appreciation of this fact. Culture frames the way people perceive, understand, experience, relate to, and respond to the social and physical worlds around them. It is a characteristic of life in smaller hunting and fishing communities all around the Greenlandic coastline that people consider the environment to be in a process of "becoming" rather than "changing." In the communities in which I have worked, acquiring personhood is a matter, in part, of growing up to be always prepared for change, for seeing the world as one of constant surprise and the environment as one of motion (Nuttall 1992). An inability to respond appropriately to this world of constant flux has much more to do with institutional, political, and social changes that provide no room to move freely in a changing world and to navigate it with reference to the experience of an intimate relationship with one's local environment.

Climate Change through Greenlandic Eyes

Greenland's climate has undergone significant periodic and often abrupt changes in the past, just as the global climate has changed historically in response to natural variability. Relatively minor variations in temperature have produced large positive feedbacks in the Greenlandic environment that have often had dramatic impacts on physical and biological systems (e.g., Vibe 1967). The successful long-term occupation of Greenland over several thousand years by various Inuit hunting and fishing societies has been possible, in part, due to their adaptive capacity (in social, economic, and cultural practices) to adjust to climate variation and change, to move around, and to see and seize opportunities in the environment. Change is a fact of life for Arctic peoples generally, and they have a rich history of culturally adaptive responses to deal with it. Many of the short-term (or coping) responses appear to be based on this tradition of flexibility and innovation (Nuttall 2005). Across the northern circumpolar world, seasonal, annual, and periodic transitions from sedentary to nomadic subsistence livelihoods and vice versa was the key to the survival and sustainability of Arctic indigenous cultures. Cultural and ecological diversity required flexibility. Resilient coping strategies during periods of extreme change and subsistence diversity were the outcome of a successful cultural and social response to climate variation and the resource instability of the Arctic (Krupnik 1993; Nuttall et al. 2005).

Resilience is often defined as "the capacity of a system to absorb disturbance and reorganize while undergoing change so as to still retain essentially the same function, structure, identity, and feedbacks" (Walker et al. 2004). Adger (2000) draws attention to the contested nature of the concept of resilience in ecology and environmental studies, and argues that social resilience is important for understanding the circumstances under which individuals and social groups respond and adapt to social change. Resilience, he points out, relates to the functioning of a system. Social and ecological resilience are clearly linked, yet merely appropriating the concept and the principles of ecological resilience and applying them to social systems "assumes that there are no essential differences in behaviour and structure between socialized institutions and ecological systems" (Adger 2000, 350). Adger defines social resilience as the ability of groups or communities to cope with external stresses and disturbances as a result of social, political, and environmental change, and he goes on to argue that it is institutionally determined and, as such, can be examined through a number of proxy indicators, such as economic structure, institutional change, and demographic change.

I would argue, however, that resilience also depends on how people perceive and conceptualize change—in short, one's world view goes some way to determine the kinds of adaptive strategies people utilize. In the small Greenlandic communities in which I have worked—in the north, south, and east of the country—I have been struck by the fact that people do not necessarily talk of the environment around them as *changing*, but of it being in a constant process of *becoming*. The environment—and each feature in it—has its own essence (*inua*), for example, *qaqqap inua*, "the essence of mountains," or *sermersuap inua*, "the essence of the great ice." Weather, or climate, is known as *sila*, and *silap inua* is "the essence of *sila*"—but its meaning is deeper, and people understand *sila* as the breath of life, the reason things move and change. *Sila* is also the word for "intelligence/consciousness," or "mind" and is understood to be the fundamental principle underlying the natural world. *Sila* is manifest in each and every person. It is an all-pervading, life-giving force connecting a person with the rhythms of the universe, and integrating the self with the natural world (Nuttall 1992). As *sila* links the individual and the environment, a person who lacks *sila* is said to be separated from an essential relationship with the environment that is necessary for human well-being. Lack of *sila* can be a temporary disorientation, such as when a person has a momentary lapse of reason, or makes an uninformed judgment about something (*silaaruppoq*). But lack of *sila* can also happen when someone goes crazy (*silanngajaarpoq*). *Silaqaraluarneq* is the state of being out of one's mind, but it can also mean "the weather is out of its mind."

Given this context, it is perhaps more understandable that when some people in Greenland experience a change in the weather, this change is experienced in a deeply personal way. And when they talk about their concerns about climate change, they articulate this in terms of how their own sense of self, personhood, and well-being is changing in relation to external climatic fluctuations. Climate change is understood as being consistent with the constant making of the world, with its uncertainty and with the environment coming into existence through continuous actualization and realization. However, the current rate of such change is faster than many people recall having experienced it in living memory.

How Greenland Inuit respond to change and remain resilient is dependent, in part, on them continuing to learn how to grow up and dwell in an environment where one is always prepared for surprise, where one is constantly challenged by uncertainty, and where one can never take anything for granted. But as I have written elsewhere, based on extensive fieldwork in northwest and south Greenland, being resilient in the face of change also depends on the strength of a sense of community, kinship, and close social associations. In a world of flux, uncertainty and unpredictability, social relationships are a source of constancy (Dahl 2000; Nuttall 1992). If a person breaks from networks of kin and social relationships, they are set adrift from the security of their social world. Loss of community is a threat to individual and social identity and, combined with loss of livelihood, exposes people to the impacts of climate change in a way that makes it difficult for them to respond effectively, if at all. To become a stranger in one's own land does not happen solely because the environment has changed, but also because political and social change threatens the social cohesion of community; endangers one's livelihood; and separates one from a fundamental relationship with people, animals, and place (Nuttall 1992).

Over the last one hundred years or so, dramatic changes in marine resources have contributed to structural changes in the Greenlandic fishing industry and have also impacted the social and economic structure of small hunting and fishing settlements. For much of the twentieth century Greenland's economy depended on a rich cod fishery, yet since the 1980s the harvest of Greenland shrimp has comprised around 75 percent of the country's total export, with a significant fishery of Greenland halibut and snow crabs representing an important diversification of marine resource exploitation.

Research on the social consequences of climate change in coastal west Greenland from the early 1900s onwards shows how people living in towns with similar social and economic settings and political and institutional structures showed a marked difference in their abilities and readiness to adapt to changing conditions (Rasmussen and Hamilton 2001). Environmental changes, particularly in climate and ocean currents, that have affected fisheries in West Greenland are well documented, as are the associated social and economic changes, especially at the beginning of the twentieth century (Hamilton, Lyster, and Otterstad 2000). As the waters of southern and western Greenland warmed, seals moved further north, making seal hunting harder for the Inuit population. Cod as well as halibut and shrimp moved into the now warmer waters and made the development of a cod fishery possible. The development of fishing in West Greenland shows how climate change can provide opportunities for some people, some local communities, and some local regions. As Thuesen (1999) argues, the political and economic changes taking place in West Greenland at the beginning of the twentieth century meant that Greenlanders were now involved in and participating in the new political structures of local municipal councils and two provincial councils, established in 1908. In 1910 experimental fisheries were taking place in West Greenland and Greenlandic fishers were learning new skills in fisheries training programs. The west coast town of Sisimiut was able to take advantage of these new developments, advantageously situated as it is at the northernmost limit of the ice-free waters on the west coast.

Greenlanders who embraced change and the opportunities it brought benefited more than the rest because they played crucial roles as local entrepreneurs and took advantage of the opportunities to diversify local economies. Thuesen (1999) argues that the development of Sisimiut as an important fishing center was due in part to a strong sense of local identity and strong dynamism in the community—in short, people had a willingness to embrace change, to diversity the economic base, and to work to develop new industries.

The development of Sisimiut into a major west coast fishing community stands in stark contrast to the development of the southwest Greenlandic town of Paamiut around the same time. Paamiut's development was based largely on plentiful resources of cod. With few other resources available in commercially viable quantities, there was little incentive to diversify the local economy (Rasmussen and Hamilton 2001). The concentration on a single resource demonstrated the vulnerability of Paamiut in the face of environmental change. The cod population began to fall, due to a combination of climate change and overfishing, and the economy and population of Paamiut declined as a result.

This highlights the importance of recognizing that in any adaptive strategy, local conditions and social and cultural differences are considerable factors in the success of a region affected by change, be it from climatic, social, economic, or political factors. The development of cod fishing in Greenland also shows, however, how climate change and social change go hand in hand. Cod fishing developed at a time when climate change was having an adverse effect on seal hunting, yet the population of Greenland was also growing, making it necessary for the Danish administration to find alternative ways for the majority of the population to make a living. Despite these studies of fisheries and climate change, the relationship between ecosystem changes, changing resource dynamics, and socioeconomic responses to these changes is poorly understood, particularly in smaller communities where marine mammal hunting (predominantly seals) and small-scale fishing provides the basis for local livelihoods.

Living in a World of Movement, Becoming, and Surprise: An Example of How to Be Adaptable

Upernavik is the most northerly municipality of West Greenland, stretching some 450 kilometers northwards from the Svartenhuk Peninsula to Melville Bay. More than half the population of roughly three thousand lives in ten settlements, with the remainder living in the town of Upernavik. People have long depended on harvesting and using marine and freshwater resources—marine mammals such as seals, walrus, narwhals, beluga, fin and minke whales, and polar bear, and fish such as Greenland halibut, salmon, and Arctic char. Land animals such as caribou and Arctic fox were of some importance until the 1960s. Many of these species are still used for food, clothing, and other products and, as in many other parts of Greenland, have long played an important and prominent role in the cash economy of local households and communities (Caulfield 1997; Dahl 2000; Nuttall 1992; Petersen 2003). Today the primary occupation providing an income is fishing for Greenland halibut, but hunting still plays a vital role in the settlements. Indeed, for hunters, fishers, and their families, movement, seasonality, and animals are the very substance of life. *Piniartoq*, the Greenlandic word for hunter, translates literally as "one who wants." To hunt is to strive for something one wants and needs. A second meaning of *piniartoq* is "provider." Ringed seals provide the main food source for much of the year. Narwhals are still caught in Melville Bay using traditional methods from kayaks, and beluga whales and polar bears are also hunted. The sea freezes over from December to June and transport to the fishing sites and hunting areas is by dogsled or snowmobile. During this time halibut fishing is carried out by using long lines under the ice, and hunters either catch seals in nets or hunt them in spring as they crawl out on the ice to bask in the sun.

The sea dominates and influences daily life, but it does not necessarily constrain it. The sea, as people in northwest Greenland understand it, is probably more akin to the Amazonian floodplain described by Harris (1998, 70) in that it is perceived and experienced as being in a "constant process of re-definition and becoming" throughout the year. This is an environment of opportunity rather than one of external environmental constraint, a place of constant unfolding of possibility. The local environment, *nuna*, is experienced not so much in spatial terms, as "a realm outside humans or their immediate living (cultural) space" (Ellen 1996, 105), but as a place in which people dwell and in which they engage in social relations of exchange between one another, and between themselves, animals, and the environment. Central to this is daily discussion of *pinngortitaq* and its significance for people and their movement around the locality. Although *pinngortitaq* is often simply translated from Greenlandic as "nature" or "creation," its literal meaning is "to come into being." *Pinngorpoq* is a process of "becoming," "to come into existence," referring to the unfolding of possibility and opportunity.

In northwest Greenland the sea is referred to locally by two names. In summer and autumn it is *imaq* (water) and fluid, becoming *siku* (ice) with the appearance of being solid in winter. In between these periods of open water and ice, the sea is no longer referred to as *imaq*, but is described as *sikuaq* ("has thin ice") in late autumn, before becoming *sikuvoq*, "frozen over in winter." Sometime between late spring and early summer, the nature of *siku* changes, becoming *imarorpoq* ("becomes open water with the breaking up of the ice"), *sikueruppoq* ("has no ice") and, eventually, *imaq* once more. It brings different opportunities and possibilities for hunting different species of marine mammals, and for harvesting different species of fish.

The experience of growing up in an environment that is also undergoing a process of becoming informs hunters and fishers that, in addition to good equipment and skill, knowledge about the movement, behavior, and habits of animals is vital to their successful capture, as is the knowledge of good hunting places, and the names and stories associated with the landscape, seascape, and icescape. Often, place names provide information about climate change and significant weather-related events. For Inuit, stories and discussions about the weather and climate are interwoven with stories and experiences of doing particular tasks like hunting, fishing, berry-picking, or traveling. Much of this is bound up with memories of past events, of local family histories, and of a strong sense of attachment to place and locality (Nuttall 2001). The weather connects people to the environment and animals, but also to their genealogical and local histories.

Moving and traveling through these memoryscapes of individual and community experience, hunters learn to identify with the hunting territories of the locality. They come to understand the movement and habits of seals and other animals, and the hunter's place in the wider social context. They also learn to appreciate the shifting nature of the environment, and to understand that *pinngortitaq* is a process of the world around them coming into existence through its actualization and through their engagement with it. As Proust once wrote, "A change in the weather is sufficient to recreate the world and ourselves." Nothing is ever fixed or certain, and success as a hunter or fisher is not just dependent on skill, but on a person's ability to be open to surprise and uncertainty. The vagaries of the weather require a certain attitude of acceptance. The sea may freeze continuously for several days (*sikujartuaarpoq*), and then the ice may suddenly be driven away with the winds and currents (*saavippoq*), only to return again.

This was driven home to me forcibly when camped on the sea ice one winter with a friend in northern Greenland, a hunter who has spent much of his life getting to know the sea and the land, the islands, bays, and headlands of his locality in all seasons and in all weathers. We had been traveling for several days by dogsled across smooth ice and looked as if we would make good progress in the days ahead. One evening, huddled around the primus stove in our small tent, I began to make a remark about how good the ice was, but he raised one finger to his lips to silence me. We awoke the following morning to find much of the ice gone and that we were camped on a rather large ice floe surrounded by open water. My companion looked around, scanned the ice through his binoculars and said calmly, "This is what I meant, *silarlukkajuppoq* (the weather is often bad)." We had no choice but to stay where we were and to wait. After two days the ice had returned sufficiently, and *imaq* had become *siku* once more, to allow us to travel and cross at an open lead and continue our journey. As Riddington (1990, 86) has written, "The essence of hunting and gathering adaptive strategy is to retain, and to act upon, information about the possible relationships between people and the environment," and I often reflect on how my companion's "adaptive strategy" in response to this situation was to accept the situation calmly and to sit and wait. We could do nothing else, of course, but it struck me that the situation was a perfect example of how Inuit grow up to expect the unexpected (Briggs 1991). I have experience of many such episodes and they are not met with hysterical responses about cataclysmic climate change. The environment is capable of surprise. When we eventually returned to the village, this became another story told about the power of the sea, the fickleness of the ice, its essence and agency, and about *sila*, the breath of life. It did not become an example of a local observation of climate change.

Victims of Modernity and Change

In May 2007 Aqqaluk Lynge, Inuit Circumpolar Council (ICC) president for Greenland, traveled to the United Kingdom to give evidence at the Stansted Airport expansion public inquiry (the British Airports Authority proposed to develop a second runway at the airport). Opposition to the plans came mainly from local residents who feared greater activity, increased noise levels, and loss of quality of life, and from environmentalists concerned that aviation is the fastest growing source of carbon dioxide emissions in the UK. At the public inquiry, Lynge spoke of the loss of sea ice and habitat critical for Arctic wildlife, of the melting of Greenland's inland ice, and of the cultural impact of climate change on Greenland Inuit. For Lynge, the anthropogenic impacts on Greenland from areas far to the south of his homeland are obvious:

> What happens in Britain affects us in the north. You may say that the expansion of London Stansted airport will play only a small part in increasing climate change, but everyone can say that about almost everything they do. It is an excuse for doing nothing. The result of that attitude would be catastrophic.
>
> The serious consequences affecting my people today will affect your people tomorrow. Most flights from Stansted are not for an important purpose. They are mostly for holidays and leisure. Is it too much to ask for some moderation for the sake of my people today and your people tomorrow? For the sake also of our wildlife and everything else in the world's precious and fragile environment that is more important than holiday flights. (Lynge 2007)

In this narrative, Lynge draws attention to both global consumption patterns and to the globalization of leisure, positioning Greenlanders as victims of modernity and environmental change. He also places his discussion of Arctic climate change in the context of global change, reiterating the familiar mantra about the Arctic as a bellwether of change, underlining the scientific argument that what will happen in the rest of the world is happening first and foremost in the Arctic. His evidence ranged across landscapes of increasing unfamiliarity to local people; to places where hunters are finding their traditional hunting grounds of ice floes, in some cases, have disappeared;

where hunting areas are impossible to get to because of eroding shorelines; a land where the weather is increasingly unpredictable, and local landscapes, seascapes, and icescapes are becoming unrecognizable.

Lynge's testimony presented to the Stansted hearings is remarkably similar to the accounts of climate change presented at international forums by other Arctic indigenous leaders. A reading of his account suggests that he drew upon and distilled key messages from the Arctic Council's Arctic Climate Impact Assessment rather than local Greenlandic examples grounded in everyday experience. Such indigenous understandings and generalized representations of a changing Arctic—and the implications for Inuit culture and livelihoods—have been expressed powerfully and emotionally by Inuit leaders and politicians in recent years, most notably by Sheila Watt-Cloutier (who was also a 2007 Nobel Peace Prize nominee) in testimony to the US Senate on the impacts of climate change in northern Canada in 2004, as well as her petition to the Inter-American Commission on Human Rights a year later.

In these accounts of climate change, where the world is told how Inuit are on the front line and experiencing the immediacy of the impacts, indigenous people's phrasings of their situation shape a simpler narrative of traditional lifestyles under threat. For indigenous peoples the Arctic is often represented as both an environment *of* risk and an environment *at* risk (Nuttall 1998, 170) and, as I discussed earlier in this chapter, climate change is an issue of cultural survival and a threat to human rights. Lynge's submission was exemplary of this in that he carefully related how Inuit have lived a sustainable lifestyle, something they have maintained despite a legacy of colonialism, rapid social change, the impacts of animal-rights campaigns and international whaling regulations, and pollutants and contaminants. For indigenous peoples climate change is another chapter in the history of how the rest of the world has reached into, explored, exploited, and influenced the Arctic for centuries. But they are portrayed (and often portray themselves) as victims of change unable to respond effectively to the environmental and social crises that the Arctic meltdown will bring.

Nation Building and Climate Change

In her petition, Watt-Cloutier (2005, 5) emphasized that "the subsistence culture central to Inuit cultural identity has been damaged by climate change." I recently heard a politician from Nuuk remark in a similar vein, as Aqqaluk Lynge did at the Stansted Airport public inquiry, that hunters in Greenland cannot hunt anymore because of climate change. While shifting ice conditions and changing animal migration routes may indeed make it difficult for hunters to secure what they need, blaming this entirely on climate change is a simplistic explanatory account, ignoring historical processes, colonial encounters, Inuit participation in the global economy, and even contemporary Inuit political attitudes towards Inuit tradition. This kind of remark, it also strikes me, is an example of the kind of taken-for-granted assumption about climate change one hears far too often, and one which climate change researchers all too enthusiastically listen to and record uncritically.

When Greenland achieved Home Rule from Denmark in 1979, it embarked on a process of nation building. Recent discussions and negotiations between Denmark and Greenland on self-government and a new form of self-governance have focused on greater autonomy within the Danish realm. A major barrier to this is Greenland's continued dependence on an annual block grant from Denmark, which essentially props up the country's economy, and Greenlandic politicians widely agree that the development of minerals and hydrocarbons is the key to financial and economic independence (Nuttall 2008). The US Geological Survey estimates that the waters off Greenland's west coast could contain more than 110 billion barrels of oil (roughly 42 percent of Saudi Arabia's reserves) are attracting interest in the territory's potential. The Home Rule administration has been involved in talks with several multinationals who covet exploration licenses for oil and gas (Rasmussen 2006), and a warmer climate, and hence easier access to exploratory sites, is seen as something positive if Greenland is to attract international investors. ExxonMobil and Chevron from the US, Husky and Encana of Canada, the UK's Cairn Energy, and Denmark's Dong Energy are among the companies that have either already won or applied for exploration licenses from Greenland's Bureau of Minerals and Petroleum for acreage.

Ironically, or perhaps just an occasion of bad timing, Aqqaluk Lynge's submission to the Stansted inquiry coincided with Air Greenland's inaugural flight on its new route between Kangerlussuaq and Baltimore/Washington International Airport the same month. Lars-Emil Johansen, Greenland's newly appointed minister of finance and foreign affairs, used the occasion of a reception at the Danish ambassador's residence in Washington, DC, held to celebrate the route, to announce:

> I am very excited about the possibilities this new airway opens up for Greenland. . . . I am also happy that I . . . can announce to you that the Greenland Home Rule government has decided to let Alcoa make a huge investment in Greenland. This investment will not only produce work places in remote areas of Greenland, it will also be a showcase for the rest of the world and [show] other American companies that Greenland has a lot to offer for companies and that it is a place you can invest in.

The ambassador himself remarked:

> It is my hope that more Americans will travel more often to Greenland, the world's biggest island. It is a place with wonderful people—Inuits [*sic*]—and a landscape you will find nowhere else in the world. In Greenland, you have the opportunity to travel on the ice cap, boat in ice fiords, climb the hills, or just experience a special culture that you only find in remote areas of the world. (Both quotes from *Diplomatic Pouch* 2007)

The excitement over Air Greenland's new North American connection was short lived. The airline's board, citing a massive loss in its first and only season, announced its cancellation in March 2008. Johansen's reference to Alcoa was to a Greenland Home Rule government decision to sign a memorandum of understanding with the world's leading producer of aluminium concerning the possible construction of the world's second largest smelter, based on the promise of major hydroelectric development. All this is consistent with current Greenlandic political thinking about economic development and aspirations for political independence. It is also another example of the way *pinngortitaq* is a process of the world coming into being—as the inland ice melts and Greenland's mineral, hydrocarbon, and hydropower potential is uncovered and revealed, the environment is mapped and described as a flexible resource that can be used to promote economic growth and political development.

The Greenlandic nation-building process denies diversity within the country, and oil and gas exploration and development projects like the one proposed by Alcoa have implications for the continuity of small villages that depend primarily on marine mammal hunting and small-scale fishing. The reality for the Home Rule authorities is that these traditional pursuits, while playing a significant part in the construction of Inuit cultural identity, do not contribute much to the economic development of the country. The future of the villages in Greenland has long been debated, with political views often divided between those who see small communities as repositories of traditional Greenlandic values and lifestyles, and those who argue that the Inuit hunting culture belongs to the past and has no place in modern Greenlandic society. Rather than Inuit hunters being prevented by climate change from catching seals, the reasons, I suggest, are rather more complicated. Long-term policies of shifting demographics, investment in a few major centers, a reluctance to introduce development policies for small villages and settlements, a redefinition of resources and rights of access to them, and a political desire to encourage the depopulation of some communities all perhaps have greater significance for changing hunting and fishing practices than climate change does.

Administratively, Greenland is being redefined as one national hunting and fishing territory, contrasted with a diversity of local hunting and fishing territories that have long characterized the social, cultural, and economic make-up of the coastal areas. Caribou, whales, seals, and fish, which have traditionally been subject to common-use rights vested in members of a local community, are becoming national and privately owned divisible commodities. The ways they are caught, used, and consumed are now subject to rational management regimes defined by the state and the interest groups of hunters and fishers (such as KNAPK, the Greenland Association of Hunters and Fishers), rather than locally understood and worked-out rights, obligations, and practices. Membership of a territorial, or place-based, community no longer gives hunters exclusive rights to harvest animals as it has done so traditionally. Hunting and fishing were largely family and community events, and kinship, locality, and territory were the mechanisms for regulating harvesting activities. Today, hunting rights are vested in people as members of social and economic associations irrespective of a local focus.

Home Rule government bodies and administrative and research institutes are increasingly charged with the task of describing and regulating access to living resources. Biologists occupy a central position in the management of resource use as primary expert advisers to the Home Rule administration. They provide advice to Home Rule agencies that then use this advice to decide upon and fix quotas for particular hunting and fishing activities. One resulting (and perhaps inevitably unsurprising) conflict is between biologists and user groups (i.e., hunters and fishers), the latter disputing the nature of this scientific expert advice because the defining of ecological sustainability ignores local knowledge and is concerned with a strict delineation and measurement of the natural world.

Climate change impacts are not universal manifestations of totalizing global transformation. In Greenland and elsewhere in the Arctic, Inuit and other indigenous peoples are facing special challenges, but are also pondering the benefits. Some are concerned over the prospect of major and irreversible impacts on indigenous communities and livelihoods. Others, as this chapter has shown, are contemplating a future of opportunities for growth and development. In Greenland, climate change is present in the unavoidable evidence of melting ice and receding glaciers. However, rather than having immediate social and economic concerns, it magnifies the threats to the cultural and economic viability of hunting livelihoods in small Greenlandic communities that come more immediately from transformations in resource-use rights and Home Rule government policy to the villages. These subvert local customary practices and knowledge systems (Dahl 2000; Nuttall 2001). How individuals, households, and communities adapt to extreme climate events will be a measure of their ability or inability to make decisions that allow them to respond effectively and with a degree of autonomy. Climate change adaptation policy is not well served by scientific knowledge alone, and discussion of such policy is hardly going on in Greenland at all. Furthermore, critics of the aluminium smelter proposal argue that it demonstrates how the Greenland Home Rule government has no vision of any kind when it comes to resource development. Opponents to the plan have also pointed out that, once in production, the Alcoa smelter will almost double Greenland's annual greenhouse gas emissions. Concerns are increasingly expressed about the absence of community and stakeholder consultation and of social and environmental impact assessment processes (Nuttall 2008).

As a researcher, I am conscious of having to understand and unravel the different kinds of meanings and implications climate change has as it directly or indirectly affects people's lives, work, and local environments. Whether sitting in people's homes in northern Greenland, or spending the summer in fishing camps, or out on long journeys by dogsled on the winter ice, I have come to appreciate that knowledge of weather and climate events grows through the experience of living in and moving through local landscapes and environments, and that uncertainty and surprise are things that people expect to encounter in a world that is always in a process of becoming. At different levels, from small villages through municipal politics and Home Rule government institutions, an anthropological challenge, as I see it, is to understand climate change within a broader context of political process and ambition, cultural specificity, and people's epistemological, social, cultural, economic, and moral relationships with the environment. As Greenlanders achieve greater autonomy over their lives, they may be forced to ask whether the process of nation building and ambitions for economic development (together with the absence of appropriate tools to manage it) will reduce the abilities of people to adapt and be flexible in coping with climate variability and change, as well as to ponder their own contributions to global climate change that may accompany such development. If all I have to offer is a comment on this, something that contributes to the development of perspectives on the political ecology of human-environment interactions, then surely that in itself is a form of anthropological action.

References

ACIA. 2005. *Arctic Climate Impact Assessment: Scientific report.* Cambridge: Cambridge University Press.

Adger, W. N. 2000. Social and ecological resilience: Are they related? *Progress in Human Geography* 23(3): 347–64.

Briggs, J. 1991. Expecting the unexpected: Canadian Inuit training for an experimental lifestyle. *Ethnos* 19(3): 259–87.

Caulfield, R. A. 1997. *Greenlanders, whales and whaling: Sustainability and self-determination in the Arctic* Hanover, NH: Dartmouth College.

Cotgrove, S. 1982. *Catastrophe or cornucopia: The environment, politics, and the future.* Chichester, UK: Wiley.

Dahl, J. 2000. *Saqqaq: An Inuit hunting community in the modern world.* Toronto: University of Toronto Press.

Diplomatic Pouch. 2007. http://www.washdiplomat.com/DPouch/2007/June/062707lifestyle.html

Ellen, R. 1996. The cognitive geometry of nature: A contextual approach. In *Nature and society: Anthropological perspectives,* eds. P. Descola and G. Palsson, 103–23. London: Routledge.

Hamilton, L. C., P. Lyster, and O. Otterstad. 2000. Social change, ecology and climate in 20th century Greenland. *Climatic Change* 47(1/2): 193–211.

Harris, M. 1998. The rhythm of life on the Amazonian floodplain: Seasonality and sociality in a riverine village. *Journal of the Royal Anthropological Institute* N.S. 4: 65–82.

Huntington, H. and S. Fox. 2005. The changing Arctic: Indigenous perspectives. In *ACIA Arctic Climate Impact Assessment: Scientific report,* 61–98. Cambridge: Cambridge University Press.

ICC Greenland. 2006. *Inuit Circumpolar Council newsletter.* December.

Krupnik, I. 1993. *Arctic adaptations: Native whalers and reindeer herders of Northern Eurasia.* Hanover, NH: University Press of New England.

Krupnik, I. and D. Jolly, eds. 2002. *The earth is faster now: Indigenous observations of Arctic environmental change.* Fairbanks, AK: ARCUS.

Lynge, A. 2007. Global warming is not just a theory to us. *The Independent*, May 30. http://www.independent.co.uk/opinion/commentators/aqqaluk-lynge-global-warming-is-not-just-a-theory-to-us-450941.html

Motzfeldt, J. 2008. Climate change in a Greenland perspective. Presentation at the Trans-Atlantic Climate Conference, Torshavn, Faroe Islands, April 7–8.

Nuttall, M. 1992. *Arctic homeland: Kinship, community and development in northwest Greenland.* Toronto: University of Toronto Press.

———. 1998. *Protecting the Arctic: Indigenous peoples and cultural survival.* New York and London: Routledge.

———. 2001. Locality, identity and memory in South Greenland. *Études/Inuit/Studies* 25(1 & 2): 53–72.

———. 2005. Inuit, marine resources and climate change: Risk and resilience in a changing Arctic. In *Indigenous use and management of marine resources*, eds. N. Kishigami and J. M. Savelle, 409–26. Osaka: National Museum of Ethnology.

———. 2008. Climate change and the warming politics of autonomy in Greenland. *Indigenous Affairs* 1-2/08: 44–51.

Nuttall, M., F. Berkes, B. Forbes, G. Kofinas, T. Vlassova, and G. Wenzel. 2005. Hunting, herding, fishing and gathering: Indigenous peoples and renewable resource use in the Arctic. In *ACIA Arctic Climate Impact Assessment: Scientific report.* 649–90. Cambridge: Cambridge University Press.

Petersen, R. 2003. *Settlements, kinship and hunting grounds in traditional Greenland. Meddelelser øm Grønland/Man and Society* vol. 27. Copenhagen: The Commission for Scientific Research in Greenland.

Rasmussen, R. O. 2006. Oil exploration in Greenland. *Indigenous Affairs* 2–3/06: 40–47.

Rasmussen, R. O. and L. C. Hamilton. 2001. The development of fisheries in Greenland. *North Atlantic regional studies research report* 53, Roskilde, Denmark: Roskilde University.

Riddington, R. 1990. *Little bit know something: Stories in a language of anthropology.* Vancouver: Douglas and McIntyre.

Thomas, R. H. 2005. Greenland ice sheet. In *Encyclopedia of the Arctic*, ed. M. Nuttall, 789–90. New York and London: Routledge.

Thuesen, S. T. 1999. Local identity and history of a Greenlandic town: The making of the town of Sisimiut (Holsteinsborg) from the 18th to the 20th century. *Études/Inuit/Studies* 23(1–2): 55–67.

Vibe, C. 1967. Arctic animals in relation to climatic fluctuations. *Meddelelser øm Grønland* Bd 170, Nr 5, Copenhagen: C.A. Reitzels.

Walker, B., C. S. Holling, S. R. Carpenter, and A. Kinzig. 2004. Resilience, adaptability and transformability in social-ecological systems. *Ecology and Society* 92(5). http://www.ecologyandsociety.org/vol19/iss2/art5.

Watt-Cloutier, S. 2005. Petition to the Inter American Commission on Human Rights seeking relief from violations resulting from global warming caused by acts and omissions from the United States. December7. Available at http://www.inuitcircumpolar.com/files/uploads/icc-files/FINALPetitionICC.pdf.

Weller, G. 2000. The weather and climate of the Arctic. In *The Arctic: Environment, people, policy*, eds. M. Nuttall and T. V. Callaghan, 143–60. New York: Taylor and Francis.

Chapter 19

Where Managerial and Scientific Knowledge Meet Sociocultural Systems: Local Realities, Global Responsibilities

P. J. Puntenney

Climate change has become the defining issue of the twenty-first century, requiring a complete readjustment of priorities on a global scale. During 2007, the debate on climate change shifted to a worldwide awareness that climate change is an ethical and moral imperative that must be negotiated within a global framework to transform rhetoric into commitments and results. While all nation-states agree upon the principle of "common but differentiated responsibilities within one's capacity," it is clear that no one country alone can confront the impacts of climate. The United Nations is at the center of this endeavor to act decisively with a "unity of purpose." The UN has declared sustainable development and climate change linked with equity and the elimination of extreme poverty as a top priority to be operationalized and mainstreamed system-wide. Member states also agreed in February 2008 to support the December 2007 Bali Roadmap to complete negotiations by 2009 for the Climate Convention's meeting in Copenhagen. The purpose is to deliver a new, comprehensive global climate change regime post-2012 for the Kyoto Protocol. Based upon the four building blocks of adaptation, mitigation, technology, and finance as the top priorities, the current agenda is driven by an emissions–climate change–impacts–responses paradigm. The human dimension priority, as building block number five, is missing. Climate change is both the cause and the effect of a much larger issue, environmental degradation.

Environmental degradation is among the most difficult and complex problems ever faced by modern society. Our current understanding of ecosystems informs us that continuing loss of biodiversity at some point will exact a large toll in ecosystem function and resilience, as is already apparent, for example, in the condition of the world's oceans. On one level, we have come to recognize that the laws of nature are not negotiable and setting a course for global sustainability is essential for the long-term survival of the planet. On another level, globalization is a human construct and sustainable development is a mediating phrase to understand better the intersection between balancing human needs and protecting the earth's life support systems. A principle challenge facing the social sciences, ecological professionals, and decision-makers in all countries is how to create initiatives that reflect the interconnection between economic prosperity, a healthy environment, and social equity. Such policies transcend traditional boundaries—those between disciplines, sectors, nationalities, cultures, and generations—requiring an ability to recognize increasingly complex and interrelated issues where attempts to ameliorate one can alter or even exacerbate the impacts of another. Within the context of regional and national sustainable development strategies, there is a shared struggle to achieve effective integration, coordination, and implementation. Under the current conditions of globalization, we are faced with the challenge that the complexity of living systems is beyond full human comprehension. So where do we go from here?

Social scientists and practitioners need to bring to bear the cultural systems dimension, addressing issues of access, equity, hope, and security in conjunction with the protection of environmental systems. Anthropologist Steve Rayner, of the University of Oxford's James Martin Institute for Science and Civilization, has pointed out that we are looking at "clumsy solutions" involving diverse stakeholder engagement with people we normally exclude from our strategic decision-making processes (Rayner 2006; Verweij et al. 2006). The larger question for anthropology in particular is how to engage and work with local communities, communities in the making, decision-makers, and the broader stakeholders to satisfy the requirements arising from adhering to the principles of sustainability as boundary constraints. Drawing upon the work of Karl-Henrik Robert, these principles "provide a practical set of design criteria used to direct social, environmental and economic actions and to transform debate into constructive discussion" (2006).

In this chapter I will first trace the intersection between environmental and human systems with growing concern over the sense of urgency to act, bringing the human-cultural dimensions to the forefront, before turning to affecting decision-making within the global context of climate impacts. I then focus on where we are in terms of constructing a sustainable vision in order to act with a unity of purpose. The most vulnerable countries that need action now from the multilateral community of nations have made it known that already the science is inadequate, technology will not solve their immediate problem of survival, and an operational strategy needs to be implemented now. This in turn focuses my discussion on how the social sciences can approach and contribute to the decision-making processes regarding the human-cultural dimensions of sustainable development and climate change. I have, interspersed throughout the chapter are recommendations for policy making.

But You Got to Know the Territory

Organized climate assessments began in the early 1970s, incorporating discourses on energy and society, and climate and society. Like the song from the musical *The Music Man*, "But You Got To Know The Territory," the human dimensions shifted from societal impacts on variability and actual local coping mechanisms to data predominately based upon mathematical models. Climate change emerged as a global policy concern near the end of the 1980s, which led to the establishment of an intergovernmental assessment mechanism, the United Nations Intergovernmental Panel on Climate Change (IPCC), and the beginning of formal negotiations. First, the IPCC produced a report to provide the technical basis for a Framework Convention on Climate Change, the UNFCCC, in preparation for the Rio Earth Summit, and later in 1997 a binding protocol negotiated at Kyoto. As then, currently the cultural dimension is framed in terms of adaptation—primarily defined in limited terms, ranging from an offsetting parameter to an abstract damage function.

Rappaport and Bennett have eloquently brought two significant understandings to the attention of anthropology as a social science: First, that it is a false logic to position the environment as equivalent or subordinate to economic and sociocultural dimensions when in actuality, economic and sociocultural dimensions are subordinate to the environment. Second, that Homo sapiens is inherently destructive and, in modern times, has the capacity to destroy the life support systems of the world. Nevertheless, as social scientists, we still formulate our problem statements, research questions, choice of methods, and recommendations employing the false logic of an outdated paradigm, that humanity is separate from and acts on the environment (human activities impacting climate), rather than humanity is an integral component of natural systems and one of many species (Bennett 2005; Puntenney 1995; Rappaport 1995; Soper 1995). The issues surrounding climate change are a good example of this paradox, as is the current state of human dimensions research and its application.

Modern science is at a crossroads moving away from an objectivity that separates the professional from the system observed, from overly specialized methods and specialized technical languages that result in little or no attention to the integration of systems. The focus now is increasingly on the whole-system level, the biosphere, where humans are key actors (Holdren 2008). The complexity and changes over time are too subtle for science to predict and beyond our comprehension. Consequently, knowledge, meaning, and understanding are of equal concern as measurement, impacts, and physical causation (Kimball 1987; Matthiessen 1987).

We have come to the realization that solving environmental problems will require broad cooperation among competing interests. While science can attempt to assess the functions and risks confronting these systems, solutions to the major environmental issues within climate change will require public choice and public responsibility. To achieve such cooperation we must draw upon many sources of knowledge, especially from the people who will be directly impacted and will have to sustain a level of commitment and adjustment over a long period of time. With an eye to mid- and long-term aims beyond the Kyoto era, decision-makers are beginning to employ an operational approach to sustaining the biosphere. We now recognize that solutions must address sustained human and natural well-being specific to the conditions needed for economic development strategies that specifically integrate regional and local realities.

Two decades ago natural resource management was primarily defined in terms of the economic value of the resource. Attempts to regulate and manage human environmental processes and solve environmental problems were primarily accomplished through legal, technical, and economic methods. Reflecting a world view based upon the assumption that humans are separate from their environment, this approach to environmental problem solving allowed valued resources to be controlled by political instruments and technical devices.

But by the 1980s, humanity was beginning to confront its environmental problems realistically in terms of interconnected systems. Advances in computer technology and satellite imagery have contributed to an increased global awareness of environmental issues and broadened our understanding of the magnitude of environmental change. By the early 1990s, the environmental outlook had shifted from a concern with problems impacting social and economic development to processes defined in terms of interrelationships and the sustainability of environmental and human systems. A significant quandary dominating the public debate about global environmental change is: "What actions might lead to unacceptable environmental, sociocultural, and economic impacts?" However the dilemma is as much sociocultural as technical and requires strategies adaptable to both realms.

As we attempt to grasp the accelerating complexity of global interdependence, creating options that will effectively protect the integrity and health of human and natural systems is daunting. The human factor operates in all of the ensuing conflicts between local, national, regional, and international environmental and socioeconomic interests. Building practical policy considerations into the development agenda requires the active participation of stakeholders in the creation of solutions. Whether cultural brokers or intermediaries, anthropologists can and must play an important part in this equation. Bringing a cross-cultural perspective that draws from a number of disciplines and areas of professional expertise in gathering concrete data, anthropologists can bridge and identify an expanded repertoire of options in the decision-making process. Anthropologists can be particularly useful in redefining and expanding potential roles for stakeholders and bringing people together in mutually beneficial ways from the local initiative to the development of a national and/or regional environmental plan of action.

Turning to the looming threat of climate change, to understand the dynamism of cultural systems and praxis (reflection plus action) -oriented policy making we must first appreciate the multidimensionality of the challenge we face. Expanding beyond emissions–climate change–impacts–responses, human-dimensions researchers and practitioners need to carve out a more meaningful role. Thus, an essential first step is understanding the basic non-negotiable principles of sustainable development as boundary constraints within the context of humans as integral components of natural systems. As scholars and practitioners, we must reexamine our notions about living systems in the context of interdependence and connectivity and within the context of our own ecological footprint writ large. We also need to be reflexive and offer our physical and natural science colleagues constructive criticism. A blueprint is needed that sets the stage to answer the question of how to integrate the human dimensions with the human condition.

The study of the interface between natural and human systems has been a major preoccupation of mine for nearly three decades. I have concluded from this work that the cocreation of a well-prepared society is the most challenging part of the environmental equation. Cultural systems are the arbitrators in the difficult trade-offs determining what choices we make to constitute our common future. The three most pressing global issues we face are climate change, loss of biodiversity, and urbanization. Increasing poverty, environmental degradation, and shortsighted decision-making are marked by the growing concentration of people in sprawling cities. As of 2008, the world urban population had reached 52-plus percent of the total global population, and a considerable portion of this population is unable to attain even the most basic necessities because of economic, social, cultural, or ethnic characteristics or because of gender or age (UN ECOSOC 2004).

It is predicted that by 2010–2012, over half of the world's population will be less than twenty-four years of age. The bad news is, many of these young people are without hope of an education, a job, decent housing, or access to the resources to marry and to start their own families. The good news is, it is the youth who can help solve the problems we face. They are a critical link to the public engagement model we are struggling to put into place within the UN system as exemplified by the Global Youth Action Network (GYAN).

Mexican anthropologist Lourdes Arizpe directed and shaped the biennial strategic series, *The World Culture Report*, which guides the work of the Cultural Division of the United Nations Educational, Scientific and Cultural Organization (UNESCO). In the preface to Part II on "Global Socialcultural Processes," Arizpe observes:

> People in many places, even when they are aware and willing to protect the natural environment, find it impossible to do so because of economic, political or cultural dislocations. Eliminating these pressures calls for good governance, organizing people through democratic processes. But it also requires proper

> conviviability or convivencia, that is, reorganizing cultural allegiances to enable human beings with different ideals of a good life to live together compatibly in a living biosphere. (Arizpe 1998)

Following her basic argument that "we are all in this together," Arizpe expands upon this conceptual framework in her recent work, stating that anthropologists must be involved in the reconstruction of cultural narratives and cross-representations in this new cultural commons (Arizpe 2004). The linking of people and nations globally through a vast information network is creating a shared understanding of the world as a whole. As a consequence, there is a growing impetus to create new platforms of dialogue from local to national to regional to global levels and more that will lead to the development of sustainable systems through mutual sharing of knowledge and mutual cooperation, albeit contentious at times.

To these ends, Peter Brosius's work is enlightening. Brosius has been building a theoretical model integrating transnational politics and conservation (Brosius 1999). By bringing anthropological knowledge to bear on the United Nations Millennium Ecosystem Assessment, he asks a series of "what if" questions that directly challenges us to think about how we do ethnographic research. His most basic question is, "Does our concept of knowledge include questions that also seek information regarding our informants' analyses of the political world, trends, tradeoffs, and response options?" How do we see our informants: are they reservoirs of local knowledge or political agents with their own ideas about the saliency and legitimacy of different forms of knowledge? Furthermore, as we think about our own fieldwork, consultancies, and engagement with stakeholders, do we systematically incorporate the data into development implementation strategies and make a more systematic effort to inform and influence alongside our other colleagues?

In 2002, ten years after the UN Conference on Environment and Development in Rio de Janeiro, Brazil, the World Summit on Sustainable Development in Johannesburg, South Africa was held to evaluate progress made by nation-states and to develop a plan of action. Moving from just "talk shops" to operationalizing action, two paradigm shifts occurred on the world stage of multilateral decision-making. The first shift put people at the center of development. The second ensured sustained impact by defining development in terms of demonstrated support for and complementary partnerships with intergovernmental commitments as spelled out in the Johannesburg Plan of Action. Within the climate change arena of decision-makers and member states of the UN, the common ground is based upon the principle of "common but differentiated responsibilities," meaning that we all have a common responsibility but it is differentiated by context, conditions, culture, and capacity (Referring to the 1987 report from the UN World Commission on Environment and Development, *Our Common Future*, and the 2004 Human Development Report (HDR), *Cultural Liberty in Today's Diverse World*).

Useful as the new paradigms are, it is still the focus on learning and learning systems that elude us amid the power struggles and the consequences of our decisions, plans, programs, and projects. We lack the necessary mechanisms to monitor and report on what is working, what is not, and how we can do better. At the global scale, we humans have never been in this position before—so there are no roadmaps. We can no longer view knowledge in terms of product and dissemination. The extensive work on multistakeholder processes demonstrates that the mediating phrase "engaging people in sustainability" is instrumental in creating dialogues leading to appropriate action.

Given the complexities of climate change, the human prospect depends to a great extent on how rapidly and effectively organizations of all kinds respond to larger trends and forces, which is to say how well and how quickly they learn (Pins and Rayner 2007). Progress in understanding sustainable systems and acting on this understanding requires collaboration and interaction between social scientists and related disciplines. It is here that the anthropologist will inevitably play a key role in the global quest for sustainability. Inventive dialogue can be an effective agent for positive change.

In 2007 this dilemma was brought to center stage with a series of events that have raised global awareness about the urgent need to act regarding climate change. The impetus for creating this global response emerged from Al Gore's film *An Inconvenient Truth* and the international press coverage of the United Nations Intergovernmental Panel on Climate Change (IPCC) reports from Working Group II on "Impacts, Adaptation and Vulnerability" and Working Group III on "Mitigation of Climate Change." In the fall, climate change was positioned on the G-8 agenda and on the UN Security Council agenda. Later in November of that year, the Fourth Assessment from the IPCC "Synthesis Report," United Nations Secretary-General Mr. Ban Ki-moon, and the UN General Assembly committed to a multilateral solution to climate change. Then, the Nobel Peace Prize was given to Al Gore and the IPCC. The UN Climate Change Conference, held in Bali, Indonesia, December 3–14, brought further attention to political mobilization around climate change and energy policy. Delegates from over 180 countries ended this momentous year by adopting the "Bali Roadmap" for a future international agreement on climate change, ended this momentous year. The media coverage of all these events and the response by professionals and the public has made an impression as people endeavor to make sense of the implications. Society has followed this lead with expectations of government being primarily responsible. What is missing is an intelligible dialogue between stakeholders and decision-makers. Without a dialogue that goes beyond "objective problem solving" relying on precise measurements, models of prediction, and accumulated data points, there is no room for process.

DESIGNING SUSTAINABLE SYSTEMS: WHOSE VISION?

The term *sustainable development* serves as a mediating phrase, in its broadest sense denoting long-term viability. To be achieved, sustainability needs to be pursued at all levels from the governmental and the corporate to the personal and the individual. Unintended consequences of development that shape climate change and energy policy have exacerbated issues of equity and poverty. They have also produced unusual patterns of land use that are disruptive of existing patterns of economic and social interaction, such as agro-landscapes extending into cities, industrial networks in desert areas, and advanced agro-industries catering to global markets from remote rural areas. As we struggle to make sense of our predicament, we attempt to interpret these realities in terms of sustainable systems.

A shift is occurring on both the global and local stages, as the scientific and policy-making communities realize that the obvious source of leadership and vision to help shape sustainable systems are broad partnerships that engage governments, the private sector, academic institutions, civil society organizations, and the public. Climate change action involves many levels of decision-making dependent upon local-level, small-scale initiatives to address the myriad of interconnected issues. There is a larger responsibility to protect and engage ordinary people. What is needed is an intelligible dialogue involving diverse partnerships that engage stakeholders (Puntenney 2003). We need a dialogue to build a global leadership network that will provide direct input into international environmental governance. We need global leadership that can support adaptation (reducing adverse impacts) to and mitigation (reducing the pace and the magnitude) of climate impacts.

The obvious candidates for providing insights, guidance, and solutions to climate change in terms of the sociocultural dimension are anthropologists. Anthropology has the ability to see beyond walls, to see how structures fit and interact with the natural and built environments. Taking the old axiom, "Think Globally, Act Locally," the cultural anthropologist's context is more often, "Think Locally, Act Globally." Wherein, the sciences focused on the human experience contribute their insights in forms meaningful to and usable by decision-makers regarding dynamic systems, with sustainability foremost in mind.

At the moment the profession is not leading but is being driven by vested interests and internal debates. The anthropologist's power has to do with asking questions, and in the process learning that a lot of attention can be drawn to such questions. The UN Millennium Development Goals [MDGs] and the UN Commission on Sustainable Development [CSD] have crafted the beginning of a global framework to redirect political realities (UN 2000). Anthropology is part of this change and will increasingly play a more significant role in major public debates. The UN Commission on Sustainable Development is preparing the foundation for the anthropologist by providing the necessary political will at the global level. These forums are also setting the stage to enable the discipline and practitioners to do what they need to do worldwide. The CSD Plan of Work 2004–2017 reads like a major issue guide to the professions and disciplines. The failure of globalization to alleviate poverty, advance sustainability goals, avert financial crisis, or stem the tide of economic and environmental refugees is creating the impetus for more integrated approaches to models of development that reflect local realities and global responsibilities. Given this present state of affairs, Arizpe (2004) points out, "people cannot manage the natural environment rationally if the 'way we live together' forces us to be hungry, greedy or destructive."

To negotiate tangible options to "Think Locally, Act Globally" or to "Think Globally, Act Locally," we need to challenge embedded assumptions by asking tough questions about what is our own ecological and cultural footprint as a profession and as a discipline. Are we creating genuine value for society or appropriating value from it? Do we want to take on the responsibilities of shaping our own nation's policies on sustainability, leading to a well-prepared society? Shaping our own local community's policies on sustainable well-being? Ultimately, in addressing issues of multidimensional analysis within a biosphere perspective, opening the interpretive process will occur to allow for diverse outside input and easy access to knowledge across sectors, across disciplines, across collaborative partnerships. This would help remove a serious obstacle to crafting potential solutions to achieve sustainable systems (Rayner 2003a, 2003b). It would also strengthen our capacity for resolving critical issues and designing institutional structures for sustaining a dialogue among those who have a stake in the outcome, especially those who will live with the outcome over the long term (Broadbent 2004; Clay 2005; Douglas 2004).

Our anthropological footprint is found throughout international policymaking debates and plans of action. Increasingly, culture is viewed as important in the discussion of policies, decision-making, and the building of sustainable enterprises. Rappaport (1995), in framing the crux of the argument and depicting where the discipline of anthropology's priorities should be, argues, "We need to understand better how humans may keep from destroying the systems upon which they depend, and themselves as well, when the complexity of those systems exceeds any hope of comprehending them." We have come to appreciate that a broad command of the environmental landscape, natural and human, is essential to understanding the continually fluctuating interpretations of reality as seen on many different scales and in many different places.

Upon closer examination of the issues surrounding climate change, we know by definition that the outcomes of international debates are the result of complex interactions of many factors, including sudden changes in the global environment or scientific understanding, macroeconomic trends, domestic and international political developments, and the presence or absence of effective leadership. On the other hand, within the context of sustainable enterprises, these dialogues have been contentious at best, struggling with how to move beyond "doing business as usual" with little interchange between members of the policy-making communities and civil society (Dodds and Middleton 2001; Dodds and Strauss 2004).

Public Debate—Private Actors

The demands on governments, nongovernmental organizations, and ordinary citizens are significantly different from what they were even a decade ago, particularly in terms of sustainability. The dilemma can be summed up pointing out the obvious: "The public agenda cannot be surrendered entirely to public institutions. . . . If global civil society is to contribute more to a sustainable future, it must come together in a more organized and decisive way." The policies of international institutions have become controversial with questions raised about equity, sustainability, and balance between economic growth, social reform, and environmental and ecosystem protection.

By 1992, the UN had expanded its original mission of preventing military conflict to include that of safeguarding the long-term health of the planet through a commitment to sustainable development. The work on sustainable development at the governmental level had moved beyond "talk shops." In response to the shortcomings and limitations of several decades of development primarily focused on the social component of international development, the United Nations sponsored a high-level intergovernmental conference in Rio de Janeiro on environment and development to create a vision of the future. The Commission on Sustainable Development (CSD) was created in December 1992 to ensure effective follow-up of the United Nations Conference on Environment and Development (UNCED), nicknamed "Rio" after the city, and to monitor and report on implementation of Earth Summit agreements, the twenty-one Agenda items (Agenda 21), mandates, and conventions at the local, national, regional, and international levels. The commission also serves to ensure the high visibility of sustainable development issues within the UN system, and helps to improve the UN's coordination of environment and development activities and to promote partnerships leading to implementation and action. It has been my observation over the last decade that one of the greatest challenges facing the CSD's monitoring of progress on Agenda 21 is staying on top of what decision-makers need to know in order to act strategically in fostering sustainable development. One of the greatest challenges facing the implementation of Agenda 21 is staying abreast of evolving, strategic approaches to sustainable development.

Knowledge, how we obtain that knowledge, and learning how to organize and use it have become increasingly important as an integral part of the outcomes of the CSD's work (Puntenney 2002; Puntenney et al. 2002). There is a growing consensus worldwide that local and regional initiatives are shaping what globalization means and directing the future of sustainable development. On the policy level, a consensus was reached and a framework for an agenda was crafted in the 2001 UN Millennium Declaration's eight goals (United Nations 2001). As they are the focal point for much of what the UN and member states have committed to in this century, it is worth noting:

1. values and principles
2. peace, security, and disarmament
3. development and poverty eradication
4. protecting our common environment
5. human rights, democracy and good governance
6. protecting the vulnerable
7. meeting the special needs of Africa
8. strengthening the United Nations

(http://www.un.org/millennium/declaration/ares552e.htm)

Ten years after Rio, the World Summit on Sustainable Development (WSSD), held in Johannesburg, South Africa in 2002, was about governments building partnerships with civil society and the private sector, action, and the implementation of agreements including the Millennium Development Goals. Sustainability is one of those goals, but it is also a prerequisite for reaching all of the others.

Science-based Decisions

The commitment to global sustainability has broadened. Here are a couple of examples that not only affect us as social scientists in how we do our work but also call into question the anthropological endeavor as we see it now. First, in May of 2002, a consensus report summarizing the findings from a two-year consultation process on "How can science and technology contribute more effectively to achieving society's goals of sustainable development?" was made public and became a part of the 2002 Global Forum Report at the WSSD in Johannesburg.

The report included global views from international science organizations, regional views grounded in grassroots efforts to harness science and technology in support of sustainable development, assessments of potential contributions from global-change science, and critical analysis of experience in designing institutions and providing financing for science and technology (S&T) directed toward solutions to sustainability problems. Those crafting the document were struck by the sense of urgency and by the potential contributions of S&T to meeting the challenge through participation in the design of solutions, and through a renewed commitment worldwide to serve as an active partner in realizing the potential. This commitment requires substantial changes in the way the S&T community conducts its work. It gives special attention to the involvement of major stakeholders affected by the outcomes of S&T research, setting agendas, and disseminating results, as well as developing conceptual frameworks of action and collaborating with civil society.

In addressing specific contributions that science and engineering can make to capacity building and global sustainability, the National Research Council study *Our Common Journey: A Transition Toward Sustainability* concludes that a transition could be achieved in the next two generations without miraculous technologies or a drastic transformation of human societies. The report stressed that advances in basic knowledge were needed in the social capacity and technological ability to use it, and in the political will to turn this knowledge into action. What is missing is the call for a dialogue and deeper analysis of the links between the human condition, culture, and the systemic issues surrounding sustainable development.

As part of the UN agenda for the new millennium, UN Secretary General Kofi Annan extended an invitation to the scientific community to bring their knowledge to the multilateral decision-making process. Peter Raven, in his presidential address to the American Association for the Advancement of Science (AAAS) on "Science, Sustainability, and the Human Prospect," singled a commitment from the AAAS to help the UN achieve the UN Millennium Development Goals, such as environmental protection and the eradication of extreme poverty. He outlined a new way of thinking about relationships—an integrated, multidimensional approach to problems of global sustainability—and challenged the scientific community to find and accept new ways for maintaining global sustainability. He further stated that the AAAS must dedicate itself to expanding the association's global leadership role on behalf of science and society, and that it must address underlying causes of concern noted above that have been ignored by the S&T community (Raven 2002). The current AAAS president, John Holden, has taken this commitment one step further, directing the S&T community to contribute their knowledge to achieving sustained well-being (Holdren 2008).

These efforts are part of a major shift in thinking towards understanding the nature of cross-scale sensitivity in global environmental issues, i.e., to relate what happens on the local scale to the regional and global scales. We have developed a fair amount of scientific and technical knowledge on one level. On another level we have made real progress in sorting out the application of practical knowledge. It is between these levels, where managerial and scientific knowledge meet in the context of political and social systems, that things are murky. Achieving sustained well-being requires democratization of expertise and involvement of local and regional communities in the analysis of issues, questions, and decision-making processes.

What is needed at this juncture, Mead-style, is that anthropology seek access to critical policy debates, be it at the organizational, community or local, national, regional, or international level of engagement. We know from the answers to the questions we ask in the communities we study that cooperation in human relations and coordination are prerequisites for sustainable systems. As Arizpe defines it, culture is "ways of living together."

The challenges of global sustainability and climate change are deeply rooted in relationships neglected too long. The anthropologist's sphere of influence is contingent upon knowing how those relationships work, and what it takes to balance the power of words and the power of action (Scott-Stevens 1987). Van Arsdale (1993; 2002) suggests a conceptual model where social scientists critically assess "Awareness, Action, and Advocacy" to ascertain a level of competence that would allow the professional to effectively engage in assessing a critical issue. Rappaport, in his seminal work, "Disorders of Our Own," cogently provides a sound theoretical framework, challenging the profession to publicly engage the discipline in what is now termed *global sustainability*. Yet the profession is strangely absent, in the larger public forums and debates surrounding climate impacts, and in partnership with decision-makers to chart a way forward.

Knowledge, Power, and Praxis

With the advent of world systems analysis and underdevelopment theory, anthropologists have chronicled the devastating impact of global economic change in the peoples of both the developing and industrialized world. It is not enough to advocate the cultural dimensions in the abstract. As social scientists our task is to combine theory and practice to demonstrate precisely how cultural pluralism and democratic participation enhances economic success, social opportunity, political stability, strategies of adaptation to global environmental change, and conflict resolution. In our research, using the prism of multidimensional analysis from a systemic approach is the essential starting point in addressing the complexities of climate change.

Anthropology has the capacity to access strategic information and in turn engage in a leadership role within the critical arena of policy debates and decision-making processes. Cultural anthropologists can also focus on second-track negotiations where, acting as transducers, we bring together the disputants, not the emissaries of the diplomatic community, to engage in a facilitated policy dialogue aimed at both relationship building and problem solving. Cultural anthropologists can also contribute to the discipline of anthropology a fuller understanding of its relationship to policy relevance in terms of methods, approaches, knowledge of ongoing processes of community, and the development of better theory to guide practice. Our challenge is to demonstrate to society at large that we can tackle the larger, multidimensional environmental-social-economic issues facing our own society and local communities regarding environmental sustainability. In addition we can enrich the debates on global sustainability nationally and internationally, creating alternative models and options through anthropological insights. The issue is what our response will be as a profession, and commitment as a discipline, to working with the UN (the American Anthropological Association has consultative status that enables AAA to accredit anthropologists to attend UN meetings) and the various policy-making arenas regarding sustainable development and climate change tied to the UN Millennium Development Goals.

Physical and natural scientists tend to gather data points, using the particular to represent the general. The human dimension of global environmental change demands not irrefutable facts alone or precise models of prediction, but a strong relationship with human experience. We are in a period of transition where there is a strong desire to comprehend and to shape knowledge to construct a strategic approach based upon cocreating a well-prepared society, and informed decision-makers. There is no silver bullet. In light of these challenges and the inevitable clumsy solutions, what is needed from an anthropological partnership with the other sciences is a strong response to the UNFCCC and Kyoto Protocol, Bali Roadmap, and post-2012 climate change regime. This inevitably implies anthropologists assuming leadership roles in the decision-making process and forums leading to action and implementation of strategic initiatives. Applying the fundamental principles of sustainability constraints as systems boundaries and the principle of "common but differentiated responsibilities within one's capacity," how can anthropological knowledge of human systems and the everyday life of human experience be summarized and presented to decision-makers who will integrate global responsibilities and local realities? How could cultural anthropology enhance collaboration, leading to long-term engagement among diverse stakeholders that will lead to a larger discussion beyond data points to action within the public debate? As professionals negotiating the cultural dimension and developing an understanding of new cultural allegiances, this is an important part in assessing our own ecological footprint.

If we were to apply the notion of praxis to our collective action to address climate impacts, Dag Hammarskjöld (1905–1961), the Swedish diplomat and second secretary general of the United Nations, would be in the place of honor in the discussion. He talks to us about his personal enquiry into his world, about cultural differences and development, and about making a difference one step at a time. He prefigures our vision of the anthropological difference in his ambition to understand. Together with his colleagues and friends, he brought the message of peace alive through his determination, strength, courage, and integrity, advising: "Never measure the height of a mountain until you have reached the top. Then you will see how low it was."

REFERENCES

Arizpe, L. 1998. *Convivencia: The goal of conviviability*. World culture report: Part two, global sociocultural processes. Paris: Unesco Publishing. http://www.unesco.org/culture/worldreport/html_eng/wcrb22.shtml (accessed April 4, 2008).

———. 2004. The intellectual history of culture and development institutions. In *Culture and public action: A cross-disciplinary dialogue on development policy*, eds. V. Rao and M. Walton, 163–84. Palo Alto: Stanford University Press.

Bennett, J. W. 2004. Applied anthropology in transition. *Human Organization* 64(sp): 1–3.

———. 2005. Malinowski award lecture 2004: Applied anthropology in transition. *Human Organization* 64(1)(sp): 1–3.

Broadbent, N. 2004. "Saami prehistory, identity and rights in Sweden." In *Conference proceedings of the third Northern Research Forum "The resilient North—Human responses to global change."* http://www.nrf.is/Publications/The%20Resilient%20North/List%20of%20authors.htm (accessed February 20, 2008).

Brosius, P. 1999. Analyses and interventions: Anthropological engagements with environmentalism. *Current Anthropology* 40(3): 277–309.

Clay, J. 2005. *Exploring the links between international business and poverty reduction.* Oxford: Oxfam Publishing.

Dodds, F. and T. Middleton. 2001. *Earth summit 2002: A new deal.* London: Earthscan.

Dodds, F. and M. Strauss. 2004. *How to lobby at intergovernmental meetings.* London: Earthscan.

Douglas, M. 2004. Traditional culture: Let's hear no more about it. In *Culture and public action: A cross-disciplinary dialogue on development policy*, eds. V. Rao and M. Walton, 85–109. Palo Alto: Stanford University Press.

Holdren, J. P. 2008. Science and technology for sustainable well-being. *Science* 319 (January 25): 424–34.

Kimball, S. T. 1987. Anthropology as a policy science. In *Applied anthropology in America*, eds. E. Eddy and W. L. Partridge, 383–97. New York: Columbia University Press.

Matthiessen, P. 1987. *The snow leopard.* New York: Viking Press.

Ny, H., J. P. MacDonald, G. Broman, R. Yamamoto, and K.-H. Robèrt. 2006. Sustainability constraints as systems boundaries: An approach to making life-cycle management strategic. *Journal of Industrial Ecology* 10(1–2): 61–77.

Pins, G. and S. Rayner. 2007. Time to ditch Kyoto. *Nature* 449(7165): 973–75.

Puntenney, P. J. ed. 1995. *Global ecosystems: Creating options through anthropological perspectives. NAPA Bulletin* 15. Arlington, VA: American Anthropological Association.

———. 2002. New policy framework: Education for our common future and Governance explained. *Outreach*, Issue IV May 30, p.5. http://www.stakeholderforum.org/news/outreach/pcIV/Thursday%2030.pdf (accessed September 27, 2008).

———. 2003. Building a sustainable enterprise: Engaging debates and debating engagements. *High Plains Applied Anthropologist* 23(fall): 182–95.

Puntenney, P. J. et al. 2002. Report to the global forum: WSSD forum on science and education: Strategic keys to sustainability. *Report of the World Summit on Sustainable Development Johannesburg, South Africa, 26 August–4 September 2002.* A/CONF.199/20, UN Publications: New York.

Rappaport, R. A. 1995. Disorders of our own: A conclusion. In *Diagnosing America: Anthropology and public engagement*, ed. S. Forman, 235–94. Ann Arbor: University of Michigan Press.

Raven, P. H. 2002. Science, sustainability, and the human prospect. *Science* 297(5583): 954–60.

Rayner, S. 2003a. Democracy in the age of assessment: Reflections on the roles of expertise and democracy in public-sector decision making. *Science and Public Policy* 30(3): 163–70.

———. 2003b. Domesticating nature: Commentary on the anthropological study of weather and climate discourse. In *Weather, Climate, Culture*, eds. S. Strauss and B. Orlove, 277–90. Oxford: Berg, Oxford.

———. 2006. The case for clumsiness. In *Clumsy solutions for a complex world*, eds. M. Verweij and M. Thompson, 1–30. Basingstoke: Palgrave Macmillan.

Scott-Stevens, S. 1987. *Foreign consultants and counterparts: Problems in technology transfer*. Boulder, CO: Westview Press.

Soper, K. 1995. *What is nature? Culture, politics and the non-human*. Oxford: Blackwell Publishing Company.

United Nations General Assembly. 2001. Fifty-sixth session. *Road map towards the implementation of the United Nations Millennium Declaration: Report of the secretary general.* http://www.un.org/documents/ga/docs/56/a56326.pdf (accessed February 20, 2008).

United Nations Department of Economic and Social Affairs, Population Division. 2004. *World urbanization prospects: The 2003 revision*. New York: United Nations.

Van Arsdale, P. W. 1993. Empowerment: A systems perspective. In *Refugee empowerment and organizational change: A systems perspective*, ed. P. W. Van Arsdale, 3–11. Arlington, VA: American Anthropological Association.

———. 2002. Awareness, action, advocacy: Mobilizing a paradigm, tackling an issue, making a difference. *High Plains Applied Anthropologist* 22(2, Fall): 119–24.

Verweij, M., M. Douglas, R. Ellis, C. Engel, F. Hendriks, S. Lohmann, S. Ney, S. Rayner and M. Thompson. 2006. Clumsy solutions for a complex world: The case of climate change. *Public Administration* 84(4): 817–43.

Chapter 20

Participatory Action Research: Community Partnership with Social and Physical Scientists

Gregory V. Button and Kristina Peterson

Introduction

With the increasing frequency of threats from natural and human-induced hazards and the early-onset effects of global warming, coastal communities around the world are facing new challenges that require innovative approaches. This chapter describes such an innovative, collaborative, and successful approach forged by the members of the Grand Bayou, Louisiana community in partnership with social and physical scientists. We argue that the project can serve as an exemplar of a non-traditional approach for other communities facing the unique challenges that now confront us. Moreover, this project illustrates how *both* social scientists and lay communities can effectively reach out to wider audiences from the local to the global to affect change.

This case demonstrates both the importance of lay knowledge and the approach of a participatory research project. We begin with an ethnographic synopsis of the people of the Grand Bayou, the described geophysical setting of their community, and show how their marginalization as an indigenous group has made them more vulnerable to environmental hazards. Then we discuss how a participatory process that validates local knowledge not only leads to an empowerment of the community but also contributes to a synthesis of traditional knowledge and science. This enhances the community's adaptation to dramatic environmental changes and it contributes to a richer understanding of climate change. The enormous successes that the Grand Bayou community has achieved in a relatively short period of time is a testament both to them and to the efficacy of such an approach for responding to climate change and other environmental challenges.

The Community of Grand Bayou

Grand Bayou is an hour and a half south of New Orleans in Plaquemines Parish, on the Bird's Foot Peninsula and almost a mile off the main highway. There are no roads or sidewalks in Grand Bayou, just fishing boats and homes on the water. After the ravages of Hurricane Katrina in August 2005, few homes remained and to date only one has been rebuilt. The community consists of twenty-five families, or 125 individuals, all of whom were displaced by the storm. They try to return for visits as often as possible and send their children and grandchildren back to the bayou during holidays and the summer.

Displaced Grand Bayou residents reside in trailers provided for storm evacuees by the Federal Emergency Management Agency (FEMA) in Plaquemines Parish, Houma, and other small Louisiana communities, as well as far north as Tennessee. Although to date no one continues to live permanently on the bayou, some members of the community can be found "at home" working during the day. Boat owners take turns living on their boats in order to protect community property against thieves. A number of the fishers, as they call themselves, continue to fish regularly. Other residents return when they can to rebuild their community, practice their cultural heritage, and spend their hours in the place that will forever be home.[1]

Contestation Over Ethnic Identity

While elements of the local natural environment contribute to the vulnerability of a given community to environmental disasters, sociopolitical and economic relations also play a significant role. In this case, the way in which outsiders and the dominant culture have perceived the residents of Grand Bayou over time contributes to this equation. The Grand Bayou community is a mixture of Native American, French, Spanish, and African heritages. Residents speak Cajun French and English. While mindful of their rich cultural ancestry, they also have a persistent collective identity. The residents identify themselves primarily as Native Americans belonging to the Atakapa tribe, many of whom are members of the Houma tribe. From the perspective of outsiders and the political infrastructure of the state, however, the residents do not conform to dominant notions of what constitutes conventional "Indianess," that is to say, do not have the romanticized look of a "noble savage." However, they have maintained their ethnic distinctiveness despite outsiders' perception of them as mixed-race nonwhites by practicing their native lifestyle and subsistence activities.

The inhabitants of Grand Bayou have occupied the site for at least three hundred years. Nearby, long inshore sand ridges, called *cheniers*, and man-made mounds reveal abundant archaeological evidence of occupation dating back one thousand years. Although there is no archaeological evidence to link the present-day community to the prehistoric inhabitants of the bayou, Grand Bayou residents trace these mounds to their ancient past and maintain an oral tradition of early residence. Present-day residents state that their parents and grandparents frequented the mounds and used the sandy ridges for raising gardens and livestock. Residents consider the *cheniers* and some of the man-made mounds as the sacred burial grounds of their ancestors and in the past objected to archaeological digging conducted without their knowledge and permission.

There is sketchy information on the Atakapa in the historical European literature of the seventeenth century. According to early European accounts, the Atakapa where thought to reside only in western Louisiana and coastal Texas and to have disappeared from the historical record (Kniffen, Gregory, and Stokes 1987). Because of their alleged extinction, any claim of being an Atakapa is dismissed as impossible. However, descendants of Atakapa are found living in both Texas and Louisiana (Leger 2007). Their "disappearance" would seem to be yet another example of the entrenched stereotype of "the myth of the vanishing Indian" (Deloria 1988).

There is a richer, if still somewhat lean, written historical record of the Houma. Early historical reports state that the main Houma village in 1700 was on bluff lands near the Mississippi and the Red Rivers (Kniffen et al. 1987). Historians have speculated that the Houma, and other indigenous people, moved into the bayous of southern Louisiana early in the nineteenth century because of pressure from the American plantation owners and the Indian removal policies of the Jacksonian era (Davis 2001). Even though there is a historical record of the Houma from the nineteenth century, they have been perceived by many over the past century not as Indians, but as a mixed ethnic group. Many other mixed Native American communities have endured similar treatment, including Haliwa and Sampson County Indians (Dane and Griessman 1972), Mashpee in Massachusetts (Campisi 1991; Plane and Button 1992), Lumbee of North Carolina (Blu 1992); and Native Americans of Louisiana (Kniffen et al. 1987). As Dane and Griessman (1992) have observed, marginalized geographical location is often a common feature of groups that some deride as "so-called Indians."

Such perceptions and attitudes have forced communities like Grand Bayou into persistent struggles for their ethnic identity and to endure a legacy of social injustice. Local whites are said to have referred to Indians in the pejorative as the "Sabine," a Spanish word for the cypress tree, which is spotted red and white (Kniffen et al. 1987). Like many natives of mixed heritage, they were suspected of being an amalgam of white, red, and black, and were persecuted at worst, or marginalized and ignored at best.[2]

As elsewhere, the members of the Grand Bayou community have consistently refused the efforts of many whites to force them into a biracial society (see Plane and Button 1993). In the 1930s the Louisiana Supreme Court (*H. L. Billiot v. Terrebone Board of Education* 143 La. 623, 79 So 78), in a case brought by the Houma, ruled individuals with "Indian Blood" were to be regarded as people of color and thus banned from the public schools (Davis 2001). For many years, then, the community lacked equal access to formal education and educational materials such as libraries. Many Native Americans in the state were reluctant to attend Negro schools because they wanted to avoid being permanently classified as Negro.

The Federal Bureau of Indian Affairs has not granted formal recognition to Houma and Atakapa Indians and other similar peoples suspected of being of "mixed blood." Houma Indians of Louisiana formally organized themselves in 1972 and are now known as the United Houma Nation. The residents of Grand Bayou were invited to join the federation but declined membership because they consider themselves to be primarily Atakapa and related to the Houma only by marriage. Even though anthropologists have acknowledged the mixed ancestry of the community, they have also long supported the Native American identity of the Houma (Speck 1943; Swanton 1911). Others, such as Davis, depict the Houma as a case of "ethnogenesis," maintaining that their modern identity must be understood in light of the state of Louisiana's post–Civil War, changing racial-classification laws and by the shifting racial views of white society (2001). Rather than resort to an analytical application of American cultural notions of racial or ethnic identity or attribute behavioral motivations, it is sufficient to conclude in this case, as Blu (1987) does in her study of the Lumbee Indians of North Carolina, that for Grand Bayou members, "identity is simply not reducible below the level at which they [Grand Bayou residents] conceive themselves."

According to current residents of Grand Bayou, their grandparents and parents, living in the shadows as marginalized people, avoided unnecessary contact with the residents of Plaquemines Parish, "minding their own business" so as to avoid conflict or controversy. Marginalized and isolated communities like this face the perpetual threats of low incomes, discrimination, human rights abuses, and a weak voice in governmental policies that determine the fate of their community and render them more powerless than the average citizen. In the process they are rendered more vulnerable to the ravages of disaster and in their ability to respond and recover from such events.

The isolation and marginalization of the community was underscored in the wake of Katrina when residents sought the same federal assistance for debris clean up that was being provided to other communities in the parish. After the fact, FEMA officials claimed they did not know the community even existed as an excuse for not having offered aid. One FEMA representative is reported to have said, "No one mentioned there was a community other than those located up and down the highway." Even after this conversation it took four months for a FEMA employee to come to the community. When the residents followed up this visit with a visit to the district council meeting, they were met with a similar response. One local elected official said, "I have heard of Grand Bayou, where exactly are you located?" (Philippe 2007). Since Grand Bayou is the oldest community in the parish, the residents are often dumbfounded when hearing such statements. This story illustrates, if nothing else, the fact that one way disaster vulnerability can be reduced is if local communities empower themselves, as the members of Grand Bayou did in this instance, and make it clear to governments that they cannot be ignored in the planning process for disaster preparedness, response, and mitigation.

Political-Ecological Setting

The political-ecological setting of a society or community is an important factor contributing to its "vulnerability index."[3] The world of today is very different for Grand Bayou residents than the world of their childhood and ancestors. Community residents began noticing the "washing away" of land in the 1970s. When present-day adults were growing up, the landmass, which stretched for miles behind their waterfront homes, was solid and firm. There they grew fruit trees and vegetables and kept animals. But erosion has affected the flora and fauna, and the land is no longer useable for such subsistence activities. Large, tall trees once stood where there is today soggy marsh. Residents' old photos of their community in earlier decades bear witness to this dramatic transformation. As the land around them sinks, due to both natural and human-induced forces, the community is increasingly vulnerable.

What are the causes of these changes? The digging of a Mississippi Gulf outlet is said to have weakened the natural defenses and the levees. Coastal erosion has also been precipitated by the clear-cutting of cypress for garden mulch and the location of hundreds of gas and oil rigs and pipelines along the coast. Moreover, transportation canals built for the petrochemical industry have contributed to the destruction of the natural protection of coastal marshes (Laska et al. 2004). These tens of thousands of cuts allow sea water to intrude into the freshwater marshes and land. According to Grand Bayou residents and USGS maps, the land was once solid but is now eroded into mini-lakes. This loss is further intensified by the sinking of the landmass of coastal Louisiana by over two feet due to natural subsidence. Before the Corps of Engineers built dikes and levees that prevent natural sedimentation and replacement of lost land, this loss was mostly replenished by the deposition of silt by the Mississippi River.

The loss of coastal marshes has accelerated the loss of coastal land and increased the vulnerability of the populace. From 1932 to 2000 Louisiana lost 1,900 miles of coastal land, an area equivalent to the state of Delaware (Sea Grant 2007). Hurricanes Katrina and Rita destroyed over 60 square miles of land in Plaquemines Parish alone. In addition to local engineering projects, global warming is contributing to coastal land loss. Like coastal Alaska, the coast of Louisiana is a harbinger of the long-term effects of global warming for coastal areas worldwide.

Climate Change, Tropical Storms, and Hurricanes

Meteorological forces such as the increasing frequency and intensity of tropical storms and hurricanes, as well as changes brought about by climate change, threaten coastal communities like Grand Bayou. Coastal Louisiana has endured 49 of the 273 hurricanes that have made landfall along the American Atlantic Coast between 1851 and 2004 (McCarragher 2007). In 1965 Hurricane Betsy hit Plaquemines Parish with wind gusts of 160 miles an hour. A testament to the community's resilience and innovative use of local, traditional knowledge occurred in the storm's aftermath when the community built a traditional palmetto lodge as temporary housing until more permanent homes could be built.

Hurricane Camille hit the coastline in 1969 with water surges of over twenty-four feet, converting the southern tip of the parish into a watery lagoon twenty miles long (MacCarragher 2007). In 2002 hurricanes Isidore and Lili flooded most of the community's homes. Six months later, many residents in this isolated community were still homeless (Peterson 2004). Because of their traditional local knowledge of the bayou, however, residents were able to protect and secure their boats and use them as temporary homes. Fishermen would seek safe harbor for their vessels along the *cheniers* of the bayou. During the ravages of Hurricane Katrina, which inundated the area with a twenty-one-foot tidal wave, they evacuated their community and took their boats to Walker Canal, where they chained them together and ran their engines full throttle during the height of the storm. None of the boats were heavily damaged and everyone on board survived. They waited in vain for FEMA to provide the community with trailers, and forty-nine people (ranging in age from eight months to ninety-three) stayed behind and lived on the boats for eight months (Peterson 2007).

According to the Union for Concerned Scientists (2007 and the IPCC Fourth Assessment Report 2007) the expected sea-level rise that will accompany climate change will have especially dramatic effects along the Gulf Coast because of the flat topography, land subsidence, and the increased shoreline development. Global warming alone could account for a sea-level rise of eight to twenty inches in the next century. Coastal Louisiana has lost an average of thirty-four miles of marshland every year for the last fifty years (Sea Grant 2007). The added influences of climate change will result in catastrophic changes. As these processes convert wetlands into water, they will increase the vulnerability of coastal communities to coastal storms. At the coast every 2.7 miles of marsh absorbs a foot of a hurricane's land surge (Tidwell 2006), reducing a storm's power, and thus its threat, as it moves inland. The importance of the marshes can be illustrated by the fact that without marshes, Hurricane Katrina's storm surge would have been approximately four-and-a-half feet higher in Plaquemines and St. Bernard Parishes (Sea Grant 2007).

THE PARTICIPATORY RESEARCH ACTION MODEL

Despite the many obstacles in their way, Grand Bayou residents remain determined to rebuild and sustain both their community and their culture. To these ends, in January 2003 the community and Kristina Peterson, a doctoral candidate at University of New Orleans, initiated a participatory action research (PAR) project (Peterson 2007). In December 2003, Shirley Laska and her team of social scientists at the Coastal Hazards Assessment Resource Technology Center (CHART) of the University of New Orleans, along with John Pine and his team of social scientists from Louisiana State University, secured funding to study how Grand Bayou residents were attempting to reorganize themselves.[4] The purpose of the project is "to test the utility of Participatory Action Research (PAR) in enhancing a coastal community's resiliency to natural disasters, namely tropical storms and hurricanes" and "to examine the reciprocal ways in which the PAR process enhances the joining of traditional ecological knowledge (TEK) and scientific knowledge to benefit the general knowledge base of all participants about the challenges [including global warming] facing coastal communities and the ways these challenges can be mitigated."

The project's research design contains a strong ethnographic focus. Oral history, preservation of local knowledge, and cultural practices have played a major role in shaping the collaboration. One emphasis of this approach has been the telling of stories to make objective and public that which formerly was informal and private. The narration of these stories has aided the community in creating a shared memory bank. In sharing these stories with themselves and outsiders they also reaffirm and celebrate their native heritage. Stories include how the residents have historically dealt with tropical storms, hurricanes, and other hazards. This process helps instill in the younger generations of Grand Bayou a validation of self and community and the importance of preserving cultural knowledge. All of these results can contribute to, among other things, the resiliency of the community in the face of the storms, hurricanes, and global warming that will continue to challenge the community in the future.

In tandem with this approach there has been a strong outreach component. Members of the community have shared their concerns regarding the preservation of the environment and their cultural heritage with groups such as the Corps of Engineers, the National Hazards Workshop, the National Academies of Science, Oxfam America, the National Council of Churches, the National Science Foundation, NOAA Coastal Services Center, the Gulf Coast Fund-Rockefeller Foundation, the Louisiana Department of Natural Resources, and many others. This outreach effort has educated these foundations and organizations about the challenges faced not only by members of the Grand Bayou community, but by many other coastal communities as well.

The Importance of Local Voices

One of the most important aspects of participatory action research is that the process ensures that the voices of the community are heard and are treated as equal to the voices of others. In our highly professionalized culture, the public debate over environmental issues and other charged topics is overwhelmed by arguments articulated by experts, whose voices, opinions, and plans for action are privileged. The lives and voices of everyday people are traditionally excluded from such conversations (Button 2002). Past research has demonstrated that, as a result of such privileging of experts over residents during the risk assessment process, a disproportionate amount of the risk is borne by local communities relative to outsiders (e.g., Button 1995). Wisner (2006) views one aspect of vulnerability as a "blockage" or "devaluation of local knowledge." He argues that we need to encourage local communities to uncover, "rediscover," and articulate the local knowledge they have of their environment. This knowledge can both be empowering to the communities vulnerable to hazards and provide great insight into the knowledge base of both lay and professionals alike. Equally important, with the training provided by the expert community, lay voices can acquire the vocabulary of the scientific community and thereby engage in discourse with the scientific community at large.

Notions of what constitutes vulnerability may be perceived very differently by local communities than by outside experts. These local perceptions can alert the larger society to vulnerabilities that experts—including anthropologists—have failed to perceive or that have been obscured by the hegemonic forces of our society. Too often our indexing of vulnerability has been obsessed with natural forces and has ignored the socioeconomic/political vulnerability that has relegated many communities to the insidious cycle of a chronic, slowly unfolding disaster that may only becomes "visible" with an event like Hurricane Katrina.

Experts tend to dismiss knowledge derived from experience as being inaccurate, biased, anecdotal, and subjective. Yet field research has demonstrated that local knowledge may sometimes be more accurate than scientific knowledge and can often provide an early warning system for environmental problems not perceived by scientists. For instance, in the wake of the *Exxon Valdez* oil spill in south Alaska in 1989, local coastal communities possessed considerable knowledge of the tides and the marine environment that was often discredited by the Coast Guard and Exxon, but which later proved to be invaluable in predicting the flow of the spill and remediating some of the spill's harmful effects. Moreover, Alaskan Native observations about the changes in marine animal behavior, and in some cases the absence of organisms typically found in the ocean's edge, proved to be important markers for toxicity that was previously undetected by scientific tests (Button 1993). In a more contemporary example, Alaskan Natives made important observations about changes in the sea ice in the 1970s, long before these changes were noted by the scientific community (NOAA 2000).

The value of the kind of local knowledge that can lessen a community's vulnerability and inform more traditional forms of scientific knowledge can be illustrated by the extensive knowledge that Grand Bayou members have about hydrologic and environmental processes. Grand Bayou residents thought it was inefficient and inaccurate to approach coastal restoration or flood/wind resiliency efforts solely through an academic/practitioner knowledge base. The scientists and practitioners on this project concurred. As one Grand Bayou resident commented, "I can tell you [about] how the water flow[s] in the bayou and canals much better than all of the models they are developing to try to describe it." In other words, community residents often know that scientific approaches are long and expensive. The community residents believe, and the scientists on this project have come to agree, that it is much more efficient in terms of timing and research costs for experts to seek local knowledge before and during the implementation of new research projects.

Amplifying the voices of the community often is not sufficient by itself. As Delica-Willison and Willison (2004) have observed, because it does not conform to the status quo, such forms of knowledge are frequently resisted and rejected by experts and decision-makers. Thus, vulnerable groups need to form alliances and partnerships with other groups that are less vulnerable, have greater access to mainstream society, and are conversant with more traditional institutionalized forms of knowledge. Such outside groups cannot speak for the more vulnerable, but they can assist them and advocate for their interests. The PAR project in alliance with the Grand Bayou community serves as an example of this kind of vital partnership. While FEMA may not have been aware, as they claim, of the existence of Grand Bayou in the wake of Hurricane Katrina, many people around the nation were, as a result of the outreach efforts mentioned above. This awareness in turn created something of a national concern about the fate of the community and its residents and gained them national attention in the media.

A long-term goal of the project is to pair the Grand Bayou community with a sister community in Alaska that historically has faced similar natural and human-induced threats. This Alaskan community has endured coastal storms and erosion caused by climate change as well as environmental degradation caused by the oil industry, and we hope that Grand Bayou residents can learn from the coping strategies of these Alaskans, the lessons they have learned, and the way in which they have empowered their local knowledge. This kind of alliance not only will help to inform and empower the Grand Bayou community, but also will, in turn, have a similar effect on the sister Alaska community.

The Confluence of Local Knowledge and Science

In rapidly changing environments, such as that along the Gulf Coast, local communities often find that their traditional coping strategies are no longer adequate. This project attests to the fact that it is the exchange and employment of both local and scientific knowledge that leads to the most robust adaptive strategies in response to new environmental challenges.

The recent dramatic meteorological changes of sea-level rise, changes in precipitation and temperature, and the increased intensity of storms show that local communities face vastly different hazards than in the past. For example, many of the safe harbor locations where community members took refuge in the past are now being destroyed by environmental alterations. Boat owners in the community, in dialogue with their science partners, are undertaking new risk assessments to locate future safe harbors.

Another way that these changes are being addressed in this participatory action research project is through an effort to improve understanding of the local risks to built infrastructure. Scientists from UNO CHART and the LSU Hurricane Center are trying to increase their understanding of wind and water dynamics, while Tulane University architecture students have been involved combining cultural practices and values with weather-resistant design and materials. During this exchange Grand Bayou community members spoke of *disaster-resilient living* as opposed to *disaster-resistant buildings* and offered by way of example the palmetto lodges that they built during Hurricane Betsy.

Science and local knowing are also working in partnership to address a problem: NOAA GIS mapping experts and researchers from UNO are meeting on site with members of the Grand Bayou community to map both ethnographic and ecological data. Historical information about the location of built infrastructure in the community (such as homes, docks, etc.) is being mapped along with information of about the transitional changes in land and water topography over the last several decades. This knowledge base can provide invaluable baseline data by which to measure future changes in the environment. Moreover, NOAA is providing GIS and map training to the community so they can create their own maps to chart future changes to their environs and the Plaquemines region.

NOAA mapping specialists not only visit the community site but also go out on boats with community members to record their knowledge of the fisheries, currents and tides, and changes that have occurred in the marine environment in recent decades. This process of discovery invokes methods of oral history, visual photographic records, and intense dialogic exchange with all members of the project. Meanwhile, members of the community are eager to learn more formal, scientific knowledge about their environment and the GIS mapping process. The listening process by all participants encourages a true cross-fertilization of knowledge that contributes to a more balanced and informed knowledge base that is richer than the sum of its parts.

Promoting Community Resiliency

As Gavanta (1988), one of the leading founders of participatory action research observes, PAR not only generates knowledge but also promotes action. One important outcome for Grand Bayou residents has been the creation of partnerships with other fishing communities in the coastal region, including collaborative efforts to strengthen marine gathering activities. Achieving this goal required overcoming historical and cultural boundaries that had in the past separated communities. The partnerships then led to the development of the first fishing cooperative in the state of Louisiana, members of which include small family fishermen (of all cultural backgrounds) of the lower Plaquemines Parish. Later, the Barataria Terrebone National Estuary Program invited Rosina Philippe, a Grand Bayou community member, well recognized for her leadership skills, to be the spokesperson for Plaquemines Parish coastal issues. The success of their work in promoting community resilience, developing economic livelihood, and forming the fishing co-op has also influenced and provided a model of "best practices" to coastal communities in nearby Lafourche and Terrebone parishes. Moreover, Grand Bayou's efforts have gained them recognition from the newly elected parish president.

Grand Bayou members have also conducted outreach and day-long educational tours of their community and the bayou environment for scores of regional, national, and international organizations and foundations. These have not only educated the public about marginalized communities and the environmental threats that coastal communities face from natural and human-induced forces, but also created an ongoing dialogue to promote a wider discourse among other sectors of society. One of the tours included people from Central America and the Caribbean and demonstrated how Grand Bayou has adapted to social, economic, and environmental challenges. Because of these communities face similar challenges, rapport was easily established. Members of the Grand Bayou community, who in the past tended to keep to themselves, have developed a new awareness and now view resiliency in the context of an entire region. They now realize the need to work in concert with other coastal communities regardless of their ethnic heritage. As the project evolved, the lay and expert collaborators became more involved in extending their focus and outreach level beyond the local to the global, eventually leading to communications with the Organization of American States, the United Nations, the International Red Cross and Red Crescent, and other international organizations. What started out as a highly localized project now has global ramifications.

Conclusion

The importance of collaboration between scientists and nonscientists has been demonstrated in the past by the experiences of communities such as those in Love Canal, NY, and Woburn, MA, who developed health survey research projects in concert with scientists in order to uncover deleterious environmental contamination in their communities (Brown and Mikkelsen 1990; Button 2002; Gibbs 1982; Levine 1982). Today, it is essential to adapt and expand the use of such collaborative approaches to the communities most affected by the increasingly frequent and more disastrous effects of climate change. Grand Bayou residents used the knowledge base that they created as a means of taking action, thereby achieving unprecedented milestones in the history of the environmental movement. The earlier successes of the twentieth century can serve as inspirational models of lay and scientific collaboration in facing the environmental challenges of the twenty-first century. What these communities have in common with that of the Grand Bayou community is what Scott (1990, xii) refers to as "the fugitive political conduct of subordinate groups" contesting the "official transcripts." Such collaborative lay/professional endeavors are in one true sense "acts of resistance" as much as they are acts of "collaboration."

In the few short years since they formed their PAR partnership, the Grand Bayou community has achieved a lot and they plan to achieve much more. PAR reduces the distinction between lay community members and experts. The process moves the voices of lay people beyond the silence imposed upon them by experts and enables them to claim a legitimate voice of their own. The process helps to insure that the environmental observations of lay people are not dismissed as merely anecdotal and shows that such observations are valid forms of traditional knowledge that can contribute to a synthesis of lay and scientific knowledge. Projects like this also enable communities to uncover and develop their capacity for self-preservation and overthrow the repressive mantle of "vulnerability' and the perception of helplessness in order to imagine a different future. Furthermore, the process cultivates and nourishes the social fabric of a community, thereby enabling them to maximize their use of their social capital in building the foundations of resiliency.

It is important to remember that PAR empowers not only communities but also experts by helping them affect policy and achieve shared goals in a concrete, real-world setting. Equally important, participatory research expands and enriches the knowledge base of local citizens and professionals alike, empowering all of us to face the seemingly overwhelming challenges of hazardous threats like global warming.

ACKNOWLEDGMENTS

While the views expressed in this chapter are those of the authors, we would like to warmly acknowledge the assistance of Shirley Laska, Rosina Philippe, Ani Philippe, and the entire community of Grand Bayou.

NOTES

1. Community members try to help one another repair their boats and rebuild their community. They also continue to bury their deceased on site and hold community meetings when possible.
2. Like the English colonists in the Carolinas, the Native Americans were a significant part of the French colony's labor market. Like the English, the French required slaves in order to create a plantation society. French inhabitants were known to incite wars

in order to purchase Indian slaves from warring tribes. This, despite the fact that the French preferred adult male African slaves because they were considered better field workers and because Native Americans were considered much more capable of escaping. When possible, the French are said to have traded Native American Slaves to the West Indies in exchange for African slaves (The native Americans for two Africans). Native American slaves' monetary value was considered lower than that of African slaves. Since the majority of African slaves imported were adult males, the French did have some preference for Native American women and children as slaves. This situation, along with African American males escapees seeking refuge in Native communities did lead to some mixed ancestry (Gallay 2002).

3. In the field of disaster studies there exist various qualitative and quanitative methodologies to measure a community's vulnerability to disasters. In an attempt to reduce the risk. An index might measure what populations are at greatest risk, how many in the population, what natural environmental features or built infrastructures, as well as features of the political economy and social orgnaiztion that might make a community vulnerable.
4. Anthropologists Anthony Oliver-Smith and Gregory Button served as advisors and consultants to the project. In addition there is an extensive list of graduate students, professors and representatives from various organizations and foundations that are in involved in the project in one form or another.

References

Blu, K. I. 1980. *The Lumbee problem: The making of an American Indian people.* Cambridge: Cambridge University Press.

Brown, P. and E. J. Mikkelsen. 1990. *No safe place: Toxic waste, leukemia, and community action.* Berkeley: University of California Press.

Button, G. V. 1993. *Social conflict and the formation of emergent groups in a technological disaster: The Exxon Valdez oil spill and the response of residents in the area of Homer, Alaska.* Unpublished diss, Department of Anthropology, Brandeis University.

———. 1995. "What you don't know can't hurt you": The right to know and the Shetland Island oil spill. *Human Ecology* 23(2): 241–57.

———. 2002. Popular media reframing of disasters: A cautionary tale. In *Catastrophe and culture: The Anthropology of disaster*, eds. S. M. Hoffman and A. Oliver-Smith. Santa Fe, NM: School of American Research.

Campisi, J. 1990. *The Mashpee Indians: Tribe on trial.* Syracuse, NY: Syracuse University Press.

Dane, J. K. and B. E. Griessman. 1972. The collective identity of marginal peoples: The North Carolina experience. *American Anthropologist* 74(3): 94–704.

Davis, D. D. 2001. A case of identity: Ethnogenesis of the new Houma Indians. *Ethnohistory* 48(3): 473–94.

Delica-Willison, Z. and R. Willison. 2004. Vulnerability reduction: A task for the vulnerable people themselves. In *Mapping Vulnerability: Disasters, Development and People*, eds. G. Bankoff, G. Ferks and D. Hilhorst. London: Earthscan.

Deloria, V. 1988. *Custer died for your sins.* Norman, OK: University of Oklahoma Press.

Gallay, A. 2002. *Indian slave trade: The rise and fall of the English Empire in the American South.* New Haven, CT: Yale University Press.

Gavanta, J. 1993. The Powerful, the powerless, and the experts: Knowledge struggles in an informational age. *In Voices of change: Participatory research in the United States and Canada*, eds. P. Park, M. Brydon-Miller, B. Hall, and T. Jackson, 21–39. Westport, CT: Bergin & Garvey.

Gibbs, L. 1982. *Love Canal: My story.* Albany, NY: State University Press of Albany.

Jeansonne, G. 2006. *Leander Perez: Boss of the delta.* Jackson: University Press of Mississippi.
Kniffen, F. B., H. F. Gregory, and G. A. Stokes. 1987. The historic Indian tribes of Louisiana: From 1542 to the present. Baton Rouge: University of Louisiana Press.
Laska, S., G. Woodell, R. Hagelman, R. Gramling, and M. Teets Farris. 2005. At risk: The human community and infrastructure of resources in coastal Louisiana. *Journal of Coastal Research* 44.
Leger, B. 2007. This isn't Cajun country. *Times of Arcadia*, July 25.
Levine, A. G. 1982. *Love Canal: Science, politics and people.* Lexington, MA: Lexington Books.
McCarragher, B. 2007. http://web.mit.edu/12.000/www/m2010/teams/neworleans1/hurricane%20history.htm. Last downloaded 9/30/08.
National Oceanic and Atmospheric Administration. 2000. Changes in Arctic sea water over the past 50 years: Bridging the knowledge gap between the scientific community and the Alaskan native community. Executive Summary, Marine Mammal Commission Workshop on the Impact of Changes in Sea Ice and Other Environmental Parameters in the Arctic, Girdwood, AK, February 15–17.
Peterson, K. 2007. A bayou community's cultural and physical survival before and after Katrina. Paper delivered at the United States-Canada Summit, Disease, Disaster, and Democracy: The Public's Stake In Health Emergency Planning, sponsored by the Center for Biosecurity. May 23.
Philippe, R. 2007. Personal interview, April.
Plane, A. M. and G. Button. 1993. The Massachusetts Indian enfranchisement act: Ethnic contest in historical context, 1849–1869. *Ethnohistory* 40(4): 587–618. Also published in *After King Philip's War: Presence and persistence in Indian New England*, ed. C. G. Calloway, 178–206. Hanover, NH: University of New England Press.
Ryan, W. 1976. *Blaming the victim.* New York: Vintage Books.
South Carolina Sea Grant Consortium. 2007. Do salt marshes function as storm surge buffers? *Coastal Heritage* 3(winter): 7.
Scott, J. C. 1990. *Domination and the arts of resistance: Hidden transcripts.* New Haven, CT: Yale University Press.
Speck, F. G. 1943. A social reconnaissance of the Creole Houma Indian trappers of the Louisiana bayous. *America Indigena* 3: 135–46.
Swanton, J. R. 1911. Indian tribes of the Lower Mississippi Valley and adjacent coast of the Gulf of Mexico. *Bureau of American Ethnology Bulletin* 43: 318–26.
Tidwell, M. 2006. *The ravaging tide.* New York: Free Press.
Union Of Concerned Scientists. 2007. Global warming: Executive summary. www.uscusa.org
United Nations International Governmental Panel on Climate Change. Fourth assessment report 2007.
Wisner, B. 2004. Assessment of capability and vulnerability. In *Mapping vulnerability: Disasters, development and people*, eds. G. Bankoff, G. Ferks and D. Hilhorst. London: Earthscan.
Zebrowski, E. and J. A. Howard. 2005. *Category 5: The story of Camille, lessons learned from America's most violent hurricane.* Ann Arbor, MI: University of Michigan Press.

Chapter 21

Terms of Engagement: An Arctic Perspective on the Narratives and Politics of Global Climate Change

Noel D. Broadbent and Patrik Lantto

Global Climate Change

It is almost universally accepted that global climate change is one of the great scientific, social, and political issues of modern times. Within the scientific community there is an unusual degree of consensus regarding the subject of global climate change. The empirical evidence clearly supports the interpretation that the planet is warming at an unprecedented rate. This consensus has, in turn, been quickly translated into political rhetoric.

The Arctic, where warming is proceeding at twice the rate as the rest of the world, has been characterized as the early warning system for the health of the planet. The Arctic regions have consequently been a central focus of climate change research. This was most recently reported in the Arctic Council's Arctic Climate Impact Assessment (ACIA 2005). The NOAA *State of the Arctic Report* (2006) updates some of the records of the ACIA report. Northern indigenous peoples have become involved as informants and collaborators in research projects and are portrayed as symbols of human survival in extreme environments. Polar bears have, by contrast, become the doomed mascots of change. Climate change has now become a regular theme in the mass media.

From an anthropological point of view, the translation from traditional environmental knowledge to Arctic systems science to public discourse and political process is problematic. With each transformation there is a loss of both substance and source criticism and the issues are often discussed entirely out of context.

While the contemporary focus of global climate change is on warming, from the geological and archaeological perspective, the overall climate context is that of glaciation. The long-term history of the earth suggests that today's climate is a geological "warm spell" in a continuing ice age (Macdougall 2004, 7). It is very likely that we will inevitably reenter a new 100,000-year cold climate phase initiated by the tilt and wobble of the Earth's axis of rotation and its distance from the sun (which is controlled by the eccentricity of its orbit) (Macdougall 2004, 121). No one can predict if anthropogenic greenhouse gases, resulting in increased precipitation, cloudiness, and changes in the Gulf Stream, will accelerate or somehow hinder this process.

What is particularly interesting in the public debate, however, is the shift in focus from the inevitability of climate change, as seen in astronomical, ice-core, and archaeological records, to the assertion of human control over climate. Whether this is achievable or not, the fact remains that humans have been responding to major environmental changes for millennia and this cultural-adaptive ability is what makes us human. Anthropology thus has a major and necessary part to play in helping to contextualize these narratives and the webs of discourse that are emerging from them. Added to this is the fact that there are major differences between first- and third-world societies and their abilities to deal with anthropogenic threats.

NARRATIVES AND VULNERABILITIES: THE CASE OF THE SWEDISH SAAMI

It can be assumed that prosperous Western countries like Sweden, one of the eight Arctic nations, are more capable of protecting their indigenous minorities in the face of rapid climate change. However, because of the ways indigenous rights have been historically defined in Sweden, the Saami, Sweden's northern indigenous people, are in a surprisingly vulnerable position. This is not only because of environmental threats, but because of the political climate and the myths that have been used to define Saami culture.

The Saami, Scandinavian, and Finnish peoples have been in contact for thousands or years. The ancestors of the Saami most probably entered the Nordic north from the east and southeast during early postglacial times (cf. Haetta 2002; Hansen and Olsen 2004). The Nordic region has thus been a long-term meeting ground of European and circumpolar peoples. Linguistic diversity, economies, and technologies reflect this pattern. This long-term interaction contrasts with most of the Arctic, where contacts with European cultures are of relatively recent date. Because of this prehistory, archaeology is of central importance for understanding the dynamics of cultural contacts in Sweden and the development of national narratives and myths associated with the formation and maintenance of the Swedish state.

The relationship of the Saami to northern lands and resources, and the impacts of global climate change, differ in fundamental ways from the Swedish majority. The Saami are culturally and ecologically part of the circumpolar world, especially with respect to landscape values (cf. Ingold 1986, 2000). By contrast, Swedish culture and landscape values have historically been aligned with the mainstreams of European history, ideology, and economy (Broadbent 2001).

Historical Narratives and the Saami Past

Myths about history and prehistory have served to perpetuate attitudes about nationalism; racial stereotypes; cultural, economic and social evolution; territoriality; and identity (Carr 1986; Hagerman 2006; Hastrup 1987; Hodder 1991; Kramer 1997; Richards 1992; Solli 1996). The Swedish nation-building process incorporated a number of myths about the origin and the nature of Saami identities. These ideas still play an important role in environmental and cultural policies. Many of the most fundamental values that led to progressive social policies in Sweden have at the same time rendered minority rights problematic. This situation, in concert with immigrant conflicts, is one of the major political dilemmas of Sweden today. Another political issue is that Saami have not been recognized by the Swedish government as indigenous people in accordance with United Nations policies (ILO nr. 169 [1989]).

The historical narrative, by definition, is a linguistic form revealing temporal significances embedded in human action (Johnson 1984, 222). The selection of a narrative is often a moral or social choice affected by beliefs about the inevitability of a given outcome, such as evolutionary and ecological ideas about cultural or economic development. According to White (1978), narrative coherence is an imposition on the historical past structured along the lines of rhetorical strategies or master allegories (tropes). Narrative "truth" thus serves as a basis for practical action and the promotion of national or community values. The situation of the Saami in Sweden has been contingent upon the view (by the majority society) that the Saami are a nomadic people in need of protection, but within a framework of conformity to Swedish social values and policies. Therein resides a basic conflict within both Swedish and Saami communities. No Swede can own reindeer, the legal basis of Saami identity, but this at the same time excludes 90 percent of the Saami people themselves.

The Saami (previously, and in a popular and sometimes derogatory sense, known as Lapps), number as many as 80,000 people in Norway, Sweden, Finland, and the Kola Peninsula in Russia. About 20,000 Saami live in Sweden, of which some 2,000 (10 percent of the total population) are involved in reindeer herding. As a result of government policies, Saami land-use rights in Sweden have been tied to reindeer ownership and herding has de facto become a key symbol of Saami identity. Lapland as an administrative region has served as a means for preserving a nomadic herding lifestyle and separating the Saami from Swedish settlers. Saami territories in Sweden are depicted as limited to the northernmost and interior regions of the country. The Saami reindeer herders have become icons for Arctic human/environment interactions and global change.

These criteria have entailed a number of assumptions about their history and the nature of their relationships to land and resources that have been defined as "traditional" by the state. Eric Hobsbawm (1983, 1) defines this type of interpretation as "invented tradition": "a set of practices, normally governed by overtly or tacitly accepted rules and of a ritual or symbolic nature, which seek to inculcate certain values and norms of behavior by repetition, which automatically implies continuity with the past."

The "past" in this case is exemplified by three competing narratives, one by a Saami and two by Swedish archaeologists. Johan Turi, a Swedish Saami reindeer herder, is the author of *Muittalus Samid Birra* (*A Book on Lappish Life,* published in 1911), one of the first indigenous ethnographies ever written. He wrote the following widely quoted text on Saami origins and settlement:

> It has not been said that the Lapp came from somewhere else. The Lapp was settled all over Lapland, and the Lapp lived on the seacoast and there were no other dwellers besides them, and that was a good time for the Lapps. And the Lapp also lived everywhere on the Swedish side and there were no settlers anywhere; the Lapps did not know there were other people besides themselves.

Oscar Montelius, the father of the typological method which was greatly influenced by evolutionary ideas, had held a similar view of Swedish settlement (Montelius 1899):

> Our forefathers have lived in Sweden for 15,000 years. When they came here what we now know as Sweden was without settlement. We possess a land which we took from no other people. We Swedes have "made" our land, established settlement areas and broken roadways.

Evert Baudou, a prominent Swedish archaeologist specialized in northern Swedish prehistory and argues, by contrast, that cultural identity cannot be ascribed to archaeological material. This text is taken from Baudou's testimony in the Reindeer Grazing Case from 1995, in which the Saami lost their traditional rights to winter grazing in Härjedalen in southern Lapland:

> Who lived in Norrland [northern Sweden] 9,000 years ago? Our definition of an ethnicity today cannot apply, and the question must remain unanswered. The very diverse material from different time periods implies that many different folk groups lived in Norrland's prehistory, as in southern Sweden. (Courtesy E. Baudou)

According to the Norwegian anthropologist Thomas H. Eriksen, the major political goal of indigenous minorities is to survive as a culture-bearing group (1993,157). This issue is therefore about minority and majority relations. The history of nationalism itself is a history of conflicts over competing narratives that seek to define a social community (Kramer 1997). The deep past has, nevertheless, been the prerogative of the Scandinavian (Germanic) majority through the field of Nordic archaeology, a narrative that often emerges as self-evident. Little archaeological research has been done on Saami prehistory. Most Swedish archaeologists, like Baudou, prefer to view northern cultures as anonymous "hunter-gatherers."

Eugenics and the Welfare State

Post–World War II Swedish and German archaeologists and anthropologists have generally avoided taking positions on ethnicity and race. This attitude is undoubtedly motivated by a need to distance oneself from the Fascist and Nazi research of the 1940s, but also to avoid the uncomfortable intellectual connections between Sweden and Germany since the 1920s. Sweden was, for instance, the first country in Europe to establish a State Institute for Racial Biology in 1922. While admitting to being "German-friendly" during World War II, Sweden was not pro-Nazi, especially following the invasion of Norway and Denmark in 1940.

The role of the Swedish welfare state, a product of the 1930s, is, however, crucial for understanding Swedish domestic policies towards minorities, particularly the Jews, the so-called Tattare, Gypsies, and the Lapps (Broberg and Tydén 2005; Lundmark 2002; Werbart 2002). Racial hygiene and education were core elements of the Swedish welfare system. The Sterilization Act, implemented from 1941, and affecting some 63,000 citizens, mostly women, was only discontinued in Sweden in 1975, 30 years after WWII (Broberg and Tydén 2005). It is very important to note that Sweden survived WWII without trauma or reparations (monetary or moral), and continued, without much debate, with social, cultural, and health policies developed before the war.

The Swedish state established special nomadic schools for Saami children, not to acculturate them, as in US Indian Schools or Canadian residential schools, but to help preserve the nomadic lifestyle and perpetuate the interpretation of the Saami as primitive, vulnerable, and consequently under the protection of the state. The Germans occupying Norway during World War II seemed to have also had a "romantic" view of the Saami, allowing them to continue with transborder herding (Lantto 2005).

The Swedish Government and the *Knowledge as Power* Report

It is completely clear that Swedish society is at a crossroads. How are we to handle the multicultural society we live in?

—Marita Ulvskog (Swedish Minister of Culture)

Swedish politicians have long been aware of the challenges of diversity, particularly regarding immigrants, who now number 12 percent of the population of Sweden. The Saami have, in fact, benefited from this concern, especially in education. Home-language programs were initiated in the 1970s.

The government-mandated report *Knowledge as Power: Action Plan for How Museums in Their Work Can Counteract Antiforeign Attitudes and Racism*, recognized the importance of history and archaeology in contemporary Swedish society (Swedish Department of Culture Report Ds. 1996, 74).

The report specifies the following social dimensions of racism and xenophobia in Sweden (1996, 15–16):

1. State uniformity under Lutheran orthodoxy has permeated Swedish social life since the sixteenth century and still steers both institutional and individual thinking and actions.
2. The concept of the Swedish folk (*folkstam*) was part of the creation of the "folk home," which led to, among other things, sterilization laws until 1975.
3. Sweden is continuing with an unexamined past from both before and after World War II.
4. Openly declared racism, which disappeared after the war, does not mean that it has disappeared from people's thoughts.
5. Norms and concepts about what it means to be Swedish have not changed along with social change.
6. Antiforeign attitudes and racism based on earlier generations' attitudes are more of an adult than a youth problem in Sweden.
7. Racism, which was originally focused on biological differences, has shifted today to mostly social, religious, and cultural differences.
8. Antiforeign attitudes are a structural problem of the majority society.
9. In order to integrate indigenous and immigrant citizens, a common base of values and principles will be necessary.

This report can be described as an attempt at applied museum studies. It goes into considerable detail about the abuses of Swedish archaeology in the past, and describes a number of museums in which the Saami have been either marginalized or represented without reference to cultural conflicts.

One of the conclusions is that museums must be active in social debates and act as forums for cultural criticism. The report was met with strong criticisms, and to date the proposed exhibit and program have not been implemented. Practicing archaeologists who have responded to indigenous/minority concerns in Sweden are still few in number and are made up predominantly of museum-based women, most notably Inger Zachrisson (1997), Ingela Bergman (1991, 1998) and Inga-Maria Mulk (1994). The *Knowledge as Power* report offers a checklist of test implications regarding Saami concerns, not least because of its focus on archaeology.

The Saami Political Movement and Swedish Law

A recent government report, *Saami Traditional Lands* (*Samernas sedvanemarker*) (2006), still equates Saami reindeer herding/identity with economic sectors like forestry and tourism. The linkage of identity and culture with economy is in this respect a serious liability that legally overrides claims based on any other criteria including continuity of settlement, self-identity, or language.

The framework for the Swedish Saami policy was established in the late 1800s, even though the image of the Saami as a reindeer-herding people has older roots. The first two Reindeer Grazing Acts of 1886 and 1898 placed reindeer herding at the center of the policy area, and other possible Saami identities were definitely excluded with the Act of 1928 (Lantto 2000;

Mörkenstam 1999). The Saami were viewed as a people only suited for reindeer herding, even endowed with a physique uniquely adapted to this industry: "The reindeer is created for the Lapp and the Lapp for the reindeer" (Bergqvist, in Lantto 2005). According to this "Lapp shall be Lapp" view, the Saami should be preserved as reindeer herders. If the Saami were to leave the industry, there was the fear that this would mean the end of them as a people, a development the Saami policy was aimed at preventing. It was, for example, against this background that the nomadic school was created in 1913. The school was organized in a way that was considered compatible with the nomadic life of reindeer herding and the Saami culture: the school year was shorter, the education less comprehensive and the living conditions more primitive than in Swedish public schools. The goal was to give the children a minimum of education, to make them religious, moral, and good citizens, but nothing more. This was seen as an adaptation to the nature of the Saami children, not threatening their future as nomadic reindeer herders (Lantto 2000).

The "Lapp shall be Lapp" view created a dualistic Swedish Saami policy; it was a policy of inclusion and exclusion, of segregation and assimilation. The reindeer herders were included in the official definition of the Saami; they were considered to be the "real" Saami and through segregation were to be protected from the threats of modern society. Non–reindeer herding Saami, on the other hand, were excluded from this context. They were not regarded as genuine Saami, and the general view was that they soon would become assimilated into the Swedish population. Since World War II, the "Lapp shall be Lapp" policy has been slowly and quietly dismantled, but despite this change, the foundation for the policy has remained the same: reindeer herding is the criterion for Saami rights and the industry remains under state regulation and control (Lantto 2000; Mörkenstam 1999).

The Saami movement started to evolve from the end of the nineteenth century as a response to the Saami policy, and the first organizations were formed in 1904. Initially, the hegemonic view of the Saami as a reindeer-herding people that dominated the Saami policy was challenged by the movement. It was argued that the Saami were one people, regardless of which profession they had. Measures were needed to support all Saami if they were to be able to preserve their separate ethnic identity. However, this strategy was unsuccessful, and as a consequence reindeer herding successively gained a more central position in the rhetoric within the Saami movement. It was still not equated with the survival of the Saami culture as in the Saami policy, but this changed from the 1950s with the foundation of the first national Saami organization, the National Union of the Swedish Saami (*Svenska Samernas Riksförbund*, SSR). The organization was primarily based on the reindeer-herding units, and consequently placed the interest of the industry at the forefront of the movement. The view of reindeer herding as a necessary precondition for the survival of Saami culture had now been completely internalized in the rhetoric of the organization. Many non–reindeer-herding Saami felt alienated by this strategy and competing organizations for these groups were formed, but the SSR held a dominant position as a representative for the Saami as a people (Lantto 2000, 2003). With the establishment of the Saami Parliament in Sweden in 1993, the heterogeneity of the Saami people and of the Saami movement finally became more visible. However, the authority of the Saami Parliament is limited, and the partition of the Saami people produced by the Saami policy has created a divided political body.

Sweden has also continually refused to acknowledge stronger Saami land rights, despite challenges from the Saami movement. In reality, the Swedish Saami policy has shown a lack of development during the last three decades. This is evident in two recent official reports concerning Saami land rights and hunting and fishing rights, both which focus solely on the reindeer-herding Saami, thus excluding all other groups. In addition, the Boundary Delimitation Committee has drawn the conclusion that the land use of modern reindeer herding is neither sufficiently intensive nor dominant to render Saami ownership and possession rights to land except for limited areas (*Samernas sedvanemarker* 2006).

The Hunting and Fishing Rights Committee does not recognize exclusive Saami rights to hunting and fishing in the core Saami area, despite strong legal arguments supporting this (*Jakt och fiske i samverkan* 2005). The state ruled against exclusive rights for Saami in 1993 (Arell 1993). This was a legal challenge by nonrural hunters and fishermen and is a close parallel to the 1990 challenge to Alaska Native subsistence in which Alaska state subsistence laws were declared unconstitutional. The new Swedish government report has examined the legal basis for hunting and fishing rights in different areas of northern Sweden. The report generally argues for shared rights, that is, that the Saami and landowners have equal rights in many areas. The foundation for argumentation is legal, rather than cultural or in terms of Saami relationships to landscapes. Most importantly, the commission does not acknowledge special Saami rights on Saami/state land. The Saami do not, in fact, have title to any land in Sweden.

Archaeology and Land Rights

Archaeology became directly involved in Swedish Saami land-rights issues in the 1995 Reindeer Grazing Case in Härjedalen in which thirty Swedish farmers sued five Saami herding villages. Two archaeologists, Emeritus Professor Evert Baudou, of Umeå University, and Dr. Inger Zachrisson, of the State Historical Museum in Stockholm, were principal witnesses on opposing sides of this case (*DOM Meddelad av Svegs Tingsrätt den 21 februari, 1996 i det så kallade renbetesmålet*).

The northern farmers challenged the customary winter reindeer grazing rights of these Saami villages, which are located in Jämtland and Härjedalen counties in southern Lapland. The animals were herded across private and state lands and wintered in farming areas. Dr. Zachrisson made seven points arguing for the Saami and recognizing that they had been in this region since the Iron Age. She relied on prior archaeological data from the region, in particular the Vivallen cemetery that dates to ca AD 600–1000. Professor Baudou rebutted her seven points regarding 1) the character of archaeology, 2) ethnicity, 3) methods, 4) continuity, 5) representativity, 6) reindeer herding, and 7) linguistic evidence.

On the issue of continuity, Baudou demanded that one must establish continuity from the historical sources on the Saami in Härjedalen since the 1600s going back to the Vivallen cemetery (AD 600–1000). Baudou went on to state that just because the Saami were in evidence at this cemetery (through grave forms and goods, physical anthropological evidence, etc.), and it was in use for as long as four hundred years, this did not mean the Saami were anywhere else in the county.

On the issue of herding, Baudou granted that the Saami practiced reindeer pastoralism during the first millennium AD, as evidenced in Norrbotten and Västerbotten farther north, but that this had not yet been demonstrated in the current region. Furthermore, the evidence of reindeer hunting in the mountains in Härjedalen and Jämtland during prehistoric and medieval times could not be ascribed to only one probable Saami group (the Vivallen Saami) but could also have been carried out by Nordic peoples. In other words, even positive evidence that the Saami were reindeer hunters and herders by this time was not deemed specific enough to be used as evidence in this case.

Baudou went on to criticize Zachrisson's methods (use of archaeological sources), which is a standard form of academic debate. Without stating it outright, Baudou went a long way towards arguing that archaeological evidence was not specific enough to be used in court cases of these types. The court agreed and this interpretation, as much as anything else, determined the outcome of the case. The state has subsequently rejected archaeological evidence in Swedish courts. This rejection of archaeological evidence is reiterated in the 2006 government report *Samernas sedvanemarker*.

Finally, Baudou cited a 1983 study by a Nordic linguist (E. Holm) of place-names. Holm concluded there was no linguistic evidence for the Saami in the region before the mid-1500s. Baudou noted, curiously enough, that in a later article Holm supported to some degree Zachrisson's reasoning on Saami and proto-Saami ethnicity. Baudou states, nevertheless, that one should keep to Holm's earlier results and follow his suggestion for more linguistic research. Since such study has not yet been carried out, "no Saami place-name continuity back to Vivallen has been proven."

While one can certainly question the probabilities of scientific evidence, as in any forensic case, the overall dismissal of archaeological evidence puts the Saami, or any people without written histories, at a major disadvantage. Ironically, historical sources and maps, unlike archaeological (i.e., physical) evidence, are nearly always biased in favor of state policies. The acceptable narrative of the Reindeer Grazing Case was, and still is, identical to that of the Swedish state.

The court accepted all of Baudou's arguments and the Saami lost the case in 1996, and then again in their appeal at the Swedish Court of Appeals in 2002. The herding villages consequently lost their customary rights to winter grazing in the region and are now expected to pay rent for such land use. These rents have proven to be too high for the Saami to pay. While the Saami recently won a court case in Umeå based on historic data, archaeology has thus far proven to be a liability for them.

The Reindeer Grazing Case is a remarkable and little-known trial in Arctic archaeology and serves as a basis for a theoretical and methodological discussion of cultural identity and rights as they are represented by archaeological data and reasoning. Legal interpretations in Norway (Bull 2004; Svensson 1999) provide highly relevant comparative material, as do court cases in the United States (cf. Hutt, Forsyth, and Tarler 2006; Roberts 2004).

New Perspectives on the Saami Past

Against the background of the narrow legal basis of Saami rights in Sweden, and the rejection of archaeological evidence by court authorities, one must consider the enormous intellectual and political significance of new archaeological research. These results cannot be ignored by a nation that has such strong international archaeological research traditions. Swedish historians have long been champions of the Saami cause for recognition and Swedish archaeologists need to similarly recognize their accountability towards people without written histories within their own borders.

The interdisciplinary *Search for a Past* project is an attempt to better define Saami territory and economy outside of Lapland and along the Bothnian coast, which is often assumed to be Swedish territory (Broadbent 2005, 2006; Broadbent and Graham 2006; Broadbent and Wennstedt Edvinger in press; Grandin, Hjärthner-Holdar, and Grönberg 2005; Wennstedt Edvinger and Broadbent 2006). The project has also helped to transform the Saami narrative from that of a passive and isolated "recipient" culture to an active and long-term force in northern cultural development. This new evidence poses substantial challenges to the Nordic archaeological narrative and the core of Swedish interpretations regarding Saami identity, rights, and acknowledgement.

Detailed mapping and archaeological excavations were carried out at nine locales along a 460-kilometer north-south transect across three counties. The northernmost investigations were carried out on Stor-Rebben Island in Norrbotten County at 65°11' N, 21°56' E. Västerbotten County is located about 80 kilometers south of Stor-Rebben. This region is as much as 200 kilometers north of any known Germanic/Scandinavian Iron Age settlement. The southernmost excavations were carried out at Hornslandsudde in Gävleborg County at 62°37' N, 17°29' E. Hornslandudde is 460 km south of Stor-Rebben, and only 300 km north of Stockholm. The full presentation of results is forthcoming, but can be summarized as follows:

1. The chronology of these coastal sites extends from ca. AD 200 to 1278. This terminus ad quem coincides with the beginnings of Swedish colonization.
2. Coastal Saami settlement was not nomadic. Bones and teeth of sheep/goats, reindeer, and even cattle have been found in the huts at five different locales, three of which are north of Germanic Iron Age agrarian settlement. The Saami practiced animal husbandry and probably evening farming.
3. A ritual bear grave of Saami type was found embedded in a hut and was radiocarbon dated to the early 11th century. Circular sacrificial features of Saami type have been found on or near six different hut sites. This is evidence of Saami shamanism.
4. Iron working is evidenced at numerous sites. Three forges indicate that the Saami practiced sophisticated iron metallurgy from as early as AD 200.
5. Of over 1,100 place-names referring to "Lapps" in Sweden, 87 percent are found in the Bothnian coastland. There are 390 of these place-names in Västerbotten.

Ironically, these sites and the archaeological evidence suggest that there was little economic difference between the Saami and other groups in the coastal region. The Saami should therefore have land entitlement on an equal basis as the "settled" Scandinavians in the Nordic north. Nomadic reindeer herding was, de facto, created by the state during historic times and should not be the sole basis for Saami traditional rights.

Seen as a whole, this evidence in Sweden suggests that the characterizing of Swedish Saami solely as nomads fails to recognize the significance of other Saami lifeways, not only today but prior to Swedish colonization. Saami territory also extended far outside of Lapland. According to the Icelandic and Norwegian Sagas from 1100s–1200s, the Saami lived as far south as Hadeland, some 20 kilometers northwest of Oslo (Zachrisson 1987, 26). They formerly lived in all but southwestern Finland (Itkonen 1947). The ancestors of the Saami were even known as far south as the Western Dvina (Daugava) River in Latvia (Eidlitz Kuolok 1991, 32). In Sweden they are recorded living as far south as the Dal River and the Mälardalen region.

Conclusion

From the archaeological perspective the fundamental problem regarding Saami acknowledgement in Sweden is that discussions have been boxed in by a narrative of Saami identity that limits indigenous history in time and space, and identity by activity. This model was originally created to protect the Saami within a prescribed economy and territory. Swedish authorities still feel that this relationship fulfills the intent of the ILO convention. This assertion is debatable, however, and is unacceptable to the majority of the Saami in Sweden, and through its lack of alignment with international Arctic policies and United Nations conventions.

Based on the facts that are emerging from archaeological research, combined with the deconstruction of biased historical narratives, a more inclusive paradigm of Saami identity can and should be realized. What is at stake, besides the loss of traditional herding, hunting, and fishing rights, is the rightful place of the Saami people and Saami culture in Sweden. Ancient or "proto-Saami" prehistory enhances our understanding of northern cultural complexity as well as of the forces that led to the formation of all northern identities in this region. These results unite rather than polarize, contextualize rather than particularize, and show that the past has much to teach us about the present.

But What About Climate?

Anthropogenic environmental impacts, like the 1986 Chernobyl nuclear disaster, disproportionately affected the Saami (Broadbent 1986). The majority of Saami were not compensated for losses in fishing, hunting, farming, and tourism due to this contamination. Only the reindeer-owning Saami were compensated for losses and even for them there was little to celebrate. Cesium 137 has a half-life of thirty years and has thoroughly worked its way into the northern ecosystem (Broadbent 1988). The environmental impacts of global climate change are likewise expected to have major long-term impacts in northern Sweden affecting the same sectors: reindeer herding, farming, forestry, and tourism (Broadbent 1988).

If we are, indeed, capable of mitigating the effects of climate change through technology, political action, and sacrifice, the question arises as to who benefits from this process. In a recent report, *Indigenous Peoples and Climate Change* (Salick and Byg 2007), a number of conclusions were drawn regarding indigenous engagement with this issue. Most obvious is that indigenous peoples have valuable knowledge and experience to share and should be involved in monitoring and information networks. There is also a need for joint actions involving anthropologists, climate researchers, and politicians. But, the report further notes, "Indigenous peoples around the world are on the front-line of climate change, experiencing the brunt: temperature rises, sea-level rises, both droughts and floods, ice sheet and snow melt, glacial retreat, increasingly violent and unpredictable weather and more."

The threats of global climate change are very real and will require considerable resources, flexibility, and adaptability. A national narrative that limits the definition of indigenous minorities through the narrow filters of statutory law based on antiquated evolutionary ideas, and thereby constrains the legal discourse over resource rights, serves only to further marginalize people and jeopardize their cultural survival. The Saami will certainly survive as fully acculturated Swedish citizens, but face serious political obstacles in their attempts to sustain the lifeways, values, and culture that make them part of the indigenous world.

References

ACIA. 2005. *Arctic Climate Impact Assessment: Scientific report.* Cambridge: Cambridge University Press.

Arell, N. 1995. Att göra en ripa av en fjäder. Reaktioner och reflektioner med anledning av jaktreform ovan odligensgränsen. In *Då, nu och sedan. Geografiska uppsatser til minnet av Ingvar Jonsson.* ed. I. Layton. Umeå: Kungl. Skytteanska Samfundets Handlingar 44.

Bergman, I. 1991. Spatial structures in Saami landscapes. *Readings in Saami History Culture and Language II* (R. Kvist, ed.). Misc. Publications 12: 85–92.

———. 1998. Visar keraminken att de var samer? *Populär arkeologi nr* 4 (1998): 26–28. Lärbro.

Broadbent, N. D. 1986. Chernobyl radionuclide contamination and reindeer herding in Sweden. *Colloquim Antropologia* 10(2): 231–42.

———. 1988. The impact of Chernobyl on the economy and cultural environment of northern Sweden. *Circumpolar Health 87. Proceedings of the 7th International Conference on Circumpolar Health. Arctic Medical Research* 47(Supplement 1): 195–98. (H. Linderholm, H. Backman, N. Broadbent, I, Joelsson, eds.).

———. 2001. Fulfilling the promise:. On Swedish archaeology and archaeology in Sweden. *Current Swedish Archaeology* 9: 25–38.

———. 2005. *Excavation report, RAÄ 78 (Fällan 1:13), RAÄ 139 (Bjurön 6:1), RAÄ 144 (Bjurön 6:1), RAÄ 70 (Bjurön 6:1), Stora Fjäderägg, Snöan, Lövångers Kyrkstad,Västerbottens län. Sweden.* National Museum of Natural History. Washington, DC: Smithsonian Institution, December.

———. 2006. The search for a past: The prehistory of the indigenous Saami in northern coastal Sweden. In: *People, material culture and environment in the north. Proceedings of the 22nd Nordic Archaeological Conference, University of Oulu, 18–23 August 2004.* ed. Vesa-Pekka Herva, 13–25. Oulu: Humanistinen tiedekunta, Oulun yliopisto.

Broadbent, N. D. and J. Graham. 2006. *Excavation report. Hornslandsudde, Rogsta Parish, Hälsingland, Sweden.* Department of Anthropology, National Museum of Natural History. Washington, DC: Smithsonian Institution, July.

Broadbent, N. D and B. Wennstedt Edvinger. In press. Sacred sites, settlements and place-names: Ancient Saami landscapes in northern coastal Sweden. In *Landscape and culture in the Siberian North*, ed. Peter Jordan. London: University College of London Press.

Broberg, G. and N. Roll-Hansen. 2005. *Eugenics and the welfare state.* East Lansing: Michigan State University Press.

Broberg, G. and M. Tydén. 2005. Eugenics in Sweden: Efficient care. In *Eugenics and the welfare state*, eds. G. Broberg and N. Roll-Hansen, 77–149. East Lansing: Michigan State University Press.

Bull, K. Strøm. 2004. Samisk forhistorie og samiske rettigheter i et juridisk perspektiv. *Samisk forhistorie.* Varanger Samiske Musuems Skrifter, eds. M. Krogh and K. Schanche, 31–40. Porsanger.

Carr, D. D. 1986. Narrative and the real world: An argument for continuity. *History and Theory* 15(1986): 17–131.

DOM Meddelad av Svegs Tingsrätt den 21 februari, 1996 i det så kallade renbetesmålet). Sveg.

Eidlitz Kuoljok-Kerstin. 1991. *På jakt efter Norrbottens medeltid.* Center for Arctic Cultural Research. Misc. Publications 10. Umeå.

Eriksen, T. H. 1993. *Etnicitet och nationalism.* Nora: Bokbolaget Nya Doxa AB.

Grandin, L., E. Hjärthner-Holdar, and E. Grönberg. 2005. *Järnhantering vid Bjuröklubb och Grundskatan. Analysrapport nummer 13–2005.* Avdelningen for arkeologiska undersökningar (UV GAL). Uppsala: Riksantikvarieämbetet.

Gullestad, M. 1991. The Scandinavian version of egalitarian individualism. *Etnologica Scandinavica* 21: 3–18.

Hagerman, M. 2006. *Det rena landet.* Om konsten att uppfinna sina förfäder. Stockholm: Prisma.

Haetta, O. M. 2002. *Samene. Nordkallottens urfolk.* Oslo: HøyskoleForlaget.

Hansen, L.-I. and B. Olsen. 2004. *Samenes historie fram til 1750.* Oslo: Cappelen Akademisk Forlag.

Hastrup, K. 1987. Presenting the past: Reflections on myth and reality. *Folk* 29: 257–77.

Hobsbawn, E. and T. Ranger, eds. 1983. The invention of tradition. Cambridge: Cambridge University Press.

Hodder, I. 1991. Archaeological theory in contemporary European societies: The emergence of competing traditions. In *Archaeological theory in Europe*, ed. I. Hodder, 1–24. London: Routledge.

Hutt, S., M P. Forsyth, and D. Tarler, eds. 2006. *Presenting archaeology in court.* Lanham, MD: AltaMira Press.

Ingold, T. 1986. *The appropriation of nature: Essays on human ecology and social relations.* Manchester, UK: Manchester University Press.

———. 2000. *The perception of the environment: Essays in livelihood, dwelling and skill.* London and New York: Routledge.

Itkonen, T. I. 1947. Lapparnas förekomst i Finland. *Ymer* 67: 43–57.

Jakt och fiske i samverkan. Slutbetänkande av jakt- och fiskerättsutredningen. Statens offentliga utredninar. SOU 2005: 116. Stockholm.

Johnsson, G. A. 1984. Historicity, narratives, and the understanding of human life. *Journal of the British Society for Phenomenology* 15(3): 216–30.

Kunskap som kraft. Handlingsprogram för hur museerna med sitt arbete kan motverka främlingsfientlighet och racism. 1996. Kulturdepartementet. Ds 1996: 64. Stockholm.

Kramer, L. S. 1997. Historical narratives and the meaning of nationalism. *Journal of the History of Ideas* 58(3): 525–45.

Lantto, P. 2000. *Tiden börjar på nytt: En analys av samernas etnopolitiska mobilisering i Sverige 1900–1950.* Umeå: Kulturgräns norr.

———. 2003. *Att göra sin stämma hörd: Svenska Samernas Riksförbund, samerörelsen och svensk samepolitik 1950–1962.* Umeå: Kulturgräns norr.

———. 2005. Raising their voices: The Sami movement in Sweden and the Swedish Sami policy, 1900–1960. In *The northern peoples and states: Changing relationships*, ed. Art Leete, 203–34. Tartu: Tartu University Press.

Lundmark, L. 1992. *Uppbörd, utarming, utveckling. Det samiska samhällets övergång till rennomadism i Lule lappmark.* Lund: Arkiv avhandlingsserie 14.

———. 2002. *"Lappen är ombytlig, ostadig och obekväm." Svenska statens samepolitik i racismens tidevarv.* Umeå: Norrlands universitetsförlag.

———. 2006. *Samerna skatteland.* Institutet för rättshistorisk forskning. Serie III. Stockholm.

Macdougall, D. 2004. *Frozen earth: The once and future story of ice ages.* Berkeley: University of California Press.

Manker, E. 1960. *Fångstgropar och stalotomter.* Acta Lapponica 15. Stockholm.

Montelius, O. 1899. Typologien eller utvecklingsläran tillämpad på det menskliga arbetet. *Svenska fornminnesföreningens tidsktift.* No. 10: 1–32. Stockholm.

Mulk, I.-M. 1994. *Sirkas. Ett fångstsamhälle i förändring Kr. f -1600 e.Kr.* Umeå: Studia Archaeologica Universitatis Umensis 6.

Mörkenstam, U. 1999. *Om "Lapparnes privilegier": Föreställningar om samiskhet i svensk samepolitik 1883–1997.* Statsvetenskapliga institutionen, Stockholms universitet.

NOAA. 2006 *State of the Arctic report.* Seattle: NOAA.

Odner, K. 2000. *Tradition and transmission.* Bergen Studies in Social Anthropology. Bergen: Norse Publications.

Olsen, B. 1994. *Boesettning og samfunn i Finnmarks forhistorie*. Oslo.
Richards, C. R. J. 1992. The structure of narrative explanation in history and biology. In *History and evolution*, eds. M. Nitecki and D. Nitecki. Albany: State University of New York Press.
Roberts, J. C. 2004. Overview of the Native American Graves Protection and Repatriation Act and related legislation; Indian Law and the federal acknowledgement process. (manuscript). Washington, DC.
Salick, J. and A. Byg. 2007. *Indigenous peoples and climate change*. University of Oxford and Missouri Botanical Garden. Oxford: Tyndall Center Publication.
Samernas sedvanemarker. Betänkande av gränsdragningskommissionen för renskötselområdet. 2006. Statens offentliga utredningar SOU 2006: 14. Stockholm.
Solli, B. 1996. Narratives of Veøy: On the poetics and scientifics of archaeology. In *Cultural identity and archaeology: The construction of European communities*, eds. P. Graves-Brown and C. Gamble, 209–26. London and New York: Routledge.
Svensson, T., ed. 1999. *On customary law and the Saami rights process in Norway*. Tromsø: University of Tromsø.
Turi, J. 1911. *En bog om Lapperness liv. Muttalus Samid Birra*. Stockholm: A.B. Nordiska Bokhandeln.
Wennstedt, B. 1989. Rituella plaster i Övre Norrlands kustland. *Oknytt* 3–4: 23–34.
Wennstedt Edvinger, B. and N. D. Broadbent. 2006. Saami Circular Sacrificial Sites in Northern Coastal Sweden. *Acta Borealia*. Vol. 23, No.1: 24–55.
Werbart, B. 2002. *De osynliga identiteterna*. Studia Archaeologica Universitatis. Umeå: Umensis.
White, H. 1978. Rhetoric and history. In *Theories of history*, eds. H. White and F. E. Manual, 1–15. Berkeley: University of California Press.
Young, O. 1992. *Arctic politics: Conflict and cooperation in the circumpolar north*. Hanover and London: University Press of New England.
Zachrisson, I. 1987. Arkeologi och etnicitet. Samisk kultur i mellersta Sverige ca 1–1500 e. Kr. *Bebyggelsehistorisk tidskrift* no.14: 24–41. Uppsala.
Zachrisson, I. et al. 1997. *Möten i gränsland. Samer och germaner i Mellanskandinavien*. Stockholm: Statens historiska museum.

Chapter 22

Shifting the University: Faculty Engagement and Curriculum Change

Peggy F. Barlett and Benjamin Stewart

> The slumped shoulders and slack jaws were impossible to ignore: these students were not responding to my lecture on climate change. My strategy was to grab students by their intellectual and ethical lapels and impress upon them the enormity of the threat that human civilization faces in the climate crisis. Despite my review of urgent reports from communities around the world affected by global warming, their eyes remained glazed.
>
> With one student-initiated shift in approach, my listeners sat up and asked engaged, clarifying questions. They willingly discussed a litany of ecological horror stories with curiosity, interest, and insight into both problems and solutions. What happened?

Educating about climate change is a huge challenge. Creating the informed populace that will support rapid and fundamental political, economic, and social change is a task that falls centrally within the purview of higher education. Not just faculty, however, but students, staff, administrators, and alumni are all part of this educational mission. Through our experiences as an anthropologist, faculty member, and campus leader for sustainability (Peggy Barlett) and theologian, graduate student, and Climate Project presenter (Ben Stewart), we have seen that opportunities to expand awareness and galvanize action exist not just in the classroom, but throughout the life of the institution (Barlett and Chase 2004; Orr 1993; Rappaport, Creighton, and Bacow 2007). How can anthropologists and others in higher education contribute effectively?

Our perspectives on educating for climate change derive from our experiences in anthropology and higher education. Peggy's work in institutional change for sustainability began in the 1990s with a neighborhood watershed alliance and then expanded to a focus on Emory University. Through dialogue, communities of practice emerged in the Ad Hoc Committee on Environmental Stewardship, the curriculum development program called the Piedmont Project, and the Sustainable Food Initiative (Barlett and Chase 2004). Activities around campus planning and construction, purchasing, forest preservation and restoration, and existing university governance and policy supported a university-wide awareness of the challenges of sustainability and climate change. This work built on an anthropological attentiveness to multiple scholarly languages, political agendas, values, and worldviews within the academy. The experiments with praxis applied ethnographic knowledge of other times and places and of other universities' efforts to Emory's early steps toward new cultural practices and meanings. This reorientation of knowledge toward the future helped guide the disparate constituencies of faculty, staff, students, and administrators toward a common commitment to sustainability (Minteer 2006; Miyazaki 2004, 2006). Ben's personal connectedness with the outdoors and with experiences of degradation in wild areas inspired him to take leadership in student environmental organizing, become involved in Al Gore's Climate Project, and pursue further graduate study into the role of religious rituals in fostering environmental sensibilities. These anthropological and theological perspectives provided the grounding for our Emory efforts: how to build cultural change and how to assess impact.

In our experience, certain intellectual and personal dispositions and their collective expression in an emerging shared vision for a more sustainable society play an important role both in basic-level learning about climate change and in longer-term engagement with its issues and movements. In this chapter, we will describe the role of three dispositions—hope, imagination, and engagement with place—in supporting climate change efforts, using illustrations from the unfolding processes at Emory University.

The Higher Education Context

The fundamental mission of higher education—educating the citizens and leaders of the future—is particularly urgent with regard to the science, technology, policy, and broader cultural shifts connected to climate change. Many universities are accepting this revised mission into traditional classroom activities. New courses and sustainability minors, majors, masters, doctorates, and certificate programs are emerging in schools at all levels (AASHE 2008). Some schools opt for a "sustainability literacy" approach that uses movies, speakers, competitions, and sustainability-themed living programs to educate about climate change and sustainability challenges. Academic leaders have been important in public education that engages with citizen groups, faith-based communities, and local governments to develop appropriate policies to reduce greenhouse gas emissions. The research function of higher education is crucial, from studies on energy-saving technologies to behavioral change strategies. Even in schools that have no researchers on climate-related issues, however, the validation of the need for urgent action offered by academic leaders can reassure innovative political leaders and support movement. Academics also help by sharing strategies and policies being enacted elsewhere—our traditional research function put to applied use.

In addition to these research and teaching roles, the operations of colleges and universities contribute to climate change efforts. As users of energy, buyers of food, and consumers of everything from computers to chalk, higher education facilities have an opportunity to walk the talk of sustainability. Nearly five hundred colleges and universities in North America, for example, have made climate change and energy conservation commitments (AASHE 2008). All types of institutions are involved: community colleges, research centers, religiously affiliated schools, liberal arts colleges, technical schools, and both public and private universities. New governance structures—energy taskforces or sustainability councils—have been established on many campuses, along with sustainability offices, centers, or institutes with full time staff. New policies to move toward climate neutrality have emerged at colleges and universities around the world, and students have voted for increased fees to support ways to address their concerns.

More sustainable practices in campus operations—buildings, grounds, energy management, and water use—are becoming better known. New "green" buildings are commonplace, including classroom buildings, libraries, residence halls, laboratory buildings, recreation centers, student unions, and stadiums. Renovations to existing buildings are important as well and teach occupants through nontoxic, renewable, and recycled materials, natural lighting, and green cleaning products. In addition to the building occupants, green construction and renovation spread awareness of climate change strategies to professional communities of architects, engineers, contractors, builders, and planners. In addition to projects in energy efficiency, colleges and universities are investing in solar panels and windmills, and other sources of renewable energy. Culture change in the use of electric lights, building temperature settings, and elevator use offer all employees and students opportunities to reduce energy use connected with daily university life.

Sustainable food initiatives also help by reducing the miles that food travels. New guidelines for purchasing locally grown and sustainably produced foods are being adopted at many schools. Waste reduction and recycling on campus are sometimes linked to changes in purchasing policies that encourage energy-efficient appliances and computers, specify paper and other products with recycled content, and award contracts to businesses with a stronger sustainability profile. Efforts to reduce the harmful effects of car use include bicycle promotions, carpooling, subsidized public transportation passes, and shuttle programs. Replacing campus fleets with hybrid vehicles, electric carts, or alternatively fueled vehicles provides significant savings, both in dollars and in emissions. Many campuses have found biofuels based on local cooking oil to be a catchy way to reuse what was previously seen as waste.

Each of these arenas of action within the educational institution requires individuals to embrace the problems of climate change and seek solutions. There are numerous circumstances in which climate change innovations may be met with slumped shoulders, slack jaws, and glazed eyes, if not with outright resistance. Our experiences with diverse constituencies suggest that in addition to accurate content, we need to focus on how to support individuals to cope with fear, despair, guilt, and confusion often generated by climate change information. We will discuss three such ways: instilling hope, supporting imagination, and building a connection to place. We have seen impact in climate change efforts without any of these three, but our experience suggests that the vitality and persistence of a movement "with legs" is fostered by such an approach. We will explore each of these issues in turn.

Hope

The vignette that opens this chapter suggests that as educators promoting campus cultural change, common climate-change teaching strategies are not as effective as they could be. Surely, we think, the electrifying photos of melting permafrost, desertified cropland, and cataclysmic hurricanes will motivate students to engage the looming environmental apocalypse. One day, after some initially sluggish discussion following such a lecture, a student asked Ben if there were any workable solutions to the climate crisis. As he briefly sketched some possible strategies—wind and solar power, compact fluorescent light bulbs, LEED-certified green buildings,[1] carbon taxes, and forest protection—the minds and bodies in the room came out of hibernation. Through attention to solutions, and the hope those solutions generated, the listeners connected with the problems and to some complex routes toward solutions.

Many anthropologists have commented to Peggy over the years how discouraged anthropology students can become when classes emphasize the massive threats to cultural integrity and economic livelihood caused by globalization, war, ecological disasters, and even sometimes government aid programs. We often seek to help our audiences take issues seriously by showing them how urgent the problems are—a logical strategy. But when it comes to global climate change, the problems are so serious that many simply shut down in the face of bad news (Moser and Dilling 2007). When listeners instead encounter real, workable solutions to the climate crisis (or other global dilemmas), they are more able to begin to engage the problems and threats. We propose, then, a paradoxical strategy: before expecting intellectual mastery of problems, try introducing some solutions. Understanding even a few small-scale solutions can create a hopeful and imaginative intellectual space in which learners may be able to engage larger-scale problems with greater thought and perseverance (Moser and Dilling 2007).

Climate change is not the only subject that can be painful to learn. Writer bell hooks argues that when learning anything authentically new, students may experience difficulty and trauma (hooks 1994). One of three key capacities necessary for "deep learning" in general is for students to "be able to handle the mental trauma that accompanies challenges to longstanding beliefs" (Bain 2004). When the seriousness of the threat entailed in climate change is added to the more basic trauma of authentic learning itself, a focus on solutions and an atmosphere of hope becomes all the more important in the classroom—as well as in other contexts. Former US vice-president Al Gore, in making presentations to audiences on the topic of climate change, has noticed this phenomenon: "An astonishing number of people go straight from denial to despair, without pausing on the intermediate step of saying, 'We can do something about this'" (Gore 2006).

The willingness to engage in action for change requires a disposition of hope in the face of serious problems. Such a disposition is based, in turn, on a sense of possibility for solving the challenges of climate change. The globally devastating problems of climate change can be experienced as disempowering, an assault on—first—individual agency. They can be perceived as an assault on collective agency as well, as political, religious, and other institutions are perceived as inadequate to the task of leadership. Resistance on the part of students and others to the "doom and gloom" of the climate change message need not be seen, then, as avoidance of responsibility, but rather an unarticulated, inchoate unwillingness to accept the implication of impotence and collective irresponsibility. It can also be seen as an inability to imagine a more sustainable future and the steps necessary to achieve it (Crapanzano 2003). Many people resist facing problems to which they can see no solution—or else become resigned to them. Glimpses of intelligent solutions, however, may stimulate further thought and curiosity.

Detailed information about the scientific issues, political debates, economic policies, and technological solutions of climate change is daunting. Especially for academics trained to value expertise, global climate change requires many of us to stretch out of our comfort zone into new territory. For some people, this new knowledge feels important and exhilarating—a chance to develop competence in an issue of great societal importance. For others, it feels like a quagmire, an opportunity to be criticized as naïve, and a hard place to take a stand. Providing information that is readily understandable, in manageable chunks, and at the appropriate level is crucial.

In addition, fully comprehending the implications of climate change requires rethinking personal values and aspirations. How much do we care about biodiversity and the survival of polar bears? Enough to sacrifice the daily comfort of air conditioning or heating? Enough to struggle with roommates, family, or office mates about building temperatures? To refuse to buy a gas-guzzling, high-status car? Or to refuse to take a vacation that requires air travel? Many complex trade-offs are required with other lifetime goals, including those of significant others. Replanting a lawn into a low-maintenance yard that requires no mowing or a home location that reduces commute time are major decisions that can emerge from new climate change information. Such rebalancing of the costs and benefits of lifestyle choices is fraught, time-consuming, and absolutely necessary. It shifts ethical priorities and "intensely local ideas about work, leisure . . . respectability, and virtue" in order to construct an alternative future (Appadurai 2004, 68).

Rethinking institutional priorities and values, as well as those of individuals, is an important response to the realities of climate change. With the massive character of the changes needed, it is easy to get discouraged. One faculty member dismissed his school's efforts to buy more local food, saying, "It's just some local apples in the dining hall." On the one hand, our training as critical observers, especially of bureaucratic shortcomings, is valuable to prevent complacency and to stimulate a realistic sense of movement. At the same time, meaningful change can sometimes begin with almost invisible shifts. More powerful change can emerge from hearing a positive story or the inspiration of tracing a thread from the past into an imagined future. One graduate student, for example, was very impressed to learn of voluntary efforts by a grounds crew to save rare wild azaleas from a campus construction site. He remarked, "I didn't know people cared so much here." Later, he led a student environmental organization that inspired scores of student activists. Hearing the story of the acts of concerned campus employees gave him hope.

At an early stage of Emory University's efforts to reduce electricity use, we promoted specific behavioral changes by making a arguments at several levels of analysis. First, globally, we emphasized that Atlanta's electricity comes mainly from coal-fired power plants that are heavy contributors to greenhouse gas emissions. A local argument pointed out that electricity reduction also improves Atlanta's air quality. An economic argument appealed to the frugal. The experiences of several schools stimulated hope by showing that others had successfully been down this path. Tulane saved $200,000 a year by turning computers off evenings and weekends, and one unit of Harvard reduced their annual expenses by $300,000 through voluntary electricity reductions (Sharp 2002).

Imagination

As anthropologists, we know how to stimulate imagination by sharing accounts of other lifeways, values, gender roles, and ecological circumstances. The realization that such alternative ways of living in the world are possible is an important beginning to fruitful anthropological study—and for many individuals, to personal growth. Just such an imaginative exercise is crucial to the teaching of climate change. As participants in privileged institutions, employees and students in higher education are members of cultures that are more or less embedded in a pattern of massive carbon dependency and waste. If the climate crisis is to be solved, correspondingly massive shifts in these cultures will need to take place. Thus, thinking through solutions to the climate crisis will require the deployment of a focused anthropological imagination: imagining a culture of the near future that intelligently and responsibly manages carbon emissions. Across the disciplines—from business to religion to architecture—students and scholars who can compellingly imagine a shift to carbon-neutral cultures will be the most effective spokespersons, and their services will be in ever-increasing demand.

Climate Change Dinners

One way that we have tried to foster imagination at Emory is through a series of climate change dinners, open to applicants from faculty, graduate students, and undergraduates. Our strategy in holding these dinners was to build a supportive group environment, share factual information and diverse perspectives on climate change, and then through small group discussions, to encourage the leap to imagining a different Emory, a different Georgia, or different national policies.

The climate change dinners emerged as an idea after several campus events highlighting climate change. There were showings of *An Inconvenient Truth* and slide shows by various faculty and graduate students trained to make Al Gore's presentation. In November 2006, we hosted a talk by Professor Eban Goodstein from Lewis and Clark College, leader of the Focus the Nation effort, followed by the development of a faculty and graduate student interest group. To promote climate change awareness and action, the group explored some curricular innovations, but decided it was more feasible to hold two dinners a semester, leading up to the Focus the Nation event on January 31, 2008.

Part of our philosophy behind the climate change dinners was a recognition that lack of familiarity with the issues and the expectations of professional expertise can work against the ability to imagine change. Relying on the willingness of individuals to self-educate on a subject as contested as climate change is too risky. Compassionate intellectual support is a first step to help faculty and students step outside their specializations. Information about climate change needs to be provided in ways that can be understood by widely divergent disciplinary languages and shared without imposing a dogma or party line. Ideally, the sharing needs to be enjoyable and compatible with the constraints of severe time pressures. A respectful collegial community helps create the intellectual risk taking that will grow into a commitment to action—whether that action is personal behavior, classroom content, or institutional service toward new policies.

Our solution, then, was to design the climate change dinners in a way that was as easy as possible to add to busy schedules, included informed lectures, but also allowed for discussion that might build connections across fields. From both research and personal experience, we know that face-to-face interaction in small groups is essential to building groups that work together effectively for deeper cultural change.

We also decided to encourage the widest possible mix of undergraduate and graduate students and faculty from all fields. We announced the dinners through an all-university email. To control numbers and balance, we asked for a short paragraph about past interests in climate change and reasons why the dinners were of interest. This small effort allowed us to pick a group of roughly fifty individuals per semester who had a serious interest in learning about the issues. Those selected were called "climate change fellows," a credential that had some value for a student resume. Table 22.1 shows that the participants came from all professional schools of Emory as well as the graduate school of arts and sciences and our two undergraduate colleges.[2]

Table 22.1: Breadth of Climate Change Dinner Participants, Spring and Fall, 2007

School	*Faculty/Staff*	*Graduate Students*	*Undergraduates*
Emory College	18	9	14
Law	2	5	–
Business	3	5	5
Theology	2	2	–
Public Health	3	11	–
Medicine	9	3	–
Oxford College	1	–	–
Nursing	1	1	–

Each dinner featured two distinct approaches to the climate change challenge, in order to attract diverse conversation partners and showcase multidisciplinary ways to engage the issues. Examples of topics and pairings of faculty presenters that were considered are:

1. The Science of Climate Change and Media Coverage (Chemistry and Journalism)
2. The Law and Politics of Climate Change (Law and the Carter Center)
3. Global Health and Ethics (Biology, Public Health, and Philosophy)
4. Business Responses and Popular Culture on Climate Change (Business and Sociology)
5. Denial and Faith (Psychology and Religion)

Lectures were limited to twenty to thirty minutes. At the first dinner, for example, a professor who teaches atmospheric chemistry presented climate change facts and debates. His presentation was followed by a journalism professor who critically reviewed the ways the media have covered (and not covered) the issues.

Guided discussions were important to build the impact of the event for participants. At each table of five to eight, one person was asked to be a discussion facilitator. Most facilitators were chosen from the faculty interest group, but well-known faculty members were also invited to increase visibility for the event. Written discussion questions were prepared in advance and provided to each table. The groups were encouraged to begin with some simple sharing of intellectual and personal backgrounds and how these informed people's interests in climate change. Tables were then adjourned to obtain dinner at the buffet, the lectures were presented, and an open question and answer session followed. Small groups at each table then turned to questions about practical implications of the lectures, both "How does this information affect my own life?" and "How does it affect our lives here at Emory?" Thus, the discussion groups moved from the personal to the didactic and then to implications for action, which allowed participants to imagine transformed practices. Some groups talked about how to overcome roadblocks, either at home or in an academic department. Tips and strategies were shared. The dinner ended with announcements that allowed further focused action.

These climate change dinners sparkled with good energy. Speakers prepared well and their information was interesting. The atmosphere combined respect for complex issues, enjoyment of learning new perspectives, and a validation of an ethic of care for the earth. The discussion guides kept tables from focusing too long on one knotty issue or being bogged down by one dominant person. The social interactions spilled over before and after the event, as people from across the university chatted, introduced each other to colleagues, and deepened their understanding of particular threads of concern. Though the experiences of participants were varied, in the language of the anthropology of hope, these dinners extended our collective horizons, affirmed a cultural heritage of possibilities, and took steps to constitute new aspirations for the future (Appadurai 2004, 61; Miyazaki 2004, 140; Moser and Dilling 2007).

Efforts like this to encourage imagination and deepen knowledge can be adapted to many different campus cultures. Our dinners were catered and paid for by the provost, but a potluck, a modest dinner fee, a shorter lunchtime gathering, or a breakfast would all work well.

The social context of learning and discourse, and especially the size of groups, is important. Nina Eliasoph, in *Avoiding Politics: How Americans Produce Apathy in Everyday Life*, observes that larger groups in North America tend to rehearse familiar social scripts and to focus on the level of individual, private concern, rather than the level of the communal, public concern. Smaller groups, when given safe opportunities, are more likely to abandon familiar social scripts and to speak generatively and hopefully to the public good (Eliasoph 1998). In the climate change dinners, the discussion guide seeks to move the small-group conversation from the individual level of concern outward toward concern for the larger community. Supportive small groups provide opportunities to express doubt, to question the need for the change, to reveal exasperation, or to confess a failure. Whether they emerge casually, such as in a workplace conversation, or more formally, such as in a faith-based study group, they can provide collegial support for the time and energy to rethink values and priorities. "The public processing of pain," even when it seems entirely to obstruct movement, can actually help name roadblocks and sometimes thereby diminish their importance.

Engagement with Place

One critical way to create a relationship with the past and optimism for the future is through a strong connection to place (Barlett 2005b). At Emory, we are in the process of building on the experiences of place-based education on campuses such as Michigan State (DeLind and Link 2004) and Grinnell College's Center for Prairie Studies (http://web.grinnell.edu/cps/). The importance of place to sustainability efforts was highlighted at Emory through woods walks in 1999. Undertaken simply to expand our knowledge of the hidden corners of the Emory campus, the walks became a valuable to community building as well as to ecological knowledge. Guided by biology professors, an interested group of faculty, staff, and students was subsequently inspired to carry out some removal of invasive kudzu, ivy, and privet and to engage in a multiyear woodlands restoration project.

Efforts to build a sense of responsibility about climate change and more sustainable campus practices require that we know something about the place where we live and work and have some sense of identification with it in order to care. Professor Barbara Patterson, leader of the Emory as Place program, argues that for us truly to inhabit the Emory campus, we have to experience it through expanded sensory awareness—through seeing the seasons come and go, hearing the birds sing, feeling the wind, and watching the creeks. The social history of the campus—the Native American "trunk road" that forms a university boundary, Emory's role in the civil rights movement—are also parts of the place we have inherited. Awareness of place raises productive questions about how we will hand this place on to future generations. Everyday campus scenes shift in meaning as they are understood as secondary forest—home to beaver, fox, and herons—and sites of past social struggle. A series of guided walks, written materials, and restoration projects spread the Emory as Place message. This kind of embodied learning supports a sense of belonging and an ethic of care and can be a useful component of climate change work.

Walking Tour Brochure

In 2000, as part of Emory's millennium year celebration, the Ad Hoc Committee on Environmental Stewardship developed a self-guided walking tour brochure to highlight some of the environmental challenges and innovations on campus (Barlett 2002). Metal signs were posted at ten sites to encourage a pause to reflect on the issues illustrated, such as a new road that reduced a campus forest, eroding stream banks after a parking lot construction, and a new certified green building. Several thousand small brochures were distributed, and the walking tour is still used in courses and by groups such as local Boy Scouts. The walking tour brochure is available online at www.environment.emory.edu/who/walkingtours.shtml.

Climate Change in the Curriculum

Climate change issues are also fostered in Emory's formal curriculum development program for sustainability, the Piedmont Project (Barlett 2005a; Eisen and Barlett 2006). This faculty development program was adapted from Northern Arizona's Ponderosa Project (Chase and Rowland 2005) and Tufts University's Tufts Environmental Literacy Program (Barlett and Rappaport 2008) and offers a supportive learning environment for twenty faculty a year who want to develop new courses or introduce a new module into an existing course in order to integrate sustainability or environmental issues.[3] Faculty applicants commit to attending a two-day workshop and then revise courses independently over the summer. In August, a field trip provides experiential learning around a sustainability issue in Atlanta and an opportunity to share syllabus progress. A final dinner the following spring provides a further moment for reflection and new connections.

The Piedmont Project does not prescribe a sustainability approach or mandate a focus on climate change. Faculty innovate in diverse content areas and also in teaching methods, inserting sustainability-related materials into Chinese, Spanish, French, German, and Portuguese language and literature courses, for example. Modules on the environmental movement, environmental justice, and environmental human rights are added to courses in sociology, history, anthropology, and political science. Green chemistry labs and environmental health issues in physical education classes are further examples of ways that sustainability is integrated across the curriculum.

Many of the 110 faculty participants in the Piedmont Project after the first six years have incorporated issues of global climate change, consciousness about consumption and energy use, and behavioral change projects into their courses. For example, some students calculate their environmental footprint, explore contrasting national energy-use patterns, assess gendered impacts of water shortages abroad, or consider coastal impacts of sea-level rise. We have found that some of the most popular innovations, however, are those that help students become more aware of Emory or Atlanta as a place.

Pedagogical Experiment on Climate Change

How we teach also effects engagement with climate change. A small experiment in teaching methods that Peggy tried in 2000 suggests that a place focus may make climate change material more meaningful. To create the experiment, colleagues in the chemistry, religion, business, and economics departments made available one class day, and two different teaching methods were used to expand climate change awareness. All classes were given a pretest on climate change knowledge, attitudes, and values, and then a posttest at the end of the semester. The first teaching method was a conventional lecture, using vivid slides on climate change and health consequences, developed by Physicians for Social Responsibility. The second method involved a shorter lecture on climate change with the same slides, followed by small group discussions. After the lecture, the class was asked to turn around in their seats to form small groups of four and to explore the following questions for fifteen minutes. These buzz-group questions brought the scientific issues into the concrete realm of students' experience and possibly with places of meaning for them:

(a) What is your hometown and what are the impacts climate change might have there over the next one hundred years?
(b) What is your favorite beach location and what might be the impact of rising water levels there?
(c) What island nations have you visited and what might be the possible effects on them of sea-level rise?
(d) Which outdoor sports or exercise do you enjoy and how will it be affected most by rising temperatures and volatility in weather events?
(e) How does your personal lifestyle contribute to global warming and what actions can your group name to reduce greenhouse gas emissions?

Post tests of information retained showed higher levels for the classes that had buzz groups; classes in which lectures lasted for forty-five minutes without discussion showed slightly lower levels of information retention when tested several months later. This result may be due to the opportunity to process the information in discussion with peers or to the exercise of translating the scientific predictions to outcomes in concrete places with personal meaning.

Given our experiences over the past seven years, we would modify both the lecture and the discussion questions to share stories that inspire hope and foster imagination of different strategies and lifeways that address climate change concerns. We would ask how overall quality of life might be enhanced by actions to reduce greenhouse gases (Nordhaus and Shellenberger 2007; Westen 2007). We would also use more concrete examples from the Emory campus, so that visible and audible cues would trigger memory. We also might make the presentation in more informal arenas—campus clubs, Greek organizations, and other groups—where people already know each other and will continue to interact. These groups might encourage the small scale needed for productive discussions and heightened community disposition to act.

Conclusion

Climate change as a phenomenon is itself reconfiguring many—perhaps most—of the earth's living systems. Simply keeping track of the physical, biological, and social changes being brought about by climate change is a challenge of immense proportions for universities and will require significant investment of educational and research energies. However, even more basically, the task of educating the citizens and leaders of the future for understanding and solving climate change calls for pedagogies that can support the deep learning necessary for large-scale cultural reorientation toward dealing with problems of such a wide scope and acute character. We have suggested that higher education can welcome the climate change challenge through traditional teaching and learning methods, but also through some counterintuitive means, such as starting with solutions before presenting problems, initiating imaginative conversations, and cultivating an on-the-ground relationship to the university's locale. Fostering the enhanced awareness of small groups and then connecting to shared institutional values around environmental and global stewardship will reorient knowledge toward the construction of a different future, building the dynamism of cultural change. Among a much wider repertoire of teaching tools, we have suggested that the capacity to learn and act is supported by teaching methods that galvanize hope, nurture imagination, and foster a relationship to place.

Notes

1. LEED stands for Leadership in Engineering and Environmental Design, a certification process developed by the US Green Building Council.
2. Emory's original location in a small Georgia town forty minutes from Atlanta is now the two-year Oxford campus. On the Atlanta campus, four-year degrees in Arts, Business, and Nursing are offered, together with professional degrees in Nursing, Medicine, Public Health, Law, Business, Theology, and Arts and Sciences.
3. Begun in 2001 with funding from a teaching innovation grant, the Piedmont Project now is funded by six deans from across the university. Faculty receive a modest stipend for their summer's work.

References

AASHE (American Association for the Advancement of Sustainability in Higher Education). 2008. AASHE digest 2007: A review of campus sustainability news. www.aashe.org/resources/pdf/aashedigest2007.pdf. Accessed 7/1/08.

Appadurai, A. 2004. The capacity to aspire: Culture and the terms of recognition. In *Culture and public action*, eds. V. Rao and M. Walton, 59–84. Stanford, CA: Stanford University Press.

Bain, K. 2004. *What the best college teachers do*. Cambridge, MA: Harvard University Press.

Barlett, P. F. 2002. The Emory University walking tour: Awakening a sense of place. *International Journal of Sustainability in Higher Education* 3(2): 105–12.

———. 2005a. Reconnecting with place: Faculty and the Piedmont Project at Emory University. In *Urban place: Reconnecting with the natural world*, ed. P. F. Barlett, 47–74. Cambridge, MA: MIT Press.

———., ed. 2005b. *Urban place: Reconnecting with the natural world*. Cambridge, MA: MIT Press.

Barlett, P. F. and G. W. Chase, eds. 2004. *Sustainability on campus: Stories and strategies for change*. Cambridge, MA: MIT Press.

Barlett, P. F. and A. Rappaport. 2008. Long-term impacts of faculty development programs: The experience of TELI and Piedmont. *College Teaching*, forthcoming.

Chase, G. W. and P. Rowland. 2004. The Ponderosa Project: Infusing sustainability in the curriculum. In *Sustainability on campus: Stories and strategies for change*, eds. P. F. Barlett and G. R. Chase, 91–106. Cambridge, MA: MIT Press.

Crapanzano, V. 2003. Reflections on hope as a category of social and psychological analysis. *Cultural Anthropology* 18(1): 3–32.

DeLind, L. B. and T. Link. 2004. Place as the nexus of a sustainable future: A course for all of us. In *Sustainability on campus: Stories and strategies for change*, eds. P. F. Barlett and G. R. Chase, 121–38. Cambridge, MA: MIT Press.

Eisen, A. and P. Barlett. 2006. The Piedmont Project: Fostering faculty development toward sustainability. *Journal of Environmental Education* 38(1): 25–38.

Eliasoph, N. 1998. *Avoiding politics: How Americans produce apathy in everyday life*. New York: Cambridge University Press.

Gore, A. 2006. *An inconvenient truth: The planetary emergency of global warming and what we can do about it*. New York: Rodale Press.

hooks, b. 1994. *Teaching to transgress: Education as the practice of freedom*. New York: Routledge.

Leal Filho, W., ed. 2002. *Teaching sustainability at universities*. Frankfurt: Peter Lang.

Minteer, B. A. 2006. *The landscape of reform: Civic pragmatism and environmental thought in America*. Cambridge: MIT Press.

Miyazaki, H. 2004. *The method of hope: Anthropology, philosophy, and Fijian knowledge*. Stanford, CA: Stanford University Press.

———. 2006. Economy of dreams: Hope in global capitalism and its critiques. *Cultural Anthropology* 21(2): 147–72.

Moser, S. C. and L. Dilling. 2007. *Creating a climate for change: Communicating climate change and facilitating social change*. New York: Cambridge University Press.

Nordhaus, T. and M. Shellenberger. 2007. *Break through: From the death of environmentalism to the politics of possibility*. New York: Houghton Mifflin.

Orr, D. W. 1993. *Earth in mind: On education, environment, and the human prospect*. Washington, DC: Island Press.

Rappaport, A., S. Hammond Creighton, and L. Bacow. 2007. *Degrees that matter: Climate change and the university*. Cambridge, MA: MIT Press.

Sharp, L. 2002. Green campuses: The road from little victories to systemic transformation. *International Journal of Sustainability in Higher Education* 3(2): 128–45.

Westen, D. 2007. *The political brain: The role of emotion in deciding the fate of the nation*. New York: Public Affairs.

Chapter 23

Car Culture and Decision-Making: Choice and Climate Change

Lenora Bohren

Human practices are changing global environmental systems and are threatening the long-term sustainability of the biosphere. In 1990 the US Committee on Global Change divided human-induced change into two principal sources: industrial metabolism and land transformation. Industrial metabolism is the flow of energy and materials through the chain of extraction, production, consumption, and disposal of an industrialized world (Ayers 1989). Land transformation is the alteration of land surfaces and its biotic cover (Meyer and Turner 1992; Turner and Meyer 1991). In this chapter I will focus on the issue of industrial metabolism in terms of the consumption and use of energy by cars and their impact on environmental systems. This impact can be best understood by examining the link between human practices and environmental change.

Anthropologists, such as Bennett (1986, 1990) have identified the locus of environmental change in the decision-making process. Decision-making, the process of making choices between alternatives that lead to a course of action, is based on available information reflecting attitudes and values embedded in a culture. These attitudes and values are linked to cultural perceptions of the environment and economic, political, technological, and socio-cultural forces and lead to choices that become integrated into the culture (Bohren 1995). Humans have adapted to varied environments around the globe by making choices in terms of available technology (Bohren 1995; Moran 1982; Netting 1977; Rappaport 1983; Stewart 1955). These choices result in activities that impact the environment often with unintended consequences such as increased levels of carbon dioxide (CO_2) and air and water pollution. The links between these activities and the resulting environmental change are well documented (Ayers 1989; Clark 1989; Meyer and Turner 1992; Miller 1991; Stern, Young, and Druckman 1992; Turner and Meyer 1991). The choice of the car as a mode of transportation is an adaptation of a human activity that has caused many unintended consequences to the environment (Bohren 1996).

Car Culture

Perhaps one of the defining characteristics of being human is the need for transport from place to place. Different societies make different choices of the type and style of technology used for transportation. These choices are made in the context of the resources available in the environment and the framework in which they are found.

The car, developed towards the end of the nineteenth century, had its beginnings in Europe where the first patent was filed in 1886. The first cars, developed by entrepreneurs such as Benz, were quite expensive and were status symbols for the wealthy. Around the turn of the twentieth century, the car was introduced to America where pioneers had been working on a similar mode of transportation but had not patented their invention. In America the car initially was adopted by wealthier individuals and was attached to social status as it had been in Europe. When this market was saturated, entrepreneurs like Henry Ford began designing a car for the average person. The new design was easier to repair than the European car and was a lighter vehicle with less attention paid to detail. The popularity of the car in America quickly surpassed the popularity of the car in Europe, mostly due to an environment characterized by a well-established mechanical process of industry (the beginnings of mass production introduced by Ford), an abundance of natural resources, no tariffs between states, and a higher per capita income and more equal income distribution (Flink 1975). The new car fit America to a "T." It was embraced by the rapidly expanding middle class and quickly changed from a symbol of social status to a symbol of the masses. It fit well with the American cultural ethos characterized by individualism, freedom of choice, and social and physical mobility.

Transportation at the turn of the twentieth century was predominantly by the horse in both Europe and America. In Europe, mass transportation such as trolleys and trains began replacing the horse. Cars were available, but they were too expensive for the general public. In America, mass transportation took a back seat to the car since mass production made the car affordable for both wealthy individuals and the masses. By 1903, it was clear that the car would quickly become the predominant form of transportation since horse transportation was problematic. Horse waste was a serious problem. In New York City, for example, 2.5 million tons of solid waste and 60,000 gallons of liquid waste had to be cleaned from the streets daily. Additionally 15,000 dead horses had to be removed from city streets each year. Germ-laden street dust from the dried waste had negative health effects, irritated noses, and had a strong odor. Fly-borne diseases also caused serious health problems. In this context, the car was considered "antiseptic" and a solution to the "horse problem." The car was cleaner, safer, more reliable, more economical, and quieter on cobblestone streets (Flink 1975).

At first it appeared that the car also solved many social problems. It provided an easy escape from the city, it helped keep rural youth on the farm by allowing them recreational access to the city, it provided faster transport of produce to city markets, and it allowed people to move to the country and commute to their jobs in the city. Thus, it was seen as a cure to the air pollution ills of the city, a cure to the transportation problems of the rural areas, and a cure to the social ills in both areas (Flink 1975, 1988). The car appealed to both men and women's desire for style, power, speed, and individual mastery. For example, styles were introduced such as the famous "duster" coats, specialized hats with ear flaps, and goggles to keep the dust from your eyes. It became a symbol of individual freedom.

United States citizens' move from the horse to the car for transportation was a key factor in shaping twentieth-century America. Planners developed cities to accommodate the automobile. Suburbs and urban sprawl emerged to accommodate the expression of individualism, freedom of choice, and mobility expressed by the car. Whatever interest there was in public transportation was quickly lost. The car—fast, comfortable and reliable—became the transportation of choice (Volti 2004).

The Problem

With ever-increasing numbers of cars, the "antiseptic" solution to horse transportation soon became a more serious problem than the horse ever was. Most of these problems were unanticipated. Problems included the costs of road construction and repair, insurance and fuel, repair and maintenance of cars, and unanticipated health issues of car exhaust; congestion of cities; and disruption of community and family life (Flink 1975). The car distorted social life and disenfranchised populations such as the disabled, youth, elderly, and women who could not afford a car and the infrastructure (gasoline, repair and maintenance, etc.) that it required (Flink 1975, 1988). Initially social life in rural areas was enhanced by the cars and the institution of Sunday drives. Eventually, this increased mobility attracted the youth to the cities, thus creating an imbalance in the population of the rural areas. In 1900 one-third of the US population was on small farms; by 1940, 23 percent were on family farms and by 1980, only 3 percent were on family farms (Kunstler 1993).

There were few alternatives. The electric trolley that was invented around the same time had only a brief life even though it was cleaner than the horse, carried more people than the car, and went to the urban fringes. Freight was originally moved by rail as were people traveling from small town to small town. However, Ford's Model T and the assembly line changed all that by making the car affordable and readily available to the middle class. City planning boards accommodated the increasing ranks of middle-class motorists. Money was allocated to roads instead of trolley systems, and automobile interests eventually killed the trolleys (Kunstler 1993). The car also brought on problems of intense resource and energy use and environmental degradation.

Solutions

Solutions to the problems caused by the car began in America. Early in the century, automobile clubs introduced rules and regulations regarding safety with a focus on speed. They wanted "a free and safe highway intended to harmonize the interests of automobilists, horse drivers and pedestrians" (Flink 1975, 25). As car numbers increased, the negative health effects of car exhaust were felt, especially in the cities. This prompted a need to regulate car exhaust. The first regulations in America fell into three categories: mandated regulations on cars, mandated regulations on fuels, and voluntary regulations on cars. The first two are technical solutions. For cars, the technical solution was emission-control devices. The Exhaust Gas Recirculation (EGR) valve, for example, was the first device to be introduced in American cars in 1963, with the catalytic converter following close behind in 1975 (not introduced in Europe until the 1990s).

For fuels, the technical solution was mandated regulations, which began with the reduction of lead in gasoline. Unleaded fuel was mandated in cars that had a catalytic converter. The introduction of alternative and reformulated fuels came later. For example, ethanol was mixed with gasoline beginning in the 1970s with the intent of reducing the dependence on foreign oil. It did not begin to be mixed with gasoline to reduce emissions until the late 1980s.

The third solution, behavioral, came later. Voluntary regulations on cars included "no drive" days, car pooling, and the establishment of high-occupancy vehicle (HOV) lanes on major access highways to congested cities. "No drive" days were days that cars with certain characteristics such as license plates ending in an even or odd number could not be driven on certain days. HOV lanes had varied applications. In Washington, DC, for example, HOV lanes are accessible by only those cars with three or four passengers. In Denver, Colorado, HOV lanes can be accessed by cars with as few as two passengers and those that pay a fee. These regulations have had varied levels of compliance in different cities.

While America stressed technological solutions, Europeans continued to enhance their behavioral solutions, especially in terms of transport diversity such as increased public transit and bicycle and pedestrian ways. Many European cities, such as Munich, Germany, have intercity areas where cars are not allowed. Other solutions are traffic calming (streets are narrowed and parking is restricted), infill (development in the cities rather than on the fringes), public education, and decreased land use for cars (such as for roads, parking, etc.). Behavioral solutions continue to be stressed more in Europe while technical solutions are stressed more in America (Bohren 1996).

The US mandated regulations on emissions were based on the Clean Air Act, signed into law in 1963 by President Lyndon B. Johnson. It was the first modern environmental law enacted by the US Congress and was the forerunner to the Clean Air Act (CAA) of 1970, amended by Congress in

1975 and 1977. The CAA was the foundation of the federal air pollution control programs and set the health-based national ambient air quality standards. The standards were to be met through the use of control technology that would reduce emissions, resulting in improved air quality. Primary standards set limits which protected human health; secondary standards protected plants and animals. The EPA set primary and secondary standards for six critical pollutants: carbon monoxide (CO), nitrogen oxides, lead, sulfur dioxides, ozone, and particulates. Standards were set on both stationary sources and mobile sources. In 1990, Clean Air Act Amendments were passed to implement more stringent standards.

Emissions reductions were required on fuels as well as from vehicles. Reformulated gasoline was introduced into cities with ozone problems in 1995 and oxygenated fuels such as "gasohol" was introduced into cities with CO problems in 1992. Air toxics (any air pollutant for which a national standard does not exist that may cause cancer or other serious health problems) were added to the list of regulated emissions in the 1990 Amendments and chlorofluorocarbons (CFCs) had to be phased out. CFCs were included since they are not destroyed in the lower atmosphere and drift into the upper air where they release chlorine that destroys the ozone layer (Commerce Clearing House 1990, 70).

Since the turn of the twenty-first century, global warming/climate change has become an issue of public and policy concern. and has changed the dynamics of emissions-control regulations. With the increased awareness of environmental issues (Kempton, Boster, and Hartley 1995) and increased concern about climate change, the focus on emissions regulations has expanded to include greenhouse gases (GHGs) especially carbon dioxide (CO_2). On April 2, 2007, the Supreme Court ruled that the EPA must take action under the Clean Air Act regarding greenhouse gas emissions from motor vehicles. CO_2 officially became a pollutant of concern. On May 14, 2007, President George W. Bush signed an Executive Order directing the EPA to develop regulations to respond to the Supreme Court's decision. Regulations will be applied to vehicles and to fuels. The initial target for vehicles is light duty cars and trucks. New regulations on fuels include a requirement to have greater quantities of renewable fuels (White 2007).

Car Culture in Mexico

Mexico City, the most populated and polluted city in the world, is a good example of the growth of car culture and its impacts on the local population and environment. When technology, such as the car, is transferred across borders, many problems ensue. These problems are often a result of an insufficient infrastructure to address the issues of vehicles, such as proper maintenance. The unintended consequence is often pollution. In 1986, the government of Mexico City under the direction of SEDUE (the environmental protection agency of Mexico in place in 1986) began addressing the problem of pollution from cars by establishing regulations. They used US regulations as a model.

The Mexican government introduced the first regulation intended to reduce the amount of lead in gasoline in the fall of 1986. In 1989 another regulation added methyl tertiary butyl ether (MTBE) to low-lead gasoline, a reformulated gasoline used in areas with high levels of ozone such as Mexico City. Also that year the government implemented a "no traffic today" program where cars were restricted to certain driving days according to the last digit of their license plates. In 1990 the government introduced reformulated gasoline with a minimum of lead (Magna Sin) for cars equipped with catalytic converters. However, catalytic converters were not introduced until 1991 and cars using Magna Sin without a catalytic converter raised the levels of volatile organic chemicals (VOCs) in the ambient air. By the time the catalytic converter was introduced, there was not enough Magna Sin gasoline to supply these cars. The use of regular gasoline in cars with catalytic converters negated the beneficial effect of reducing exhaust pollution.

The intention of the regulations was to reduce 20 percent of vehicular emissions. The results, however, were different. A number of unintended consequences resulted from these regulations. The "no traffic today" program encouraged citizens to buy a second older car, which was usually more polluting, to drive to work on the day that their "clean" car's license plate was not allowed to be driven. The addition of MTBE (now phased out due to its toxicity) increased the levels of VOCs such as formaldehyde. The results of these regulations were increased emissions of the precursors of ozone and increased numbers of days exceeding of the Mexican Ozone Air Quality Standards (Bravo 1992).

The Mexican example stressed the need to factor in the cultural context when building a regulatory framework. Mexico copied the air quality regulations from the US, which resulted in many unintended consequences that increased rather than decreased air pollution. Today Mexico is making more culturally appropriate decisions for the regulation of car emissions. Clean cars, for example, are exempted from the "no traffic today" and the incentive to buy older, more polluting cars was eliminated. There is enough Magna Sin available to refuel the cars equipped with catalytic converters, and emissions programs are in place to check on the condition and emissions of these vehicles.

Global Issues

In 1927, the US had 80 percent of the world's vehicles. The car, a worldwide symbol of personal freedom, was a technology desired around the world. By the mid-twentieth century there were 2.6 billion people on earth and there were 50 million cars (WorldWatch 1996). Soon, the unanticipated consequence of exhaust emissions spread around the world. The World Bank projected that pollution from cars would increase fourfold from 1990 to 2030 in developing countries (Bohren 1996).

This projection is already becoming a reality. In Bangkok, Thailand, for example, pollution levels are so high that health and productivity is down. In China the government has been promoting the dream of "a car for every family" (which is becoming possible with increased affluence) without addressing the infrastructure questions of land use (parking and roads), fuel, and repair. With the increasing population of cars in China, the resulting emissions could increase carbon concentrations to an extent that could offset emission reductions in other countries (WorldWatch 1996). Bilateral agreements (such as the United Nations Partnership for Clean Fuel and Vehicles) and multilateral programs encouraging low-sulfur fuels, robust vehicle and fuel compliance programs, etc. are in effect in China in an attempt to address public health concerns, especially before the upcoming Beijing Olympics. Regional organizations and the global Partnership for Clean Fuels and Vehicles are focused on completing a global lead phase-out and adoption of cleaner vehicles. This will be done by partnerships with local organizations, i.e., policy support promoted by the US EPA (Leaf 2007).

With the increased concern about global climate change, many countries in Europe are addressing the concern and initiating actions. Vaxjo, Sweden, for example, decided to wean itself from fossil fuels in 1996 by using wood chips from sawmills to replace oil at power plants. The water that is heated up in the process of cooling the plant is used to heat homes and offices in Vaxjo. Their goal is to cut emissions by 50 percent by 2010 and 70 percent by 2025. Barcelona required new buildings in 2006 to install solar panels to supply at least 60 percent of energy for hot water. Copenhagen and Paris both introduced a public bicycle service in 1995. London has recently instituted a congestion charge. Many European cities continue to encourage an increase in transport diversity choices such as increased public transit, increased bicycle ways, and increased pedestrian ways. Other European cities, such as Munich and Groningen, Netherlands, continue to promote intercity areas where cars are not allowed; traffic calming where streets are narrowed and parking is restricted; infill development rather than on the fringes, thus reducing sprawl; public education; and decreased land use for cars. Belgium, for example, has reduced the number of traffic lanes into some cities from three lanes to two, thus encouraging the use of public transportation in terms of buses or trains. The European Union has taken notice of these trends and is encouraging cities to adopt their own emissions targets. As a result, Stockholm, Copenhagen, and London have set targets to cut CO_2 emissions by 60 percent by 2025 (Reporter-Herald 2007; WorldWatch 1996).

Other areas of the world are also initiating actions. Bogotá, Colombia has reduced emissions with the increased use of a municipal bus system and an extensive network of bicycle paths. Brazil has been promoting the use of renewable fuels in terms of ethanol produced from sugar cane, an abundant natural resource, for many years. Actions in the US, including Austin, TX; Portland, OR; and Seattle, WA have addressed the reduction of GHGs at a local level.

Conclusion

This discussion of "car culture" illustrates the impact of decision-making in relation to the car as a transportation choice and the environmental consequences of this choice. The car, the symbol of the American ethos of individualism, personal freedom, and mobility, shaped twentieth-century America. The decisions by Americans to choose cars as the primary mode of transportation led to a large population of cars, urban sprawl, little mass transportation, and air pollution. Regulations to control air pollution were primarily technological, in terms of emissions control systems. Emissions control technology in newer vehicles is very sophisticated and successful. However, the emissions reductions are often counteracted by the increasing number of cars on the road and the increased vehicle miles traveled (VMT).

Very recently the control of GHGs has been added to pollution control regulations. The use of more fuel-efficient cars is being promoted and is adding a new dimension to car choices. New technology vehicles that are more fuel efficient, such as hybrid cars, are entering the market. The choice of vehicles on the market is being expanded by regulations requiring a greater quantity of renewable fuels. Flex-fuel vehicles are now available that use a mixture of gasoline and up to 85 percent ethanol (a renewable fuel) and diesel vehicles that can use between 5 to 100 percent biodiesel (a renewable fuel).

These choices are gaining popularity, yet there seems to be a trade-off in the choices. Smaller, lighter, fuel-efficient vehicles and hybrids reduce GHGs. Renewably fueled vehicles, like the flex-fuel vehicles or biodiesel vehicles, reduce our dependence on foreign oil and help local economies. The choice of renewable fueled vehicles is being encouraged by many local, state, and federal governments, but these vehicles are not as fuel efficient as the hybrids or smaller, lighter cars and do not reduce GHGs as much. Very few choices are being made to reduce VMT by increasing the use of transportation diversity.

In order to truly reduce emissions we must reduce VMT, increase fuel efficiency (miles per gallon) in vehicles, and increase the supply of renewable fuels in an environmentally sound manner (by looking at the full life cycle of renewable fuels including production, transportation, and use). We must choose cars that are fuel efficient, fueled with locally produced renewable fuels, and that reduce VMT. We can reduce VMT by choosing to car pool; to make fewer, shorter trips; and to utilize and promote mass transportation. We must look at both the technological and behavioral components of emissions reduction. The cultural challenge for the US is to transform the American ethos of individualism, personal freedom, and mobility away from reliance on the car. Rappaport (1993) stressed the importance of the role of anthropologists in this challenge. Anthropologists have extensive experience in the study of culture and adaptation. Julian Stewart (1955) coined the term *cultural ecology* to describe how human activities, driven by cultural values, have utilized available tools to adapt to local environments. Culture has been the intermediary between the environment and human activity (Altman and Chemers 1980); environmental change has always been the result (Netting 1977). With the anthropological understanding of the interface between culture, technology, and the environment, anthropologists can help policy makers assist countries (including our own) to make choices that create culturally appropriate, balanced systems that are environmentally sustainable.

This will be a difficult challenge. Many cultures around the world are quickly adopting the car as a mode of transportation, and pollution, especially in larger cities, is increasing to levels that are dangerous to the health of both the citizens and the natural environment. The example of Mexico City demonstrates the need to understand the use of the car and its place in culture before appropriate regulations can be made to address the unintended consequences. Anthropologists' holistic approach can help policy makers with appropriate regulations that can help develop the needed Infrastructure for the car. For example, a project I worked on in Sri Lanka, before the tsunami, was designed to help develop not only an emissions testing program but also an infrastructure that would establish an emissions program to monitor pollution and train administrators and mechanics to repair and maintain vehicles with the goal of reducing pollution. It is well known that a properly maintained vehicle will reduce pollution and increase the reliability of the vehicle. The training went well and the emissions program was almost established before the tsunami prevented its completion.

Sociocultural changes are evident everywhere the car has gained a toehold (Volti 2004). Miller (2001) discusses how cultures quickly integrate technology into meanings that are embedded into social and cultural relations. The technology, in our case the car, quickly becomes part of a personal and social system of values. Mimi Sheller (2004) takes this a little further by stressing the importance of the "deep" affective dimensions, i.e., the role of emotions and senses in making decisions about cars. With the prediction that world car travel will triple between 1990 and 2050 (Hawken, Lovins, and Lovins 1999) and the fact that cars produce one-third of worldwide CO_2 emissions (Urry 2004), it is essential that we address these "deep" cultural issues. Change is needed in the human sociocultural practices that have accompanied the adaptation to the car in order to reduce its contribution to global climate change.

References

Altman, I and M. Chemers. 1980. Culture and environment. Monterey, CA: Brooks/Cole Publishing Co.

Ayers, R. U. 1989. Industrial metabolism and global change. *International Social Science Journal* 121(41): 363–73.

Bennett, J. W. 1986. Summary and critique: Interdisciplinary research on people-resources relations. In *Natural resources and people: Conceptual issues in interdisciplinary research*, eds. K. A. Dahlberg and J. W. Bennett, 342–72. Boulder, CO: Westview Press.

———. 1990. Ecosystems, environmentalism, resource conservation and anthropological research. In The ecosystem approach in anthropology from concept to practice, ed. E. F. Moan, 435–57. Ann Arbor: Univ. of Michigan Press.

Bohren, L. 1995. Socio-cultural factors in land use/management decisions. PhD diss., Colorado State University. Fort Collins.

———. 1996. Car culture: Society and mobility, Keynote address, First International Clean Air Conference, March 5, Munich, Germany.

Bravo, H. A. 1992. Mexico's strategy toward cleaner air. Paper presented at the 8th Annual Clean Air Conference, September 2, Steamboat Springs, CO.

Clark, W. C. 1989. Human ecology of global change. *International Social Science Journal* 121(41): 315–45.

Commerce Clearing House, Inc. 1990. Clean air act: Law and explanation, Chicago, Illinois.

Flink, J. J. 1975. *The car culture*. Cambridge, MA: MIT Press.

———. 1988. *The automobile age*. Cambridge, MA: MIT Press.

Hawken, P., A. Lovins, and L. H. Lovins. 1999. *Natural capitalism*. New York: Little, Brown and Company.

Kempton, W., J. S. Boster, and J. A. Hartley. 1995. Environmental values in American culture. Cambridge, MA: MIT Press.

Kunstler, J. H. 1993. *The geography of nowhere*. New York: Simon & Schuster.

Leaf, D. 2007. EPA international clean fuels and vehicles programs. Keynote address, 23rd Annual Clean Air Conference, September 25, Breckenridge, CO.

Meyer, W. B. and Turner, B. L. 1992. Human population growth and global land use/cover change. *Annual Review Ecological Systems* 23: 39–61.

Miller, D. 2001. Driven societies. In *Car cultures*, ed. D. Miller, 1–34. Oxford: Berg.

Miller, R. 1991. Social science and the challenge of global environmental change. *International Social Science Journal* 130: 609–17.

Moran, E. F. 1982. Human adaptability: An introduction to ecological anthropology. Boulder, CO: Westview Press.

Netting, R. 1977. Cultural ecology. Menlo Park, CA: Cummings Publishing.

Rappaport, R. A. 1893. Distinguished lecture in general anthropology: The anthropology of trouble. *American Anthropologist* 95(2): 295–303.

Reporter-Herald. 2007. Pushing green. Loveland, CO, Oct. 14.

Sheller, M. 2004. Automotive emotions: Feeling the car. *Theory, Culture and Society* 21: 221–42.

Stern, P. C., O. R. Young, and D. Druckman, eds. 1992. Global environmental change: Understanding the human dimensions. Washington, DC: National Academy Press.

Stewart, J. 1955. The concept and method of cultural ecology: Theory of culture change. Urbana: University of Illinois Press.

Turner, B. L. and Meyer, W. B. 1991. Land use and land covering global environmental change: Considerations for study. *International Social Science Journal* 130: 669–79.

Urry, J. 2004. The system of automobility in *Theory, Culture and Society* 21: 25–39.

Volti, R. 2004. *Cars and culture: The life story of a technology*. Westport, CT: Greenwood Press.

White, R. 2007. EPA's GHG Rule. Paper presented at the 23rd Annual Clean Air Conference, September 27, Breckenridge, CO.

WorldWatch report. 1996. The billion-car accident waiting to happen. February 19. Washington, DC: The WorldWatch Institute.

Chapter 24

Anthropologists Engaging in Climate Change Education and Outreach: Curating *Thin Ice—Inuit Traditions within a Changing Environment*

A. Nicole Stuckenberger

In the most fundamental ways, places are made (Raffles 2002, 329).

Introduction

Arctic temperatures are now rising at nearly twice the rate seen for the rest of the world (ACIA 2005). The major impacts of global climate change on the Arctic environment are evident in reductions in the extent and thickness of sea ice, melting permafrost, changes in local flora and fauna, changes in the character of the seasons, and increasingly erratic weather patterns (ACIA 2005; Krupnik and Jolly 2002). The cultures of many Arctic peoples, such as the Inuit, Iñupiat, and Yup'ik groups of Canada, Greenland, and Alaska are closely interconnected with their environment through their subsistence practices, and these kinds of changes in the weather and climate have far-reaching social, cultural, and economic implications.

The polar regions have become the focus of major research efforts worldwide under the umbrella of the International Polar Year (IPY) 2007/2008. The John Sloan Dickey Center's Institute of Arctic Studies at Dartmouth College participates in the IPY with project #160, "Arctic Change: An Interdisciplinary Dialog between the Academy, Northern Peoples, and Policy Makers." A major part of this project has been the exhibition *Thin Ice: Inuit Traditions within a Changing Environment*. I was responsible for curating this from January 12 to May 31, 2007, at the Hood Museum of Art at Dartmouth College, Hanover NH,[1] and as an anthropologist I was concerned that the exhibition should approach climate change and its impacts from the perspective of culture. In this chapter, I discuss the exhibition as an example of anthropological action to communicate and educate about climate change and its significance for local livelihoods and indigenous traditions in a global context.

Exhibition Backgrounds

The development of policies on adaptation to climate change is a growing concern among Arctic nations, stakeholders, scientists, policymakers, and indigenous groups (ACIA 2005; Community of Kangiqsujuaq 2003; Fox 2002; Gearheard and Shirley 2007, 64; Huntington et al. 2004; IPCC 2007; Krupnik and Jolly 2002; McCarthy and Martello 2005; Nilsson et al. 2004; Nickels et al. 2002; Nuttall 1998; Nuttall and Callaghan 2000; Nuttall et al. 2005; Oozeva et al. 2004).[2] Many of these efforts in policy development draw on scientific findings and increasingly aim to consider local knowledge alongside the pure science. The inclusion of local knowledge has itself become a subject of policymaking with regard to indigenous empowerment (Agrawal 2002; Berkes, Colding, and Folke 2000; Dybbroe 1999; Ellis 2005; Huntington 2000; Huntington and Fox 2005; Nadasdy 1999; Raffles 2002; Sillitoe 1998).[3] Sharing of knowledge and a mutual understanding of the distinct systems of scientific and local knowledge traditions and practices are keys to the success of such discourses.

Generally, most scientific research on Arctic indigenous understandings of climate change have focused on inventorying local observations of change rather than on analyzing the cultural framings that inform these observations (Krupnik and Jolly 2002; SEARCH SSC 2001; Community of Kangiqsujuaq et al. 2005). Likewise, indigenous political leaders aid national and international policy-makers by adopting Western discourses on climate change. Given that both scientific and policymaking discourses are largely foreign to Arctic indigenous communities, the *Thin Ice* exhibition aimed to underline the relevance of research into the diversity of concepts and perceptions.

Taking the position that 1) successful local adaptation is a crucial factor in community well-being and for national security; and 2) policies impacting the local level should be based on both a high degree of literacy on climate change science, economics, and politics, and on local lifeways among policymakers and the public (Mehdi, Mrena, and Douglas 2006, 8f), the exhibition focused on indigenous knowledge of the Arctic to argue the case.

Inuit Indigenous Knowledge

Among the largest groups of the Arctic indigenous peoples are the related Inuit, Iñupiat, and Yup'ik of Greenland, Canada, and Alaska.[4] Historically, these groups maintained a subsistence lifestyle, and their social lives, economic practices, and spirituality were closely interconnected with the seasonal cycle and daily weather conditions of the Arctic. Today, Inuit live in centralized permanent communities, are linked to the global market economy, and are citizens of countries with national and international policies on resource development, trade, biodiversity, and animal rights. However, local politics and the local economy often operate at least partially on principles different from those of mainstream institutions. For example, Inuit in Nunavut developed a political discourse based on consensus rather than the opposition governance principle; and people interconnect the profit-oriented market economy with the older economy of sharing. While the social conditions of Inuit community life changed drastically, Inuit managed to maintain much continuity in their lifeways. For the Inuit, it is not so much *what* a person does but *how* he or she does it that is important (Nuttall et al. 2005; Omura 2002, 107; Stuckenberger 2005; Young and Einarsson 2004).

Knowledge

From an anthropological point of view, all knowledge, including scientific knowledge, is constructed and shaped according to values of validity and relevance that are accepted within a community. As Agrawal has pointed out: "Nothing even makes sense without at least an imaginable context. The only choice one possesses about context is which context to highlight. This choice exists whether one talks about indigenous or modern knowledge systems" (1995, 425).

A main distinguishing feature between scientific and indigenous knowledge is that the latter often includes the subjective, cultural, and situational rather than extracting it from the data. However, it is important to keep in mind that the distinction between knowledge systems is strict only within a theoretical discourse. In daily life, people tend to make pragmatic use of various knowledge systems, and this includes Western scientific knowledge as much as local stories, observations, and religious beliefs (Agrawal 1995; Berkes 1999; Berkes et al. 2000; Ellis 2005; Huntington 1998; Huntington and Fox 2005; Nadasdy 1999; Raffles 2002; Sillitoe 1998, 230; Zamparo 1996). Dybbroe (1999, 15) states that "indigenous knowledge is the knowledge that Inuit . . . possess right here and now: values and insights transmitted in oral or other form, experiential knowledge acquired from living on the land, knowledge acquired through formal schooling or the media, etc." And as Ingold and Kurtilla (2000, 195) point out, science is itself a form of local traditional knowledge—the difference between local and scientific knowledge, they suggest, "lies not in the epistemological status of the knowledge itself, but in the nature of the skilled practices through which it is generated."

While heterogeneous in its specific items, indigenous knowledge is tied to a cultural context that generates a particular understanding. It is probably this connection that is the foundation for the strong feature of continuity maintained by Inuit communities like Qikiqtarjuaq in Canada's Nunavut, amidst drastic changes. If indigenous knowledge is removed from its cultural framework of reference its significance in contributing to the science and policy results is limited to being data rather than information (Agrawal 1995; Ellis 2005, 73; Fazey et al. 2006; Nadasdy 1999; Roué and Nakashima 2002; Scott 1996; Sillitoe1998; Simpson 2001; Stevenson 1996; Stuckenberger 2005; Wenzel 1999; White 2006).

Inuit Knowledge, Hunting, and Sila: the 'Outside, the Weather'

In 1998, a Nunavut working group introduced the concept of *Inuit Qaujimajatuqangit* (literally translated as "that which has long been known to Inuit"), noting its holistic character: "the Inuktitut term encompasses [...] the notion of Inuit knowledge [of the workings of humans, nature, and animals], social and cultural values, practices, beliefs, language and world view" (Fienup-Riordan 1999; Laugrand 2002; Wenzel 1999; Working Group on Traditional Knowledge 1998, 5).

The Inuit logic of human-environment relations interconnects the moral, physical, social, religious, aesthetic, technological, and the economic domains. The holistic study of insiders' perspectives is one of the major projects of sociocultural anthropology and its qualitative methodology. Anthropology is therefore particularly suited to generate relevant insights into climate change through a focus on local observations and lifeways. An exhibition such as *Thin Ice* that is based on the anthropological study of Inuit human-environment relations, through the lens of hunting, gives the audience an opportunity to view these and their own communities more clearly through a point of comparison, and through an understanding that the Arctic is not a region that is distant from their own lives, but is a place where people's lives are affected by what happens elsewhere. The audience is invited to ponder, reflect and perhaps appreciate different ways of knowing and relating to the environment, as well as to understand the interlinkages between Arctic ecosystems and the rest of the globe.

For the Inuit, hunting had always been more than a technical, social, or economic enterprise, as people believed that humans, animals, and various natural forces shared spiritual qualities. Today, many Inuit still formulate their self-perceptions and views on weather and climate in such holistic cosmological terms, emphasizing relations among people, God, the land, and animals (Dorais 1997; Fienup-Riordan 1996, 2000; Laugrand and Oosten 2007; Nuttall and Callaghan 2000; Stuckenberger 2005, 2006). For example in Qikiqtarjuaq, the seasons and the actual weather conditions continue to form communal and religious life and subsistence practices (Stuckenberger 2006).

The Inuktitut concept of *sila* translates as "universe", "outside", "weather", "intelligence", "understanding" and captures the relationship between an individual and the environment through knowledge (Stuckenberger 2007). It is a notion that resembles the most the scientific concept of climate. However, in its full scope of meaning, *sila* integrates nature and the human mind and lifeways (see also the chapter by Nuttall in this volume). Iñupiat and Inuit believe that the acquisition of knowledge and understanding is closely related to exposure to the Arctic environment through subsistence practices. As people thus learn, they grow into a state of *silatujuq* ("gifted with *sila*," or, more generally, "having understanding") (Nuttall et al. 2005; Oosten 1989; Saladin d'Anglure 1980, 37; Stuckenberger 2007). How do impacts of climate change affect and how are they affected by people's self-perception and understandings of the environment and of change?

Continuity and Change in Inuit Life

In the pre-Christian worldview, nature and human society were interconnected expressions of a moral universe that evolved during mythic times from the misdeeds of the first human beings who were born of the union between *sila* and *nun*a ("earth") (Boas 1901; Rasmussen 1929, 1931). While *sila* was perceived as a spirit being and moral agent in the past, anthropologists know little about contemporary interpretations of *sila.* In the last century, Inuit traditions and lifeways have been drastically changed due to conversion to Christianity and the often forced relocation of small nomadic hunting camps into centralized permanent settlements by outside nation-states. These settlements were conceived on Western models of community.

While the new facilities and technologies created a greater independence from Arctic climate and weather, and while indigenous groups embrace scientific knowledge and Western education, there is evidence that people also continue to embrace notions of a moral universe. Elder Nutaraq, for example, suggested to students of Nunavut Arctic College that the disappearance of the right snow conditions for igloo building is associated with changes in lifestyle. He thought that the preference of hunters for readily available hunting cabins and ownership thereof would make them unreasonably impatient (*silaittuq*, "lacking *sila*"). This example shows that Inuit understandings of climate change may be quiet different from a Western scientific perspective (Kappianaq and Nutaraq 2001, 153).

Taking its lead from the concept of *sila* and Inuit hunting practices, *Thin Ice* was designed to explore historic and contemporary Inuit perceptions of human-environment relationships in order to underline the importance of cultural understandings for developing processes of adaptation to impacts of climate change.

THE EXHIBITION DESIGN

In contrast to a text, an exhibition generally speaking organizes information in a non-linear designed space and communicates information through objects and contextual labels and panels. *Thin Ice* was built in three adjacent galleries, the center gallery being very small and more like a niche. The Hood Museum being an art museum favored an object-oriented style of display. The exhibition team[5] aimed to find a balance between an autonomous display of objects and binding them to context. A thematic combination of objects, and titles and brief panels and labels provided the context while a generous use of space and spotlighting gave objects room to express themselves. We wanted the audience to experience an aesthetically welcoming space that invited exploration and engagement.

Aiming at introducing the holistic Inuit insider perspective on the environment, we faced the problem of either convoluted displays with dense descriptions continuously cross-referencing between the various domains of life (such as religion, social organization, technology), or fragmenting the holistic perspective by dealing with each domain of life in isolation. Overlaying both approaches was a combination of the indigenous teaching method of storytelling with sectional thematic foci provided the necessary clarity in conveying information and a framework that emphasized their interconnection. The exhibition narrative told about the different stages of a hunt and zoomed into relevant themes that were then dealt with in detail.

While each section showed theme-specific objects, objects also appeared repetitively and formed part of the organizational structure of the space. A harpoon or spear was mounted onto each wall, and a class cases with kayak and boat miniatures was placed in each section. While the details of each of the repetitive objects spoke to the specific theme of the section it was assigned to, the repetition in itself emphasized similarity. For example, the miniature of a kayak frame relates to technological know-how, but also to the gender relations that shaped its production, the environment for which it was designed, and local hunting practices (as well as trade relationships with outsiders etc.) The fact that the same kind of object can be used to talk about the various themes expressed the interconnection between those themes in the holistic Inuit peoples' lifeways. The multi-dimensionality of Arctic indigenous objects together with the simultaneous visibility of next to all objects in the exhibition was our attempt to translate the holistic character of Inuit lifeways into design.

Walk-through

Upon entering the exhibition space, but before stepping through a small gateway into the galleries, a large introductory panel informed the visitor about the context, content, and limitations of the exhibition. Right opposite the gateway, the visitor encountered a large video display of *Eyewitness to Change: Inuit Observations on Climate Change* introducing climate change and contemporary Canadian Inuit life.[6] Walking around the video-wall, the vista opened to the three successive galleries displaying Inuit objects. The focal point was a shaman's *nepcetaq* mask (St. Michael, Alaska, early twentieth century) depicting the physical and spiritual nature of hunting placed onto the far back wall of the third gallery.

The walk around the galleries was then guided by wall titles and sub-titles telling about the process of hunting and highlighting associated themes. Starting with the preparation, the hunt continues with the kill and proper treatment of the game. This is followed by the transition from the land to the community as the game is brought home and the animal materials shared and used. The titles of the two larger galleries 1 and 3 depicted these practices as follows: (1) *Knowing When* dealt with the knowledge required to decide

Figure 24.1: The entrance to the exhibition. © Jeff Nitzel. Courtesy Hood Museum of Art.

on when and where to go hunting and with the increasing unfamiliarity of the environment due to impacts of global warming; (2) *Being Hunter - Being Game* dealt with the killing of and respect for game, (3) *Cosmology* explored human-animal relationships, mythology, cosmological relationships within the framework of Christianity; (4) *Being Social* elaborated on the relevance of animals for social life by focusing on the practices of sharing and cooperation, and on gender relations in Inuit societies; and (5) *Knowing How* dealt with indigenous knowledge of various kinds of technologies, skills in producing items, and the transfer of knowledge to the younger generations.

While following this general narrative of a hunting trip, the visitors passed by the small niche gallery. The niche was entitled *Facing Climate Change* and sub-titled with a quote by Aqqaluk Lynge, President, Inuit Circumpolar Council, Greenland: "*We are all in this together. But our perceptions are different. This is the challenge.*" (Lynge 2007, 10). Setting up the larger discourse on climate change related policy-making, the knowledge perspectives of three stakeholders—indigenous peoples, scientists, and politicians—were presented in posters and associated objects.

The Objects

The Dartmouth College Hood Museum of Art has more than 65,000 objects, of which about 3,000 relate to the art and material culture of Greenland, Alaska, Canada, and other regions of the circumpolar North. Much of the Hood Museum's material was collected in the 1950s and 1960s in the wake

Figure 24.2: Third gallery with mask. © Jeff Nitzel. Courtesy Hood Museum of Art.

of a lively period of northern activity at Dartmouth. Dartmouth faculty, students, and friends of the College conducting research and traveling in the Arctic purchased items, such as, of clothing, tools, art, and trade goods and later donated them to the museum (Woodward 2007, 22). Additional exhibition material was taken from the Rauner Special Collections Library holding the Stefansson Collection on Polar Exploration at Dartmouth College. Additional items were received on loan from the Canadian Museum of Civilization and the Hudson Bay Archives, Manitoba.

The exhibition displayed 52 objects, including boat miniatures, harpoons, masks, a hunting model, clothing, household goods, tools, and contemporary art prints and sculptures. Accompanying object labels and text panels

Figure 24.3: Part of section "Being Social." © by Jeff Nitzel. Courtesy Hood Museum of Art.

were written in an active voice and closely connected the objects to indigenous practices (for labels see the exhibition catalogue, Stuckenberger 2007).

The items originated from various geographic areas within the Arctic; although Inuit groups, including Iñupiat and Yup'ik, Greenlanders, and Canadian Inuit, share many fundamental beliefs and practices, they are also quite culturally distinct. This exhibition emphasized their commonalities. Furthermore, most of the objects in the Hood's collection reflect on practices conducted largely by men, such as hunting, rather than women, so men's voices were more prominent.

The question arises as to what is there to learn from historic material culture for the present situation that evolved from social drastic changes and that is at the onset of increasingly severe environmental shifts? Many of the historic objects, such as the *ulu* (woman's knife) are still in use; others refer to aspects of human-environment relations, such as respect, that continue to be perceived as relevant among Inuit peoples (Stuckenberger 2005). The use of present-day prints and sculptures, photographs, quotes by indigenous people, anthropological comment, video and interactive displays show continuity while also showing change.

The theme of continuity and change was further explored through two multimedia displays. Dartmouth student Sasha Earnhart-Gold conducted interviews with Inuit from Nunavut for the Inuit Circumpolar Council on local observations of environmental change. The video documentation of those interviews was shown in the section *Knowing How* to elaborate on the

Inuit notion of *uggianaqtuq*—the increasing unfamiliarity of the weather. The interactive program *Anijaarniq: Inuit Land Skills and Wayfinding* produced by Claudio Aporta with the Igloolik Oral History Project provided an opportunity to explore the complex and dynamically evolving indigenous knowledge of navigation in the following section *Being Hunter—Being Game*.

Reception of the Exhibition[7]

Most visitors perceived the exhibition as "timely" both with respect to climate change in general and with regard to the focus on the rapidly changing Arctic in specific. The exhibition was accompanied by an extensive educational program for children and adults for whom a total of 47 tours were scheduled for the larger community.

Children visitors from first through fifth grades had been introduced to issues of climate change in their school curricula and they had discussed issues of sustainability. Many of the children had visited the earlier exhibition *El Anatsui: Gawu* that pre-dated *Thin Ice* and also had seen the two other Arctic exhibitions shown together with *Thin Ice*, namely *Our Land*, displaying art form Nunavut (on loan from the Peabody Essex Museum) and *American Resource Wars*, a photographic exhibition by Subhankar Banerjee. The combination of four exhibitions seemed to have a cumulative value that opened discourses on and critical reviews of resource use and ethics.

Partaking in a guided tour of *Thin Ice*, children easily accepted that there are different lifeways—even when confronted with the realities of killing "cute" animals.[8] For most children, the Arctic clothing display was a favored spot and may have been a key-experience. Connecting closely to their own livelihoods (warm clothing in winter), the garments proved to invite a culturally non-judgmental reflection on the killing and use of animals and an interest in how Inuit relations to animals are different from their own.

Guides noted that adults were often highly interested in exploring the interconnections between technology, social life, and spirituality comparing them to their own lives. It seemed that for adults and children alike, the curator's choice to avoid a critical or alarmed stance and to create a friendly and aesthetically engaging environment supported intensive investigation, reflections, and the development of questions regarding ways, and costs and benefits of adaptation to climate change and responsibility for one's own community and the world community at large.

Thin Ice also attracted a large number of college class visits. According to the Hood Museum's Collection Use Reports, twice as many classes visited *Thin Ice* as visited the entire museum during the same term (winter) in 2006. The Hood Museum especially appreciated that classes from environmental studies and geography programs at Dartmouth also made use of the exhibition, which are courses that have not made as much use of the museum in the past.

Climate change and its human dimension is clearly recognized as a global issue that requires an interdisciplinary approach between different disciplines and ways of knowing, but one that also demands that anthropologists and others engaged in seeking to understand it involve themselves in strategies that communicate greater awareness to a range of diverse audiences. This understanding, generated through the anthropological approach to climate change, is of great relevance in approaching the questions of if and how local lifeways and adaptive capacities should be integrated into larger-scale perspectives of climate change policy-making. As *Thin Ice* shows, the curation and exhibiting of the cultural dimensions of climate change is not intended as mere display for an uncritical public gaze. It is one crucial way that anthropologists, in partnership with people experiencing the immediate effects of climate change, can take anthropological action and engage, communicate and educate in an effective, creative and powerful way.

NOTES

1. See http://hoodmuseum.dartmouth.edu/exhibitions/thinice/index.html.
2. See presentations by Mary Simon, Anne Hanson, and Peter Irniq at Dartmouth College in 2006 and 2007.
3. See for example Inuit Tapiriit Kanatami and Inuit Circumpolar Council-Canada: Inuit Action Plan, February 2007.
4. For the sake of brevity, I will henceforth use the (not entirely correct) term *Inuit* when referring to all three groups.
5. Thanks to Katherine Hart, Patrick Dunfey, Kellen Haak, and their teams.
6. The video footage was selected from a documentary made in 2000 by the International Institute for Sustainable Development and the US National Oceanic and Atmospheric Administration and adapted by Random Video for the exhibition *A Friend Acting Strangely* at the Smithsonian Institution, National Museum. The video documents observations by Inuvialuit and scientists about changes in weather and ice conditions.
7. I would like to thank Kristin Berquist, Adrienne Kermond, Cynthia Gilliland, Kristin Garcia, and Lesley Wellman—all members of the staff of the Hood Museum of Art—for data and observations.
8. Cognitive processes vary with each individual and with respect to the various educational activities associated with the museum visit. As we did not conduct a formal survey of changes in perception and knowledge, this chapter remains somewhat vague, presenting informal observations rather than hard data.

REFERENCES

ACIA. 2005. *Arctic Climate Impact Assessment: Scientific report*. Cambridge: Cambridge University Press.

Agrawal, A. 1995. Dismantling the divide between indigenous and scientific knowledge. *Development and Change* 26: 413–39.

———. 2002. Indigenous knowledge and the politics of classification. *International Social Science Journal* 54: 287–97.

Berkes, F. 1999. *Sacred ecology: Traditional ecological knowledge and resource management*. Philadelphia: Taylor and Francis.

Berkes, F., J. Colding, and C. Folke. 2000. Rediscovery of traditional ecological knowledge as adaptive management. *Ecological Applications* 10(5): 1251–62.

Boas, F. 1901. The Eskimo of Baffin Land and Hudson Bay: From notes collected by Capt. George Comer, Capt. James S. Mutch, and Rev. E. J. Peck. *Bulletin of the American Museum of Natural History* 15.

Community of Kangiqsujuaq, C. Furgal, M. Qiisiq, B. Etidloie, and P. Moss-Davies. 2005. *Unikkaaqatigiit—putting the human face on climate change: Perspectives from Kangiqsujuaq, Nunavik.* Ottawa: Inuit Tapiriit Kanatami, Nasivvik Centre for Inuit Health and Changing Environments at Université Laval, and the Ajunnginiq Centre at the National Aboriginal Health Organization.

Dorais, L.-J. 1997. *Quaqtaq: Modernity and identity in an Inuit community.* Toronto: University of Toronto Press.

Dybbroe, S. 1999. Researching knowledge: The terms and scope of a debate. *Topics in Arctic Social Sciences* 3: 13–26. Keynote address at the International Arctic Social Sciences Association (IASSA), Copenhagen, May 21–23, 1998.

Einarsson, N., J. N. Larsen, A. Nilson, and O. R. Young, eds. 2004. *Arctic human development report (AHDR).* Akureyri, Iceland: Stefansson Arctic Institute.

Ellis, S. C. 2005. Meaningful consideration? A review of traditional knowledge in environmental decision making. *Arctic* 58(1): 66–78.

Fazey, J., J. A. Fazey, J. G. Salisbury, D. B. Lendenmayer, and S. Dovers. 2006. The nature and role of experiential knowledge for environmental conservation. *Environmental Conservation* 33(1): 1–10.

Fienup-Riordan, A. 1983. *The Nelson Island Eskimo: Social structure and ritual distribution.* Anchorage: University of Alaska Press.

———. 1996. *The living tradition of Yup'ik masks.* Seattle: University of Washington Press.

———. 1999. Yaqulget qaillun pilartat (What birds do): Yup'ik Eskimo understanding of geese and those who study them. *Arctic* 52(1): 1–22.

———. 2000. *Hunting traditions in a changing world: Yup'ik lives in Alaska today.* New Brunswick, NJ: Rutgers University Press.

Fox, S. 2002. These are things that are really happening: Inuit perspectives on the evidence and impacts of climate change in Nunavut. In *The earth is moving faster now: Indigenous observations of arctic environmental change*, eds. I. Krupnik and D. Jolly, 12–53. Fairbanks, AK: Arctic Research Consortium of the United States.

Gearheard, S. and J. Shirley. 2007. Challenges in community-research relationships: Learning from natural sciences in Nunavut. *Arctic* 60(4): 62–74.

Huntington, H. 1998. Observations on the utility of the semi-directive interview for documenting traditional ecological knowledge. *Arctic* 51(3): 237–42.

———., ed. 2000. *Impact of changes in sea ice and other environmental parameters in the Arctic.* Bethesda, MD: Marine Mammal Commission.

Huntington, H., T. V. Callaghan, S. Fox, and I. Krupnik. 2004. Matching traditional and scientific observations to detect environmental change: A discussion on Arctic terrestrial ecosystems. *Ambio* Nov.(13): 18–23.

Huntington, H., and S. Fox. 2005. The changing Arctic: Indigenous perspectives. In *ACIA, Arctic Climate Impact Assessment.* 61–98. Cambridge: Cambridge University Press.

Ingold, T., and T. Kurttila. 2000. Perceiving the environment in Finnish Lapland. *Body and Society* 6(3–4): 183–96.

Intergovernmental Panel on Climate Change (IPCC). 2007. *Climate change 2007: The physical science basis. Summary for policy makers.* http://www.ipcc/ch.

Kappianaq, G., and C. Nutaraq. 2001. *Traveling and surviving on our land: Inuit perspectives on the 20th century.* Iqaluit: Language and Culture Program of Nunavut Arctic College.

Krupnik, I. and D. Jolly, eds. 2002. *The earth is faster now: Indigenous observations of Arctic environmental change.* Fairbanks, AK: Arctic Research Consortium of the United States.

Laugrand, F. 2002. Écrire pour prendre la parole: Conscience historique, memoires d'aines et regimes d'historicité au Nunavut. *Anthropologie et Sociétés* 26(2–3): 91–116.
Laugrand, F. and J. Oosten. 2007. Reconnecting people and healing the land: Inuit Pentecostal and evangelical movements in the Canadian Eastern Arctic. *Numen* 54: 229–69.
Lynge, A. 2007. Whose climate is changing? In *Thin ice: Inuit traditions within a changing environment,* ed. A. N. Stuckenberger, 10–11. Hanover, NH and London: University Press of New England.
McCarthy, J. and M. L. Martello. 2005. Climate change in the context of multiple stressors and resilience. In *Arctic Climate Impact Assessment: Scientific report,* 945–88. Cambridge: Cambridge University Press.
Mehdi, B., C. Mrena, and A. Douglas. 2006. *Adapting to climate change: An introduction for Canadian municipalities.* C-CIARN (Canadian Climate Impacts and Adaptation Research Network).
Nadasdy, P. 1999. Politics of TEK: Power and the "integration" of knowledge. *Arctic Anthropology* 36(1–2): 1–18.
Nickels, S., F. Christopher, J. Castleden, P. Moss-Davies, M. Buell, B. Armstrong, D. Dillon, and R. Fonger. 2002. Putting the human face on climate change through community workshops. In *The earth is faster now: Indigenous observations of Arctic environmental change,* eds. I. Krupnik and D. Jolly, 300–33. Fairbanks, AK: Arctic Research Consortium of the United States.
Nilsson, A., J. N. Larsen, and N. Einarsson. 2004. Arctic human development report (AHDR). Akureyri: Stefansson Arctic Institute.
Nuttall, M. 1998. *Protecting the Arctic: Indigenous peoples and cultural survival.* Amsterdam: Harwood Academic Publishers.
Nuttall, M., and T. V. Callaghan, eds. 2000. *The Arctic: Environment, people, policy.* Amsterdam: Harwood Publishers.
Nuttall, M., F. Berkes, B. Forbes, G. Kofinas, T. Vlassova, and G. Wenzel. 2005. Hunting, herding, fishing and gathering: Indigenous peoples and renewable resource use in the Arctic. In *ACIA Arctic Climate Impact Assessment: Scientific report,* 649–90. Cambridge: Cambridge University Press.
Omura, K. 2002. Construction of *inuinnaqtun* (real Inuit-way): Self-image and everyday practices in Inuit society. Self- and other-images of hunter-gatherers. Paper presented at the 18th International Conference on Hunting and Gathering Societies (CHAGS 8) National Museum of Ethnology, October 26–30, Osaka, Japan.
Oosten, J. 1989. Theoretical problems in the study of Inuit shamanism. In *Shamanism: Past and present,* eds. M. Hoppal and O. J. von Sadovsky, 331–48. Budapest and Los Angeles: Fullerton.
Oozeva, C., C. Noongwook, G. Noongwok, C. Alowa, and I. Krupnik. 2004. *Watching ice and weather our way./Sikumengllu eslamengllu eghapalleghput.* Washington, DC: Arctic Studies Center, Smithsonian Institution.
Raffles, H. 2002. Intimate knowledge. *International Social Science Journal* 54(173): 325–35.
Rasmussen, K. 1929. *Intellectual culture of the Iglulik Eskimos. Report of the fifth Thule expedition, 1921–24.* Vol. 7. Copenhagen: Nordisk.
———. 1931. *The Netsilik eskimo: Social life and spiritual culture. Report of the fifth Thule expedition, 1921–1924.* Vol. 8. Copenhagen: Gyldendalske Boghandel.
Roué, M. and D. Nakashima. 2002. Knowledge and foresight: The predictive capacity of traditional knowledge applied to environmental assessment. *International Social Science Journal* 173: 337–47.
Saladin d'Anglure, B. 1980. "Petit-ventre", l'enfant-geant du cosmos inuit: ethnographie de l'enfant et enfance de l'ethnographie dans l'arctique central inuit. *L'Homme* 20(1): 7–46.

Scott, C. 1996. Science for the west, myth for the rest? The case of James Bay Cree knowledge construction. In *Naked science: Anthropological inquiry into boundaries, power, and knowledge*, ed. L. Nader, 69–86. New York: Routledge.
SEARCH SSC. 2001. *Study of environmental Arctic change, science plan.* Seattle: Polar Science Center, Applied Physics Laboratory, University of Washington.
Sillitoe, P. 1998. The development of indigenous knowledge: A new applied anthropology. *Current Anthropology* 39(2): 223–35.
Simpson, L. 2001. Aboriginal peoples and knowledge: Decolonizing our processes. *Canadian Journal of Native Studies* 21(1): 137–48.
Stevenson, M. G. 1996. Indigenous knowledge in environmental assessment. *Arctic* 49(3): 278–91.
Stuckenberger, A. N. 2005. *Community at play: Social and religious dynamics in the modern Inuit community of Qikiqtarjuaq*. Amsterdam: Rozenberg Publishers.
———. 2006. Sociality, temporality and locality in a contemporary Inuit community. *Études/Inuit/Studies* 30(2): 95–111.
Stuckenberger, A. N., with contributions by W. Fitzhugh, A. Lynge, and K. H. Woodward. 2007. *Thin ice: Inuit traditions within a changing environment.* Exhibition catalogue, Hood Museum of Art, Dartmouth College. Hanover, NH and London: University Press of New England.
Wenzel, G. W. 1999. Traditional ecological knowledge and Inuit: Reflections on TEK research and ethics. *Arctic* 52(2): 113–24.
White, G. 2006. Cultures in collision: Traditional knowledge and Euro-Canadian governance processes in northern land-claim boards. *Arctic* 59(4): 401–14.
Woodward, K. H. 2007. Arctic, northwest coast, and polar exploration collection of Dartmouth College. In *Thin ice: Inuit traditions within a changing environment*, A. N. Stuckenberger, 21–27. Hanover, NH and London: University Press of New England.
Working Group on Traditional Knowledge. 1998. *Presentations.* Iqaluit, Canada.
Young, O. R., and N. Einarsson. 2004. Introduction: Human development in the Arctic. *Arctic Human Development Report*, 15–26.
Zamparo, J. 1996. Informing the fact: Inuit traditional knowledge contributes another perspective. *Geoscience Canada* 23(4): 261–66.

Epilogue: Anthropology, Science, and Climate Change Policy

Susan A. Crate and Mark Nuttall

We began this book with the idea of producing a comprehensive assessment of anthropology's engagement with climate change, but also to map out where the discipline can head as it carves out new research and policy-oriented approaches and examines the dynamics of various epistemologies and practices. Although anthropologists talk about the weather with the people with whom they live, research, and work, and even though climate enters into our considerations of ecology, culture, and human-environment interactions, climate change research has long been regarded the domain of practitioners of various disciplines in the natural sciences. Indeed, identifying the causes and nature of climate change, and assessing its impacts, is something physical scientists still claim as their legitimate prerogative (Duerden 2004), leaving social scientists alone in their considerations of how to explain how the physical manifestations of change are perceived, experienced, interpreted, and negotiated at community levels.

Scientists talk about climate, and they seek to develop sophisticated tools to study variations in climate that occur on time scales from seasonal to many millions of years; presumably anthropologists (like their field and research partners) tend to only talk about the weather. What we claim as evidence for climate change is often regarded politely by our colleagues in the physical sciences as useful, but anecdotal, information that some find occasionally corroborates or enhances scientific understanding.

Scientific concern with climate change is about abstraction, generalization, replication, measurement, and quantification, with the refinement of models and the elimination of uncertainty. Like all science it marks out its territory in bold and ambitious ways, and with the dazzling language of global circulation models. But in talking about the weather we seek to understand lived worlds, symbolic forms, and human well-being. The anthropological concern with how climate change is apparent in local weather patterns, how rain and wind feel on one's face, how snow now feels differently underfoot than it did in one's childhood, how places no longer resemble what they did a generation or two ago, what people notice in their gardens, and how human-environment relations are redefined and people are displaced allows us to shed light on the complexities of real life situations as well as the interconnections between global and local place. In a sense we bring climate change down to earth, allow it to assume and be understood in its many human faces, on an earth experienced by more and more people as a highly shifting ground. Yet while science has provided the foundation for our understanding of climate change, now consolidated with the reports of the IPCC's fourth assessment, anthropology is well-equipped to contribute to this understanding, to highlight the complexities of what are assumed to be simple distinctions between scientific and social scientific approaches, and also between scientific knowledge and indigenous/local knowledge, and to taking a lead role in policy-focused discussions about adaptation strategies.

The last few years have seen an increase (albeit a modest one) in the number of anthropologists involved in climate change assessments, and in policy-oriented conference and workshop contexts.[1] Climate change is a variation in climatic parameters attributed directly or indirectly to human activity, and a major goal of the IPCC's work is assessing anthropogenically driven climate change. Anthropologists are well placed to help identify the possible role of humans, their needs, their wants, and their choices, and their cultures of production and consumption in influencing the recent variations in global climate. Yet we still lag behind our physical science colleagues in the contributions we make. While scientists admit to uncertainty and concede the limitations of models and scenarios, particularly when it comes to issues of scale, socioeconomic scenarios for understanding the nature of future climate change remain poorly developed and inadequate. The challenge is one of bridging scales and of revealing their interconnections (Magistro and Roncoli 2001), and disentangling natural variability and change from the consequences of human action.

Anthropologists encounter climate change in situ, in the field, by being in place and encountering and pondering the physical and social evidence for and effects of climate change. Anthropology's attention to human being means that our points of entry into data collection are people and communities. Our special skills are in interpreting and translating how a changing environment interfaces, transforms, underpins, and undermines human communities. We gather climate change data by talking and listening to people who feel they are vulnerable to the changes they either experience personally, or who worry about the possible impacts of the shifts in weather and climate they hear and read about. Additionally a growing number of anthropologists seek and are involved in understanding climate change through participation in scientific assessments and by working with scientists on interdisciplinary projects. A handful of anthropologists, including one coeditor of this volume (Nuttall), have collaborated on a number of recent major international initiatives that seek to understand the human dimensions of climate change impacts, including the Arctic Climate Impact Assessment (ACIA), the IPCC fourth assessment, the Canadian national assessment of climate change and adaptation, the 2007 Bali Climate Declaration, and the Millennium Ecosystem Assessment. Environmental anthropologists and other social scientists have combined remote sensing techniques with an understanding of the human use of resources (Finan and Nelson 2008; Liverman et al. 1998). Anthropologists have also worked in multidisciplinary teams with other social scientists (such as sociologists, political scientists, and economists) and natural scientists to understand human interactions with the carbon cycle (see, for example Stern 2002; Willett Kempton's[2] website: http://www.ocean.udel.edu/people/profile.aspx?willett#ResearchInterests). Such work focuses on questions concerning the underlying causes of fossil fuel consumption, agricultural intensification, and energy production, as well as human activities that respond to the carbon cycle (including environmental management regimes).

Much of the early effort put into these initiatives goes into forging effective means of communication between social scientists and natural scientists. Anthropologists and other social scientists who have worked in these multidisciplinary contexts have similar stories to tell of how they are often compelled to legitimate their participation in climate change studies and assessments that are framed by paradigms and approaches from the natural sciences. While this participation has advanced our understanding of climate change and its impacts, it often prompts anthropologists to reflect on their involvement and to critique climate change models, scientific scenarios, and storylines for the future (Nuttall 2001). Our responsibility is to interrogate and challenge prevailing scientific views and perspectives, rather than to bring a perspective on social, cultural, and economic life that merely confirms them. But we also know that critique is not our main role, and that in going beyond it our task is both epistemological and ontological. We have become well versed in explaining that climate change, loss of biodiversity, atmospheric pollution, and a host of other issues of pressing contemporary global change are of concern to social scientists because they are social as well as environmental issues. And in doing so, we aim to demonstrate that science has a limited reach. Our effort goes into arguing that the social sciences complement scientific research on global environmental change and offer a vital perspective on understanding the relationship between climate, environment, politics, economy, society, culture, and human behavior (e.g., Rayner and Malone 1998; Redclift and Benton 1994). Human-environment perspectives in anthropology contribute to understanding both the causes of global environmental change (by focusing on processes connecting human activity and environmental change) and the human responses (in terms of policy solutions, public opinion and action, social movements, and environmental management systems). As a discipline concerned with (among other things) understanding social complexity, cultural diversity, and the interrelationships between society and environment, anthropology is well suited to make significant and finely tuned contributions to integrated assessments of climate change. This is particularly so when studies of climate change are contextualized within the broader perspective of global change (Rayner and Malone 1998).

The reality is that climate change challenges researchers in both the natural and social sciences to forge strong working partnerships across disciplines, as well as with various stakeholders. The development of innovative multidisciplinary forms of collaboration between the natural and social sciences seems the only real way forward if we are to identify and understand the processes driving climate change, for instance, or if we are to assess, analyze, and evaluate the impacts of climate change within the broader context of social and cultural change for indigenous peoples and local communities. To these ends, integrated assessments of global change potentially offer advantages by being both inter- and multidisciplinary. For example, while scientific research on climate change in the Arctic or the Amazon rainforest that focuses only on the physical aspects may greatly enhance our understanding of the climatic and environmental consequences of anthropogenic greenhouse gas emissions, it remains limited unless it also considers the social, economic, and political processes that both affect the emissions and are affected by climate change (Nuttall and Callaghan 2000, xxvi). Rotmans and Dowlatabadi (1998, 357–58) state that integrated assessment modelers face challenges in building a relevant knowledge base and designing models that reflect reality. In particular, they argue that knowledge of key dynamics of social systems is extremely limited and this creates problems in the modeling of socioeconomic aspects. Understanding the environmental, social, and economic consequences of climate change and seeking to design appropriate policy responses is difficult without knowledge of local and regional socioeconomic conditions. To rectify the dearth of basic socioeconomic data, integrated climate change impact studies will need to build up a socioeconomic profile of each locality under scrutiny, examine settlement patterns and demographic changes, and reveal the extent to which social and cultural understandings of resources and the environment have been shaped by socioeconomic development (as well as human choices and decisions) at local, regional, and national levels.

Social scientists have often been seen as contributing to or merely complementing the "pure scientific" work of climate change scientists, rather than being integral to such research, or even influencing its direction and scope. Anthropologists, as we have shown in this volume, have many roles to play. As Nuttall and Callaghan (2000) have written with respect to work on climate change in the Arctic, research on the causes and consequences of climate change needs to focus on a range of issues, such as resource use practices and the human modification of the environment, and the roles that cultural, political, and institutional processes play in local and regional decisions about the use and management of resources. A thorough analysis of the complex and multifaceted interactions and feedback mechanisms involved in climate change needs to place the scientific quantifications of, for example, changes in greenhouse gas emissions and their impacts into wider social, economic, and political perspectives. In many parts of northern Europe, for example, local land use and resource management practices have resulted in major changes to local and regional natural ecosystems, which have affected trace gas exchanges. Socioeconomic studies concerned with trace gases and climate change in Northern wetlands would be concerned with gaining an understanding of the choices and decisions people make in relation to how they modify their local environments, in both historical and contemporary perspective. One methodological and theoretical challenge for integrated assessment studies, therefore, is how to gain a greater understanding of the relationship between land and resource use practices, resource management choices, and climate change.

We also have a role to play in strengthening the reliability of models and storylines for the future in terms of their representation of social, economic, and cultural processes. Integrated assessment models may provide thorough descriptions of cause and effect situations and of the interactions between social and physical processes, yet these are by necessity simplified representations of complex relationships (Nuttall 2001). An interdisciplinary attempt at refinement of these models not only enhances understanding of the complex interface between human societies and the environment—it also has greater policy relevance (Nuttall and Callaghan 2000, xxvi).

Climate change science contributes to the process of consensus decision-making, influencing what people come to think about what is "real" about the world and how it is changing. International processes and climate change policymaking require a "sound scientific basis" and, in providing this, science assumes social authority over environmental and global arenas (Wynne 1994). But scientific knowledge, as the case of climate change science shows us, is contested and subject to major indeterminacies. As Wynne argues, scientific uncertainty remains the main obstacle to more consensual and authoritative policies (Wynne 1994, 175). Science is far from being able to offer clear-cut accounts of the world and its physical processes, and scientists struggle with uncertainties generated by climate models (and in any case, the kinds of problems encountered by science such as climate change do not fit within neat disciplinary boundaries).

Carvalho (2007) has examined the representation of climate change in the British "quality press," and shows that the discursive (re)construction of scientific claims in the media is strongly entangled with ideological standpoints. Ideology, she argues, works in a powerful way to define and select what is scientific news, what the relevant "facts" are, and who the authorized "agents of definition" of matters of scientific importance actually are. The representation of scientific knowledge has significant implications and consequences for both public understandings of climate change and the political responsibility for addressing it. Public perceptions of, understandings of, and attitudes to climate change are greatly influenced, for example, by the way the media convey information about scientific research on climate change, and how media images are deployed to depict vulnerable lives and fragile ecosystems on the edge of a changing world. This fact emphasizes the key roles anthropologists have in communicating with a variety of media and doing an appropriate job of translating their messages into ones that their public audiences will grasp and feel mobilized to act upon.

We as anthropologists have a critical role in analyzing the socioeconomic, human, and political processes in industrial countries and need to understand the cultural knowledge of the groups that affect global climate change, not only the affected. We need to understand how consumers, policymakers, environmentalists, industry representatives, and the like use cultural knowledge to frame and interpret scientific climate change information. Yes, anthropologists must continue to study marginalized peoples most impacted by the local effects of climate change. However, "it is crucial to also overcome our marginality in scholarship and policy arenas related to global change research to study all types of relevant 'locals', and especially those populating institutions of power" (Lahsen 2008: 587). This point is relevant especially since one of the central drivers of climate change is consumption and policymaking in industrialized and industrializing countries linked to fossil fuel use and environmental degradation (e.g., deforestation).

Some anthropologists engaging in this pursuit argue that we need to overcome our abated but continued aversion to study those who are much more important in shaping climate change and associated knowledge and policies than are the marginal populations we are accustomed to studying—namely power brokers such as scientists, governmental decision-makers, industry leaders, journalists, and financial elites (Lahsen 2007, 2008). Lahsen poignantly states, "Climate-related cultural dynamics among these population segments have long been rampant, but few anthropologists have engaged them in their research, despite pioneer Steve Rayner's now nearly two-decade-long provocation that we 'stop fiddling while the global warms.' We have marginalized ourselves and lost important opportunities to do research and intervene further 'upstream' in arenas where knowledge and policy is produced because we have been insufficiently inclined to revise our own cultural (disciplinary) inclinations" (Lahsen 2008).

Policy *will* change—whether we as anthropologists actively engage or not. The question is how quickly, of what character, and with what level of response. As we hope to have shown in this volume and also to have begun a wider and more active dialogue on, anthropologists have a multitude of key roles to play in the issue of climate change that can both help to move the policy process along more quickly and bring policy results that are more effective and long lasting exactly because they engage the human dimension.

NOTES

1. For example: Thirteenth conference of parties to the UNFCCC (UN Framework Convention on Climate Change) in Bali, Indonesia, December 2007; the Tyndall Centre's "Living with climate change: are there limits to adaptation," http://www tyndall.ac.uk/research/programme3/adaptation2008/index_outputs.html; the World Bank's "Social Dimensions of Climate Change" workshop, http://web.worldbank.org/WBSITE/EXTERNAL/TOPICS/EXTSOCIALDEVELOPMENT/0,contentMDK:21659919~pagePK:210058~piPK:210062~theSitePK:244363,00.html.
2. http://www.ocean.udel.edu/people/profile.aspx?willett#ResearchInterests.

REFERENCES

Carvalho, A. 2007. Ideological cultures and media discourses on scientific knowledge: Rereading news on climate change. *Public Understanding of Science* 16(2): 223–43.

Duerden, F. 2004. Translating climate change impacts at the community level. *Arctic* 57(2): 204–12.

Finan T. J. and D. R. Nelson. 2008. "Decentralized planning and adaptation to drought in rural Northeast Brazil: An application of GIS and participatory appraisal toward transparent governance." In *Adapting to climate change: Thresholds, values, governance*, eds. W. N. Adger, I. Lorenzoni, and K. O'Brien. Cambridge: Cambridge University Press.

Lahsen, M. 2007. Anthropology and the trouble of risk society. *Anthropology News*. December: 10–11.

———. 2008. Comments on S. Crate, Gone the bull of winter? Grappling with the cultural implications of and anthropology's role(s) in global climate change. *Current Anthropology* 49(4): 569–95.

Liverman, D., E. F. Moran, R. R. Rindfuss, and P. C. Stern eds. 1998. *Peoples and pixels: Linking remote sensing and social science*. Washington, DC: National Academy Press.

Magistro, J. and C. Roncoli. 2001. Anthropological perspectives and policy implications on climate change research. *Climate Research* 19: 91–96.

Nuttall, M. 2001. Indigenous peoples and climate change research in the Arctic. *Indigenous Affairs* 4/01: 26–33.

Nuttall, M. and T. V. Callaghan. 2000. Introduction. In *The Arctic: Environment, people, policy*, eds. M. Nuttall and T. V. Callaghan. New York: Taylor and Francis.

Rayner, S. and E. L. Malone, eds. 1998. *Human choice and climate change*. Columbus, OH: Batelle Press.

Redclift, M. and T. Benton, eds. 1994. *Social theory and the global environment*. London: Routledge.

Rotmans, J. and H. Dowlatabadi. 1998. Integrated assessment modeling. In *Human choice and climate change*, vol. 3: *Tools for policy analysis*, eds. S. Rayner and E. L. Malone. Columbus, OH: Battelle Press.

Stern, P. 2002. Human interactions with the carbon cycle: Summary of a workshop. Washington, DC: National Academy Press.

Wynne, B. 1994. Scientific knowledge and the global environment. In *Social theory and the global environment*, eds. M. Redclift and T. Benton, 169–89. London: Routledge.

About the Contributors

Peggy F. Barlett, Goodrich C. White Professor of Anthropology at Emory University, is a specialist in agricultural systems and sustainability programs in higher education. Recent works include *Sustainability on Campus: Stories and Strategies for Change*, edited with Geoffrey W. Chase, (MIT Press 2004) and *Urban Place: Reconnecting with the Natural World* (MIT Press 2005). Co-founder of the Piedmont Project at Emory, she leads workshops on curriculum development and faculty engagement with sustainability around the country. She leads Emory's Sustainable Food Initiative, which builds on her past work on the U.S. farm crisis of the 1980s. Her current work seeks to combine the science behind sustainability with the joy and satisfaction of deeper engagement with the animate earth.

Lenora Bohren is a senior research scientist and the Director of the National Center for Vehicle Emissions Control and Safety (NCVECS) and the Director of Research for the Institute of the Built Environment (IBE) at Colorado State University (CSU). She has been working on environmental issues with organizations such as the USEPA and the USDA for over twenty years. While working with these issues, she has helped develop methodologies to assess the socioeconomic characteristics of managing tropical soils and has conducted air pollution studies throughout the United States and Mexico. In addition, she has supervised the development of an environmental education curriculum for middle and junior high school students to teach pre-drivers how to become environmentally responsible drivers. More recently, she has worked with the Natural Resource Ecology Laboratory (NREL) at CSU interviewing farmers and ranchers assessing their attitudes toward adaptation to climate change and global warming.

Inge Bolin has been professor of anthropology and research associate at Malaspina University College in Nanaimo, Canada. Her areas of expertise are Andean studies, ethnology of South America, applied and medical anthropology, and human ecology. She has published two books, *Rituals of Respect: the Secret of Survival in the High Peruvian Andes* (1998) and *Growing up in a Culture of Respect: Child Rearing in Highland Peru* (2006). In 1992 she founded the volunteer sustainable development NGO Yachaq Runa in highland Peru, where she continues to work.

Kenneth Broad is associate professor in marine affairs and policy and fellow at the Abess Center for Ecosystem Science and Policy at the University of Miami. He also is co-director of the Center for Research on Environmental

Decisions based at Columbia University. His research focuses on climate impacts and human perception, the (mis)use of scientific information, decision-making under uncertainty, and ecosystem based management. He works closely with hydrologists, oceanographers, economists, ecologists, climatologists, and other strange creatures. Broad is also active in underwater cave exploration and was awarded the 2006 National Geographic Society Emerging Explorer Award.

Noel D. Broadbent's interests include historic and prehistoric archaeology, anthropology, indigenous research, and international science policy. He is a founding member of the International Arctic Social Sciences Association, and was founding director of the Center for Arctic Cultural Research at Umeå University, founding director of the Arctic Social Sciences Program at NSF, and chair of the Department of Archaeology and Saami Studies at Umeå University. He is currently completing the NSF-funded "Search for a Past" research project on Saami prehistory at the Department of Anthropology at the Smithsonian Institution in Washington, D.C.

Gregory V. Button is a faculty member in the Department of Anthropology at the University of Tennessee, Knoxville. He has been conducting research on disasters for nearly three decades. His work has focused on issues of social conflict, vulnerability, scientific uncertainty, forcible displacement, post-disaster recovery, and disaster policy. His most recent publication, co-authored with Anthony Oliver-Smith, is "Disaster, Displacement, and Employment: Distortion of Labor Markets During Post-Katrina Reconstruction" in *Capitalizing on Disaster: Neoliberal Strategies in Disaster Reconstruction*, edited by Nandini Gunewardena & Mark. Schuller (2008). He is currently writing a book about unnatural disasters.

Benedict J. Colombi is assistant professor in American Indian studies and anthropology and a member of the Institute for the Study of Planet Earth at the University of Arizona. He conducts research in ecological and environmental anthropology and is writing several articles and developing a book based on recent research with the Nez Perce Tribe about large dams and Pacific salmon in the Columbia River Basin.

Todd Crane is a lecturer in the Technology and Agrarian Development chair group at Wageningen University in the Netherlands. His research interests include cultural adaptations to climate variability, political ecology and participatory rural development. He has research experience in West Africa and the American southeast.

Susan A. Crate, assistant professor in the Environmental Policy and Social Sciences Department at George Mason University, is an interdisciplinary scholar specializing in environmental and cognitive anthropology. She has worked with indigenous communities in Siberia since 1988 and specifically with Viliui Sakha since 1991, and is author of *Cows, Kin and Globalization: An Ethnography of Sustainability* (2006). Her current research focuses on understanding local perceptions, adaptations, and resilience of Viliui Sakha communities in the face of unprecedented climate change. She trained with Al Gore in the Climate Project in 2007, and since has given over 20 community presentations.

Tim Finan is a research anthropologist and the director of the Bureau of Applied Research in Anthropology (BARA) at the University of Arizona. His research activities have spanned five continents and three decades. He is a student of socioeconomic change particularly under environmental stress, and his published work has contributed to the understanding of livelihood vulnerability and adaptation, to the impacts of climate change, and to the theory and practice of development. Since 2001, he has carried out multiple projects in Bangladesh that include the impacts of sea-level rise on coastal livelihoods, the nature of indebtedness among vulnerable households, and the determinants of economic change in the poorest regions of the country.

Shirley J. Fiske is adjunct professor in the Department of Anthropology, University of Maryland, and a consultant in environmental anthropology. She has over 20 years of experience in natural resources policy and programs in executive and legislative branches of the federal government, including the National Oceanic and Atmospheric Administration (NOAA) and the United States Senate. As an anthropologist she promoted the value of social sciences and anthropology for common pool resources, ocean governance, marine fisheries, coastal communities, and the human dimensions of global change under the US Global Change Research Program. Most recently she was legislative staff in the Senate, working on energy, environment, natural resources (including oceans), and global climate change legislation. Dr. Fiske is co-editor of a casebook in applied anthropology, *Anthropological Praxis: Translating Knowledge into Action* (1987), among other publications.

Donna Green is a research scientist in the Climate Change Research Centre, University of New South Wales, Australia, where she leads a programme that uses indigenous and non-indigenous knowledge to understand climate impacts on remote communities in northern Australia. Her research focuses on human-environment interactions, specifically on social and economic vulnerability, adaptation, and risk. Her current work builds on ten years' experience working in the areas of energy, environment, and sustainable development in the Asia-Pacific region. She was a contributing author to the Nobel Prize–winning IPCC's *Fourth Assessment Report* WGII.

Fekri Hassan is Petrie Professor of Archaeology at the Institute of Archaeology and Department of Egyptology at University College London. His work focuses on North Africa and the Middle East, and he has conducted fieldwork in Egypt, Algeria, Lebanon and Jordan. His research interests include climate change, water and civilization, cultural heritage management, the origins of civilization and state societies, and the relevance of archaeology to contemporary human issues. He is editor of *Droughts, Food and Culture: Ecological Change and Food Security in Africa's Later Prehistory* (2002) and co-editor of *Damming the Past: Dams and Cultural Heritage Management* (2008).

Anne Henshaw is a marine conservation program officer with the Oak Foundation. Before joining Oak, she was a visiting professor in the Sociology and Anthropology Department at Bowdoin College and director of Bowdoin's Coastal Studies Center. Her community-based research activities in Arctic Canada focus on linking Inuit experiential knowledge and land use with western science using geographic information systems. The results of her work have been published in a variety of peer reviewed journals and international venues including the Arctic Climate Impact Assessment. She currently serves on the Advisory Committee for the Office of Polar Programs at the National Science Foundation.

Robert K. Hitchcock is professor and chair of the Anthropology Department at Michigan State University. He also serves on the board of the Kalahari Peoples Fund (KPF), a non-profit organization that provides assistance to the peoples of southern Africa. Hitchcock has carried out research, development, and human rights work among the San of southern Africa and other groups for the past three decades. Some of his recent work focuses on impacts of resettlement and environmental change in southern and eastern Africa and the Great Plains of the United States.

Jerry Keith Jacka is assistant professor of environmental anthropology at North Carolina State University. His work focuses on the intersections of development, religion, and environmental change in Papua New Guinea. He has just completed a monograph entitled *Alchemy in the Rain Forest: Mining, Christianity, and the Environment in Highlands Papua New Guinea.*

Patrik Lantto is associate professor of history and senior lecturer at the Centre for Sami Research, Umeå University. He is author of several articles and book chapters, as well as *The time begins again: An analysis of the Sami etnopolitiska mobilisation in Sweden 1900–1950* (2000) and *To make its voice heard: Swedish Sami Riksförbund, samerörelsen and American samepolitik 1950–1962* (2003).

Heather Lazrus is a PhD Candidate at the University of Washington where she specializes in environmental anthropology. Her dissertation research focuses on the political ecology of climate change impacts in Tuvalu, South

Pacific. Heather has consulted with organizations active in the Pacific including the World Bank. She has also been involved in several projects with NOAA's National Marine Fisheries Service which examine impacts of environmental and policy change on fishing communities in Alaska and the Pacific Northwest.

Elizabeth Marino is a PhD student at the University of Alaska Fairbanks. She is currently researching the possible displacement of indigenous villages in Northwestern Alaska due to increasing erosion, storm activity, and rising water, with a particular focus on government response to climate crisis. Recent publications include a coauthored agency report entitled: "Shishmaref co-location cultural impact assessment" (2005) and "Defending the invisible: a biographical sketch of an Inupiaq soldier" to appear in the IPSSAS conference journal (2008).

Mark Nuttall holds the Henry Marshall Tory Chair in the Department of Anthropology at the University of Alberta and has worked extensively in Greenland, Canada, Alaska, Scotland and Finland. He was involved in the Arctic Council's Arctic Climate Impact Assessment (ACIA), the Intergovernmental Panel on Climate Change (IPCC) fourth assessment, the national Canadian assessment of climate change and adaptation, and the UN's Millennium Ecosystem Assessment. He is author or editor of several books, including the *Encyclopedia of the Arctic* (Routledge 2005). He is also Academy of Finland Distinguished Professor at the University of Oulu, Finland, and was elected to a Fellowship of the Royal Society of Canada in 2008.

Anthony Oliver-Smith is professor emeritus of anthropology at the University of Florida. He also held the Munich Re Foundation Chair on Social Vulnerability at the United Nations University Institute on Environment and Human Security in Bonn, Germany, for 2007–2008. He has done anthropological research and consultation on issues relating to disasters and involuntary resettlement in Peru, Honduras, India, Brazil, Jamaica, Mexico, Japan, and the United States. He is the author of *The Martyred City: Death and Rebirth in the Andes* (1986; 1992) and the co-editor of *Catastrophe and Culture* (2002) and *The Angry Earth* (1999).

Ben Orlove is professor of environmental science and policy at the University of California Davis, adjunct senior research scientist at the International Research Institute for Climate and Society at Columbia University, and a Fellow of the American Association for the Advancement of Science. He has conducted fieldwork in Latin America, Africa, and Australia. His work on climate began with studies of indigenous knowledge and forecast use, and developed into the study of climate-forecasting organizations and their relations with many social actors. In recent years, he has carried out research on climate change, particularly adaptation to glacier retreat. His recent books include *Lines in the Water: Nature and Culture at Lake Titicaca* (2002); *Weather, Culture, Climate* (2003); *Darkening Peaks: Glacier Retreat, Science and Society* (2008).

Rev. Kristina Peterson has worked on disaster relief issues for over three decades. She is currently a doctoral student at the University of New Orleans and a Research Fellow at the Center for Hazard Assessment Response and Technology. Her passion for sustainable equitable development is built on the belief that radical democracy is primary in addressing human and environmental rights. Her work is focused on participatory action as a tool for the preservation and restoration of the Louisiana wetlands and for the resilience of traditional and indigenous, at-risk coastal communities.

Nicole D. Peterson is a postdoctoral researcher at the Center for Research on Environmental Decisions at Columbia University. Her current research focuses on the role of local and international institutions in the construction of and adjustment to risk, examining decision-making processes in marine resource management and the role of agricultural development projects in mitigating climate risks.

Pamela J. Puntenney is founder and executive director of Environmental & Human Systems Management, a private sector consulting firm that seeks to facilitate and support the further implementation of sustainability strategies through learning processes, research and influencing policy, provision and distribution of key information, and training and capacity building. She serves as co-chair of the UN Commission on Sustainable Development Education Caucus, a senior foreign policy advisor advising senior management on creating inter-linkages, and has presented keynote addresses to ministries, business, scientific, technology, and education conferences throughout the world on a variety of environmental and sustainable development policy issues.

Carla Roncoli is an economic and environmental anthropologist and a research scientist in the Department of Biological and Agricultural Engineering at the University of Georgia, working with interdisciplinary research projects on climate applications in agriculture. Her interests focus on the human dimension of climate variability and change, particularly on the interactions of climate perceptions and knowledge and agricultural decision-making. For over twenty years she has carried out research among agriculturalists and pastoralists in dryland Africa, and, recently, among family farmers of the Southeast United States. She has mostly published in interdisciplinary journals, such as *Climate Research*, *Agricultural Systems*, and *Society and Natural Resources*.

Peter Schweitzer is professor of anthropology at the University of Alaska Fairbanks and director of Alaska EPSCoR (Experimental Program to Stimulate Competitive Research). Apart from numerous article publications, he is co-author of *Russian Old-Settlers of Siberia: Social and Symbolic Aspects of Identity* (2004), editor of *Dividends of Kinship: Meanings and Uses of Social Relatedness* (2000) and a co-editor of *Hunters and Gatherers in the Modern World: Conflict, Resistance, and Self-Determination* (2000). Schweitzer served as president of the International Arctic Social Sciences Association (IASSA) from 2001 to 2004.

Benjamin Stewart is assistant professor of worship at the Lutheran School of Theology at Chicago. His research interests include the role of religious ritual in shaping ecological sensibilities. He is completing his doctoral dissertation at Emory University's Graduate Division of Religion, where he was awarded a Piedmont Project Fellowship in Teaching and the Environment.

Sarah Strauss is associate professor in the Department of Anthropology at the University of Wyoming in Laramie and project coordinator for the Societal-Environmental Research and Education Laboratory at the National Center for Atmospheric Research in Boulder, Colorado. She has conducted ethnographic research in India, Switzerland, and the United States on topics related to health and the environment; current work focuses on water, weather, and climate change. Recent books include *Positioning Yoga* (2004) and *Weather, Climate, Culture* (2003; co-edited with Benjamin S. Orlove).

Nicole Stuckenberger is visiting assistant professor in the Department of Anthropology and research fellow at the John Sloan Dickey Center for International Understanding at Dartmouth College. She is a specialist in Inuit and Inupiat religion, society, and perceptions of the environment. She was curator of the exhibition *Thin Ice: Inuit Traditions within a changing Environment* at the Hood Museum of Art (January 27–May 14, 2007), and her recentbooks include *Community at Play: Social and Religious Dynamics in the Modern Inuit Community of Qikiqtarjuaq* (2005) and *Thin Ice: Inuit Traditions within a Changing Environment* (2007).

Richard Wilk is professor of anthropology and gender studies at Indiana University. He has done research with Mayan people in the rainforest of Belize, in West African markets, and in the wilds of suburban California on topics relating to consumption. His most recent books are *Home Cooking in the Global Village* (2006) and the edited *Fast Food/Slow Food* (2006).

Index